Springer Handbook
of Nanotechnology

Bharat Bhushan (Ed.)

3rd revised and extended edition

Springer Handbook of Nanotechnology

Since 2004 and with the 2nd edition in 2006, the Springer Handbook of Nanotechnology has established itself as the definitive reference in the nanoscience and nanotechnology area. It integrates the knowledge from nanofabrication, nanodevices, nanomechanics, nanotribology, materials science, and reliability engineering in just one volume. Beside the presentation of nanostructures, micro/nanofabrication, and micro/nanodevices, special emphasis is on scanning probe microscopy, nanotribology and nanomechanics, molecularly thick films, industrial applications and microdevice reliability, and on social aspects. In its 3rd edition, the book grew from 8 to 9 parts now including a part with chapters on biomimetics. More information is added to such fields as bionanotechnology, nanorobotics, and (bio) MEMS/NEMS, bio/nanotribology and bio/nanomechanics. The book is organized by an experienced editor with a universal knowledge and written by an international team of over 145 distinguished experts. It addresses mechanical and electrical engineers, materials scientists, physicists and chemists who work either in the nano area or in a field that is or will be influenced by this new key technology.

"The strong point is its focus on many of the practical aspects of nanotechnology... Anyone working in or learning about the field of nanotechnology would find this an excellent working handbook."

IEEE Electrical Insulation Magazine

"Outstandingly succeeds in its aim... It really is a magnificent volume and every scientific library and nanotechnology group should have a copy."

Materials World

"The integrity and authoritativeness... is guaranteed by an experienced editor and an international team of authors which have well summarized in their chapters information on fundamentals and applications."

Polymer News

List of Abbreviations
1 Introduction to Nanotechnology

Part A Nanostructures, Micro-/Nanofabrication and Materials

2 Nanomaterials Synthesis and Applications: Molecule-Based Devices
3 Introduction to Carbon Nanotubes
4 Nanowires
5 Template-Based Synthesis of Nanorod or Nanowire Arrays
6 Templated Self-Assembly of Particles
7 Three-Dimensional Nanostructure Fabrication by Focused Ion Beam Chemical Vapor Deposition
8 Introduction to Micro-/Nanofabrication
9 Nanoimprint Lithography-Patterning of Resists Using Molding
10 Stamping Techniques for Micro- and Nanofabrication
11 Material Aspects of Micro- and Nanoelectromechanical Systems

Part B MEMS/NEMS and BioMEMS/NEMS

12 MEMS/NEMS Devices and Applications
13 Next-Generation DNA Hybridization and Self-Assembly Nanofabrication Devices
14 Single-Walled Carbon Nanotube Sensor Concepts
15 Nanomechanical Cantilever Array Sensors
16 Biological Molecules in Therapeutic Nanodevices
17 G-Protein Coupled Receptors: Progress in Surface Display and Biosensor Technology
18 Microfluidic Devices and Their Applications to Lab-on-a-Chip
19 Centrifuge-Based Fluidic Platforms
20 Micro-/Nanodroplets in Microfluidic Devices

Part C Scanning-Probe Microscopy

21 Scanning Probe Microscopy-Principle of Operation, Instrumentation, and Probes
22 General and Special Probes in Scanning Microscopies
23 Noncontact Atomic Force Microscopy and Related Topics
24 Low-Temperature Scanning Probe Microscopy
25 Higher Harmonics and Time-Varying Forces in Dynamic Force Microscopy
26 Dynamic Modes of Atomic Force Microscopy
27 Molecular Recognition Force Microscopy: From Molecular Bonds to Complex Energy Landscapes

Part D Bio-/Nanotribology and Bio-/Nanomechanics

28 Nanotribology, Nanomechanics, and Materials Characterization
29 Surface Forces and Nanorheology of Molecularly Thin Films
30 Friction and Wear on the Atomic Scale
31 Computer Simulations of Nanometer-Scale Indentation and Friction
32 Force Measurements with Optical Tweezers
33 Scale Effect in Mechanical Properties and Tribology
34 Structural, Nanomechanical, and Nanotribological Characterization of Human Hair Using Atomic Force Microscopy and Nanoindentation
35 Cellular Nanomechanics
36 Optical Cell Manipulation
37 Mechanical Properties of Nanostructures

Part E Molecularly Thick Films for Lubrication

38 Nanotribology of Ultrathin and Hard Amorphous Carbon Films
39 Self-Assembled Monolayers for Nanotribology and Surface Protection
40 Nanoscale Boundary Lubrication Studies

Part F Biomimetics

41 Multifunctional Plant Surfaces and Smart Materials
42 Lotus Effect: Surfaces with Roughness-Induced Superhydrophobicity, Self-Cleaning, and Low Adhesion
43 Biological and Biologically Inspired Attachment Systems
44 Gecko Feet: Natural Hairy Attachment Systems for Smart Adhesion

Part G Industrial Applications

45 The *Millipede*-A Nanotechnology-Based AFM Data-Storage System
46 Nanorobotics

Part H Micro-/Nanodevice Reliability

47 MEMS/NEMS and BioMEMS/BioNEMS: Materials, Devices, and Biomimetics
48 Friction and Wear in Micro- and Nanomachines
49 Failure Mechanisms in MEMS/NEMS Devices
50 Mechanical Properties of Micromachined Structures
51 High-Volume Manufacturing and Field Stability of MEMS Products
52 Packaging and Reliability Issues in Micro-/Nanosystems

Part I Technological Convergence and Governing Nanotechnology

53 Governing Nanotechnology: Social, Ethical and Human Issues

Subject Index

使 用 说 明

1.《纳米技术手册》原版为一册，分为A～I部分。考虑到使用方便以及内容一致，影印版分为7册：第1册—Part A，第2册—Part B，第3册—Part C，第4册—Part D，第5册—Part E，第6册—Part F，第7册—Part G、H、I。

2.各册在页脚重新编排页码，该页码对应中文目录。保留了原书页眉及页码，其页码对应原书目录及主题索引。

3.各册均给出完整7册书的章目录。

4.作者及其联系方式、缩略语表各册均完整呈现。

5.主题索引安排在第7册。

6.目录等采用中英文对照形式给出，方便读者快速浏览。

材料科学与工程图书工作室

联系电话　0451-86412421
　　　　　0451-86414559
邮　　箱　yh_bj@yahoo.com.cn
　　　　　xuyaying81823@gmail.com
　　　　　zhxh6414559@yahoo.com.cn

Springer 手册精选系列

纳米技术手册

扫描探针显微镜

【第3册】

Springer
Handbook *of*
Nanotechnology

〔美〕Bharat Bhushan 主编

（第三版影印版）

哈尔滨工业大学出版社
HARBIN INSTITUTE OF TECHNOLOGY PRESS

黑版贸审字 08-2013-001号

Reprint from English language edition:
Springer Handbook of Nanotechnology
by Bharat Bhushan
Copyright © 2010 Springer Berlin Heidelberg
Springer Berlin Heidelberg is a part of Springer Science+Business Media
All Rights Reserved

This reprint has been authorized by Springer Science & Business Media for distribution in China Mainland only and not for export there from.

图书在版编目（CIP）数据

纳米技术手册：第3版. 3, 扫描探针显微镜 =Handbook of Nanotechnology. 3, Scanning–Probe Microscopy：英文 / (美) 布尚 (Bhushan,B.) 主编. —影印本. —哈尔滨：哈尔滨工业大学出版社,2013.1
（Springer手册精选系列）
ISBN 978-7-5603-3949-8

Ⅰ. ①纳… Ⅱ. ①布… Ⅲ. ①纳米技术 – 手册 – 英文②纳米技术 – 应用 – 扫描电子显微镜 – 手册 – 英文 Ⅳ. ①TB303-62②TN16-62

中国版本图书馆CIP数据核字(2013)第004304号

责任编辑　杨　桦　许雅莹　张秀华
出版发行　哈尔滨工业大学出版社
社　　址　哈尔滨市南岗区复华四道街10号　邮编 150006
传　　真　0451-86414749
网　　址　http://hitpress.hit.edu.cn
印　　刷　哈尔滨市石桥印务有限公司
开　　本　787mm×960mm　1/16　印张 16
版　　次　2013年1月第1版　2013年1月第1次印刷
书　　号　ISBN 978-7-5603-3949-8
定　　价　48.00元

（如因印刷质量问题影响阅读，我社负责调换）

Foreword by Neal Lane

In a January 2000 speech at the California Institute of Technology, former President W.J. Clinton talked about the exciting promise of *nanotechnology* and the importance of expanding research in nanoscale science and engineering and, more broadly, in the physical sciences. Later that month, he announced in his State of the Union Address an ambitious US$ 497 million federal, multiagency national nanotechnology initiative (NNI) in the fiscal year 2001 budget; and he made the NNI a top science and technology priority within a budget that emphasized increased investment in US scientific research. With strong bipartisan support in Congress, most of this request was appropriated, and the NNI was born. Often, federal budget initiatives only last a year or so. It is most encouraging that the NNI has remained a high priority of the G.W. Bush Administration and Congress, reflecting enormous progress in the field and continued strong interest and support by industry.

Nanotechnology is the ability to manipulate individual atoms and molecules to produce nanostructured materials and submicron objects that have applications in the real world. Nanotechnology involves the production and application of physical, chemical and biological systems at scales ranging from individual atoms or molecules to about 100 nm, as well as the integration of the resulting nanostructures into larger systems. Nanotechnology is likely to have a profound impact on our economy and society in the early 21st century, perhaps comparable to that of information technology or cellular and molecular biology. Science and engineering research in nanotechnology promises breakthroughs in areas such as materials and manufacturing, electronics, medicine and healthcare, energy and the environment, biotechnology, information technology and national security. Clinical trials are already underway for nanomaterials that offer the promise of cures for certain cancers. It is widely felt that nanotechnology will be the next industrial revolution.

Nanometer-scale features are built up from their elemental constituents. Micro- and nanosystems components are fabricated using batch-processing techniques that are compatible with integrated circuits and range in size from micro- to nanometers. Micro- and nanosystems include micro/nanoelectro-mechanical systems (MEMS/NEMS), micromechatronics, optoelectronics, microfluidics and systems integration. These systems can sense, control, and activate on the micro/nanoscale and can function individually or in arrays to generate effects on the macroscale. Due to the enabling nature of these systems and the significant impact they can have on both the commercial and defense applications, industry as well as the federal government have taken special interest in seeing growth nurtured in this field. Micro- and nanosystems are the next logical step in the *silicon revolution*.

Prof. Neal Lane
Malcolm Gillis University Professor,
Department of Physics and Astronomy,
Senior Fellow,
James A. Baker III Institute for Public Policy
Rice University
Houston, Texas

Served in the Clinton Administration as Assistant to the President for Science and Technology and Director of the White House Office of Science and Technology Policy (1998–2001) and, prior to that, as Director of the National Science Foundation (1993–1998). While at the White House, he was a key figure in the creation of the NNI.

The discovery of novel materials, processes, and phenomena at the nanoscale and the development of new experimental and theoretical techniques for research provide fresh opportunities for the development of innovative nanosystems and nanostructured materials. There is an increasing need for a multidisciplinary, systems-oriented approach to manufacturing micro/nanodevices which function reliably. This can only be achieved through the cross-fertilization of ideas from different disciplines and the systematic flow of information and people among research groups.

Nanotechnology is a broad, highly interdisciplinary, and still evolving field. Covering even the most important aspects of nanotechnology in a single book that reaches readers ranging from students to active researchers in academia and industry is an enormous challenge. To prepare such a wide-ranging book on nanotechnology, Prof. Bhushan has harnessed his own knowledge and experience, gained in several industries and universities, and has assembled internationally recognized authorities from four continents to write chapters covering a wide array of nanotechnology topics, including the latest advances. The authors come from both academia and industry. The topics include major advances in many fields where nanoscale science and engineering is being pursued and illustrate how the field of nanotechnology has continued to emerge and blossom. Given the accelerating pace of discovery and applications in nanotechnology, it is a challenge to capture it all in one volume. As in earlier editions, professor Bhushan does an admirable job.

Professor Bharat Bhushan's comprehensive book is intended to serve both as a textbook for university courses as well as a reference for researchers. The first and second editions were timely additions to the literature on nanotechnology and stimulated further interest in this important new field, while serving as invaluable resources to members of the international scientific and industrial community. The increasing demand for up-to-date information on this fast moving field led to this third edition. It is increasingly important that scientists and engineers, whatever their specialty, have a solid grounding in the fundamentals and potential applications of nanotechnology. This third edition addresses that need by giving particular attention to the widening audience of readers. It also includes a discussion of the social, ethical and political issues that tend to surround any emerging technology.

The editor and his team are to be warmly congratulated for bringing together this exclusive, timely, and useful nanotechnology handbook.

Foreword by James R. Heath

Nanotechnology has become an increasingly popular buzzword over the past five years or so, a trend that has been fueled by a global set of publicly funded nanotechnology initiatives. Even as researchers have been struggling to demonstrate some of the most fundamental and simple aspects of this field, the term nanotechnology has entered into the public consciousness through articles in the popular press and popular fiction. As a consequence, the expectations of the public are high for nanotechnology, even while the actual public definition of nanotechnology remains a bit fuzzy.

Why shouldn't those expectations be high? The late 1990s witnessed a major information technology (IT) revolution and a minor biotechnology revolution. The IT revolution impacted virtually every aspect of life in the western world. I am sitting on an airplane at 30 000 feet at the moment, working on my laptop, as are about half of the other passengers on this plane. The plane itself is riddled with computational and communications equipment. As soon as we land, many of us will pull out cell phones, others will check e-mail via wireless modem, some will do both. This picture would be the same if I was landing in Los Angeles, Beijing, or Capetown. I will probably never actually print this text, but will instead submit it electronically. All of this was unthinkable a dozen years ago. It is therefore no wonder that the public expects marvelous things to happen quickly. However, the science that laid the groundwork for the IT revolution dates back 60 years or more, with its origins in fundamental solid-state physics.

By contrast, the biotech revolution was relatively minor and, at least to date, not particularly effective. The major diseases that plagued mankind a quarter century ago are still here. In some third-world countries, the average lifespan of individuals has actually decreased from where it was a full century ago. While the costs of electronics technologies have plummeted, health care costs have continued to rise. The biotech revolution may have a profound impact, but the task at hand is substantially more difficult than what was required for the IT revolution. In effect, the IT revolution was based on the advanced engineering of two-dimensional digital circuits constructed from relatively simple components – extended solids. The biotech revolution is really dependent upon the ability to reverse engineer three-dimensional analog systems constructed from quite complex components – proteins. Given that the basic science behind biotech is substantially younger than the science that has supported IT, it is perhaps not surprising that the biotech revolution has not really been a proper revolution yet, and it likely needs at least another decade or so to come into fruition.

Prof. James R. Heath

Department of Chemistry
California Institute of Technology
Pasadena, California

Worked in the group of Nobel Laureate Richard E. Smalley at Rice University (1984–88) and co-invented Fullerene molecules which led to a revolution in Chemistry including the realization of nanotubes. The work on Fullerene molecules was cited for the 1996 Nobel Prize in Chemistry. Later he joined the University of California at Los Angeles (1994–2002), and co-founded and served as a Scientific Director of The California Nanosystems Institute.

Where does nanotechnology fit into this picture? In many ways, nanotechnology depends upon the ability to engineer two- and three-dimensional systems constructed from complex components such as macromolecules, biomolecules, nanostructured solids, etc. Furthermore, in terms of patents, publications, and other metrics that can be used to gauge the birth and evolution of a field, nanotech lags some 15–20 years behind biotech. Thus, now is the time that the fundamental science behind nanotechnology is being explored and developed. Nevertheless, progress with that science is moving forward at a dramatic pace. If the scientific community can keep up this pace and if the public sector will continue to support this science, then it is possible, and even perhaps likely, that in 20 years we may be speaking of the nanotech revolution.

The first edition of Springer Handbook of Nanotechnology was timely to assemble chapters in the broad field of nanotechnology. Given the fact that the second edition was in press one year after the publication of the first edition in April 2004, it is clear that the handbook has shown to be a valuable reference for experienced researchers as well as for a novice in the field. The third edition has one Part added and an expanded scope should have a wider appeal.

Preface to the 3rd Edition

On December 29, 1959 at the California Institute of Technology, Nobel Laureate Richard P. Feynman gave at talk at the Annual meeting of the American Physical Society that has become one of the 20th century classic science lectures, titled *There's Plenty of Room at the Bottom*. He presented a technological vision of extreme miniaturization in 1959, several years before the word *chip* became part of the lexicon. He talked about the problem of manipulating and controlling things on a small scale. Extrapolating from known physical laws, Feynman envisioned a technology using the ultimate toolbox of nature, building nanoobjects atom by atom or molecule by molecule. Since the 1980s, many inventions and discoveries in fabrication of nanoobjects have been testament to his vision. In recognition of this reality, National Science and Technology Council (NSTC) of the White House created the Interagency Working Group on Nanoscience, Engineering and Technology (IWGN) in 1998. In a January 2000 speech at the same institute, former President W.J. Clinton talked about the exciting promise of *nanotechnology* and the importance of expanding research in nanoscale science and technology, more broadly. Later that month, he announced in his State of the Union Address an ambitious US$ 497 million federal, multi-agency national nanotechnology initiative (NNI) in the fiscal year 2001 budget, and made the NNI a top science and technology priority. The objective of this initiative was to form a broad-based coalition in which the academe, the private sector, and local, state, and federal governments work together to push the envelop of nanoscience and nanoengineering to reap nanotechnology's potential social and economic benefits.

The funding in the US has continued to increase. In January 2003, the US senate introduced a bill to establish a National Nanotechnology Program. On December 3, 2003, President George W. Bush signed into law the 21st Century Nanotechnology Research and Development Act. The legislation put into law programs and activities supported by the National Nanotechnology Initiative. The bill gave nanotechnology a permanent home in the federal government and authorized US$ 3.7 billion to be spent in the four year period beginning in October 2005, for nanotechnology initiatives at five federal agencies. The funds would provide grants to researchers, coordinate R&D across five federal agencies (National Science Foundation (NSF), Department of Energy (DOE), NASA, National Institute of Standards and Technology (NIST), and Environmental Protection Agency (EPA)), establish interdisciplinary research centers, and accelerate technology transfer into the private sector. In addition, Department of Defense (DOD), Homeland Security, Agriculture and Justice as well as the National Institutes of Health (NIH) also fund large R&D activities. They currently account for more than one-third of the federal budget for nanotechnology.

European Union (EU) made nanosciences and nanotechnologies a priority in Sixth Framework Program (FP6) in 2002 for a period of 2003–2006. They had dedicated small funds in FP4 and FP5 before. FP6 was tailored to help better structure European research and to cope with the strategic objectives set out in Lisbon in 2000. Japan identified nanotechnology as one of its main research priorities in 2001. The funding levels increases sharply from US$ 400 million in 2001 to around US$ 950 million in 2004. In 2003, South Korea embarked upon a ten-year program with around US$ 2 billion of public funding, and Taiwan has committed around US$ 600 million of public funding over six years. Singapore and China are also investing on a large scale. Russia is well funded as well.

Nanotechnology literally means any technology done on a nanoscale that has applications in the real world. Nanotechnology encompasses production and application of physical, chemical and biological systems at scales, ranging from individual atoms or molecules to submicron dimensions, as well as the integration of the resulting nanostructures into larger systems. Nanotechnology is likely to have a profound impact on our economy and society in the early 21st century, comparable to that of semiconductor technology, information technology, or cellular and molecular biology. Science and technology research in nanotechnology promises breakthroughs in areas such as materials and manufacturing, nanoelectronics, medicine and healthcare, energy, biotechnology, information technology and national security. It is widely felt that nanotechnology will be the next industrial revolution.

There is an increasing need for a multidisciplinary, system-oriented approach to design and manufacturing of micro/nanodevices which function reliably. This can only be achieved through the cross-fertilization of ideas from different disciplines and the systematic flow of information and people among research groups. Reliability is a critical technology for many micro- and nanosystems and nanostructured materials. A broad based handbook was needed, and the first edition of Springer Handbook of Nanotechnology was published in April 2004. It presented an overview of nanomaterial synthesis, micro/nanofabrication, micro- and nanocomponents and systems, scanning probe microscopy, reliability issues (including nanotribology and nanomechanics) for nanotechnology, and industrial applications. When the handbook went for sale in Europe, it was sold out in ten days. Reviews on the handbook were very flattering.

Given the explosive growth in nanoscience and nanotechnology, the publisher and the editor decided to develop a second edition after merely six months of publication of the first edition. The second edition (2007) came out in December 2006. The publisher and the editor again decided to develop a third edition after six month of publication of the second edition. This edition of the handbook integrates the knowledge from nanostructures, fabrication, materials science, devices, and reliability point of view. It covers various industrial applications. It also addresses social, ethical, and political issues. Given the significant interest in biomedical applications, and biomimetics a number of additional chapters in this arena have been added. The third edition consists of 53 chapters (new 10, revised 28, and as is 15). The chapters have been written by 139 internationally recognized experts in the field, from academia, national research labs, and industry, and from all over the world.

This handbook is intended for three types of readers: graduate students of nanotechnology, researchers in academia and industry who are active or intend to become active in this field, and practicing engineers and scientists who have encountered a problem and hope to solve it as expeditiously as possible. The handbook should serve as an excellent text for one or two semester graduate courses in nanotechnology in mechanical engineering, materials science, applied physics, or applied chemistry.

We embarked on the development of third edition in June 2007, and we worked very hard to get all the chapters to the publisher in a record time of about 12 months. I wish to sincerely thank the authors for offering to write comprehensive chapters on a tight schedule. This is generally an added responsibility in the hectic work schedules of researchers today. I depended on a large number of reviewers who provided critical reviews. I would like to thank Dr. Phillip J. Bond, Chief of Staff and Under Secretary for Technology, US Department of Commerce, Washington, D.C. for suggestions for chapters as well as authors in the handbook. Last but not the least, I would like to thank my secretary Caterina Runyon-Spears for various administrative duties and her tireless efforts are highly appreciated.

I hope that this handbook will stimulate further interest in this important new field, and the readers of this handbook will find it useful.

February 2010   Bharat Bhushan
Editor

Preface to the 2nd Edition

On 29 December 1959 at the California Institute of Technology, Nobel Laureate Richard P. Feynman gave at talk at the Annual meeting of the American Physical Society that has become one of the 20th century classic science lectures, titled "There's Plenty of Room at the Bottom." He presented a technological vision of extreme miniaturization in 1959, several years before the word "chip" became part of the lexicon. He talked about the problem of manipulating and controlling things on a small scale. Extrapolating from known physical laws, Feynman envisioned a technology using the ultimate toolbox of nature, building nanoobjects atom by atom or molecule by molecule. Since the 1980s, many inventions and discoveries in the fabrication of nanoobjects have been a testament to his vision. In recognition of this reality, the National Science and Technology Council (NSTC) of the White House created the Interagency Working Group on Nanoscience, Engineering and Technology (IWGN) in 1998. In a January 2000 speech at the same institute, former President W. J. Clinton talked about the exciting promise of "nanotechnology" and the importance of expanding research in nanoscale science and, more broadly, technology. Later that month, he announced in his State of the Union Address an ambitious $497 million federal, multiagency national nanotechnology initiative (NNI) in the fiscal year 2001 budget, and made the NNI a top science and technology priority. The objective of this initiative was to form a broad-based coalition in which the academe, the private sector, and local, state, and federal governments work together to push the envelope of nanoscience and nanoengineering to reap nanotechnology's potential social and economic benefits.

The funding in the U.S. has continued to increase. In January 2003, the U. S. senate introduced a bill to establish a National Nanotechnology Program. On 3 December 2003, President George W. Bush signed into law the 21st Century Nanotechnology Research and Development Act. The legislation put into law programs and activities supported by the National Nanotechnology Initiative. The bill gave nanotechnology a permanent home in the federal government and authorized $3.7 billion to be spent in the four year period beginning in October 2005, for nanotechnology initiatives at five federal agencies. The funds would provide grants to researchers, coordinate R&D across five federal agencies (National Science Foundation (NSF), Department of Energy (DOE), NASA, National Institute of Standards and Technology (NIST), and Environmental Protection Agency (EPA)), establish interdisciplinary research centers, and accelerate technology transfer into the private sector. In addition, Department of Defense (DOD), Homeland Security, Agriculture and Justice as well as the National Institutes of Health (NIH) would also fund large R&D activities. They currently account for more than one-third of the federal budget for nanotechnology.

The European Union made nanosciences and nanotechnologies a priority in the Sixth Framework Program (FP6) in 2002 for the period of 2003-2006. They had dedicated small funds in FP4 and FP5 before. FP6 was tailored to help better structure European research and to cope with the strategic objectives set out in Lisbon in 2000. Japan identified nanotechnology as one of its main research priorities in 2001. The funding levels increased sharply from $400 million in 2001 to around $950 million in 2004. In 2003, South Korea embarked upon a ten-year program with around $2 billion of public funding, and Taiwan has committed around $600 million of public funding over six years. Singapore and China are also investing on a large scale. Russia is well funded as well.

Nanotechnology literally means any technology done on a nanoscale that has applications in the real world. Nanotechnology encompasses production and application of physical, chemical and biological systems at scales, ranging from individual atoms or molecules to submicron dimensions, as well as the integration of the resulting nanostructures into larger systems. Nanotechnology is likely to have a profound impact on our economy and society in the early 21st century, comparable to that of semiconductor technology, information technology, or cellular and molecular biology. Science and technology research in nanotechnology promises breakthroughs in areas such as materials and manufacturing, nanoelectronics, medicine and healthcare, energy, biotechnology, information technology and national security. It is widely felt that nanotechnology will be the next industrial revolution.

There is an increasing need for a multidisciplinary, system-oriented approach to design and manufacturing of micro/nanodevices that function reliably. This can only be achieved through the cross-fertilization of ideas from different disciplines and the systematic flow of information and people among research groups. Reliability is a critical technology for many micro- and nanosystems and nanostructured materials. A broad-based handbook was needed, and thus the first edition of Springer Handbook of Nanotechnology was published in April 2004. It presented an overview of nanomaterial synthesis, micro/nanofabrication, micro- and nanocomponents and systems, scanning probe microscopy, reliability issues (including nanotribology and nanomechanics) for nanotechnology, and industrial applications. When the handbook went for sale in Europe, it sold out in ten days. Reviews on the handbook were very flattering.

Given the explosive growth in nanoscience and nanotechnology, the publisher and the editor decided to develop a second edition merely six months after publication of the first edition. This edition of the handbook integrates the knowledge from the nanostructure, fabrication, materials science, devices, and reliability point of view. It covers various industrial applications. It also addresses social, ethical, and political issues. Given the significant interest in biomedical applications, a number of chapters in this arena have been added. The second edition consists of 59 chapters (new: 23; revised: 27; unchanged: 9). The chapters have been written by 154 internationally recognized experts in the field, from academia, national research labs, and industry.

This book is intended for three types of readers: graduate students of nanotechnology, researchers in academia and industry who are active or intend to become active in this field, and practicing engineers and scientists who have encountered a problem and hope to solve it as expeditiously as possible. The handbook should serve as an excellent text for one or two semester graduate courses in nanotechnology in mechanical engineering, materials science, applied physics, or applied chemistry.

We embarked on the development of the second edition in October 2004, and we worked very hard to get all the chapters to the publisher in a record time of about 7 months. I wish to sincerely thank the authors for offering to write comprehensive chapters on a tight schedule. This is generally an added responsibility to the hectic work schedules of researchers today. I depended on a large number of reviewers who provided critical reviews. I would like to thank Dr. Phillip J. Bond, Chief of Staff and Under Secretary for Technology, US Department of Commerce, Washington, D.C. for chapter suggestions as well as authors in the handbook. I would also like to thank my colleague, Dr. Zhenhua Tao, whose efforts during the preparation of this handbook were very useful. Last but not the least, I would like to thank my secretary Caterina Runyon-Spears for various administrative duties; her tireless efforts are highly appreciated.

I hope that this handbook will stimulate further interest in this important new field, and the readers of this handbook will find it useful.

May 2005   Bharat Bhushan
                                                                                         Editor

Preface to the 1st Edition

On December 29, 1959 at the California Institute of Technology, Nobel Laureate Richard P. Feynman gave a talk at the Annual meeting of the American Physical Society that has become one classic science lecture of the 20th century, titled "There's Plenty of Room at the Bottom." He presented a technological vision of extreme miniaturization in 1959, several years before the word "chip" became part of the lexicon. He talked about the problem of manipulating and controlling things on a small scale. Extrapolating from known physical laws, Feynman envisioned a technology using the ultimate toolbox of nature, building nanoobjects atom by atom or molecule by molecule. Since the 1980s, many inventions and discoveries in fabrication of nanoobjects have been a testament to his vision. In recognition of this reality, in a January 2000 speech at the same institute, former President W. J. Clinton talked about the exciting promise of "nanotechnology" and the importance of expanding research in nanoscale science and engineering. Later that month, he announced in his State of the Union Address an ambitious $ 497 million federal, multi-agency national nanotechnology initiative (NNI) in the fiscal year 2001 budget, and made the NNI a top science and technology priority. Nanotechnology literally means any technology done on a nanoscale that has applications in the real world. Nanotechnology encompasses production and application of physical, chemical and biological systems at size scales, ranging from individual atoms or molecules to submicron dimensions as well as the integration of the resulting nanostructures into larger systems. Nanofabrication methods include the manipulation or self-assembly of individual atoms, molecules, or molecular structures to produce nanostructured materials and sub-micron devices. Micro- and nanosystems components are fabricated using top-down lithographic and nonlithographic fabrication techniques. Nanotechnology will have a profound impact on our economy and society in the early 21st century, comparable to that of semiconductor technology, information technology, or advances in cellular and molecular biology. The research and development in nanotechnology will lead to potential breakthroughs in areas such as materials and manufacturing, nanoelectronics, medicine and healthcare, energy, biotechnology, information technology and national security. It is widely felt that nanotechnology will lead to the next industrial revolution.

Reliability is a critical technology for many micro- and nanosystems and nanostructured materials. No book exists on this emerging field. A broad based handbook is needed. The purpose of this handbook is to present an overview of nanomaterial synthesis, micro/nanofabrication, micro- and nanocomponents and systems, reliability issues (including nanotribology and nanomechanics) for nanotechnology, and industrial applications. The chapters have been written by internationally recognized experts in the field, from academia, national research labs and industry from all over the world.

The handbook integrates knowledge from the fabrication, mechanics, materials science and reliability points of view. This book is intended for three types of readers: graduate students of nanotechnology, researchers in academia and industry who are active or intend to become active in this field, and practicing engineers and scientists who have encountered a problem and hope to solve it as expeditiously as possible. The handbook should serve as an excellent text for one or two semester graduate courses in nanotechnology in mechanical engineering, materials science, applied physics, or applied chemistry.

We embarked on this project in February 2002, and we worked very hard to get all the chapters to the publisher in a record time of about 1 year. I wish to sincerely thank the authors for offering to write comprehensive chapters on a tight schedule. This is generally an added responsibility in the hectic work schedules of researchers today. I depended on a large number of reviewers who provided critical reviews. I would like to thank Dr. Phillip J. Bond, Chief of Staff and Under Secretary for Technology, US Department of Commerce, Washington, D.C. for suggestions for chapters as well as authors in the handbook. I would also like to thank my colleague, Dr. Huiwen Liu, whose efforts during the preparation of this handbook were very useful.

I hope that this handbook will stimulate further interest in this important new field, and the readers of this handbook will find it useful.

September 2003

Bharat Bhushan
Editor

Editors Vita

Dr. Bharat Bhushan received an M.S. in mechanical engineering from the Massachusetts Institute of Technology in 1971, an M.S. in mechanics and a Ph.D. in mechanical engineering from the University of Colorado at Boulder in 1973 and 1976, respectively, an MBA from Rensselaer Polytechnic Institute at Troy, NY in 1980, Doctor Technicae from the University of Trondheim at Trondheim, Norway in 1990, a Doctor of Technical Sciences from the Warsaw University of Technology at Warsaw, Poland in 1996, and Doctor Honouris Causa from the National Academy of Sciences at Gomel, Belarus in 2000. He is a registered professional engineer. He is presently an Ohio Eminent Scholar and The Howard D. Winbigler Professor in the College of Engineering, and the Director of the Nanoprobe Laboratory for Bio- and Nanotechnology and Biomimetics (NLB²) at the Ohio State University, Columbus, Ohio. His research interests include fundamental studies with a focus on scanning probe techniques in the interdisciplinary areas of bio/nanotribology, bio/nanomechanics and bio/nanomaterials characterization, and applications to bio/nanotechnology and biomimetics. He is an internationally recognized expert of bio/nanotribology and bio/nanomechanics using scanning probe microscopy, and is one of the most prolific authors. He is considered by some a pioneer of the tribology and mechanics of magnetic storage devices. He has authored 6 scientific books, more than 90 handbook chapters, more than 700 scientific papers (h factor – 45+; ISI Highly Cited in Materials Science, since 2007), and more than 60 technical reports, edited more than 45 books, and holds 17 US and foreign patents. He is co-editor of Springer NanoScience and Technology Series and co-editor of Microsystem Technologies. He has given more than 400 invited presentations on six continents and more than 140 keynote/plenary addresses at major international conferences.

Dr. Bhushan is an accomplished organizer. He organized the first symposium on Tribology and Mechanics of Magnetic Storage Systems in 1984 and the first international symposium on Advances in Information Storage Systems in 1990, both of which are now held annually. He is the founder of an ASME Information Storage and Processing Systems Division founded in 1993 and served as the founding chair during 1993–1998. His biography has been listed in over two dozen Who's Who books including Who's Who in the World and has received more than two dozen awards for his contributions to science and technology from professional societies, industry, and US government agencies. He is also the recipient of various international fellowships including the Alexander von Humboldt Research Prize for Senior Scientists, Max Planck Foundation Research Award for Outstanding Foreign Scientists, and the Fulbright Senior Scholar Award. He is a foreign member of the International Academy of Engineering (Russia), Byelorussian Academy of Engineering and Technology and the Academy of Triboengineering of Ukraine, an honorary member of the Society of Tribologists of Belarus, a fellow of ASME, IEEE, STLE, and the New York Academy of Sciences, and a member of ASEE, Sigma Xi and Tau Beta Pi.

Dr. Bhushan has previously worked for the R&D Division of Mechanical Technology Inc., Latham, NY; the Technology Services Division of SKF Industries Inc., King of Prussia, PA; the General Products Division Laboratory of IBM Corporation, Tucson, AZ; and the Almaden Research Center of IBM Corporation, San Jose, CA. He has held visiting professor appointments at University of California at Berkeley, University of Cambridge, UK, Technical University Vienna, Austria, University of Paris, Orsay, ETH Zurich and EPFL Lausanne.

List of Authors

Chong H. Ahn
University of Cincinnati
Department of Electrical
and Computer Engineering
Cincinnati, OH 45221, USA
e-mail: *chong.ahn@uc.edu*

Boris Anczykowski
nanoAnalytics GmbH
Münster, Germany
e-mail: *anczykowski@nanoanalytics.com*

W. Robert Ashurst
Auburn University
Department of Chemical Engineering
Auburn, AL 36849, USA
e-mail: *ashurst@auburn.edu*

Massood Z. Atashbar
Western Michigan University
Department of Electrical
and Computer Engineering
Kalamazoo, MI 49008-5329, USA
e-mail: *massood.atashbar@wmich.edu*

Wolfgang Bacsa
University of Toulouse III (Paul Sabatier)
Laboratoire de Physique des Solides (LPST),
UMR 5477 CNRS
Toulouse, France
e-mail: *bacsa@ramansco.ups-tlse.fr;
bacsa@lpst.ups-tlse.fr*

Kelly Bailey
University of Adelaide
CSIRO Human Nutrition
Adelaide SA 5005, Australia
e-mail: *kelly.bailey@csiro.au*

William Sims Bainbridge
National Science Foundation
Division of Information, Science and Engineering
Arlington, VA, USA
e-mail: *wsbainbridge@yahoo.com*

Antonio Baldi
Institut de Microelectronica de Barcelona (IMB)
Centro National Microelectrónica (CNM-CSIC)
Barcelona, Spain
e-mail: *antoni.baldi@cnm.es*

Wilhelm Barthlott
University of Bonn
Nees Institute for Biodiversity of Plants
Meckenheimer Allee 170
53115 Bonn, Germany
e-mail: *barthlott@uni-bonn.de*

Roland Bennewitz
INM – Leibniz Institute for New Materials
66123 Saarbrücken, Germany
e-mail: *roland.bennewitz@inm-gmbh.de*

Bharat Bhushan
Ohio State University
Nanoprobe Laboratory for Bio- and
Nanotechnology and Biomimetics (NLB^2)
201 W. 19th Avenue
Columbus, OH 43210-1142, USA
e-mail: *bhushan.2@osu.edu*

Gerd K. Binnig
Definiens AG
Trappentreustr. 1
80339 Munich, Germany
e-mail: *gbinnig@definiens.com*

Marcie R. Black
Bandgap Engineering Inc.
1344 Main St.
Waltham, MA 02451, USA
e-mail: *marcie@alum.mit.edu;
marcie@bandgap.com*

Donald W. Brenner
Department of Materials Science and Engineering
Raleigh, NC, USA
e-mail: *brenner@ncsu.edu*

Jean-Marc Broto
Institut National des Sciences Appliquées
of Toulouse
Laboratoire National
des Champs Magnétiques Pulsés (LNCMP)
Toulouse, France
e-mail: *broto@lncmp.fr*

Guozhong Cao
University of Washington
Dept. of Materials Science and Engineering
302M Roberts Hall
Seattle, WA 98195-2120, USA
e-mail: *gzcao@u.washington.edu*

Edin (I-Chen) Chen
National Central University
Institute of Materials Science and Engineering
Department of Mechanical Engineering
Chung-Li, 320, Taiwan
e-mail: *ichen@ncu.edu.tw*

Yu-Ting Cheng
National Chiao Tung University
Department of Electronics Engineering
& Institute of Electronics
1001, Ta-Hsueh Rd.
Hsinchu, 300, Taiwan, R.O.C.
e-mail: *ytcheng@mail.nctu.edu.tw*

Giovanni Cherubini
IBM Zurich Research Laboratory
Tape Technologies
8803 Rüschlikon, Switzerland
e-mail: *cbi@zurich.ibm.com*

Mu Chiao
Department of Mechanical Engineering
6250 Applied Science Lane
Vancouver, BC V6T 1Z4, Canada
e-mail: *muchiao@mech.ubc.ca*

Jin-Woo Choi
Louisiana State University
Department of Electrical
and Computer Engineering
Baton Rouge, LA 70803, USA
e-mail: *choi@ece.lsu.edu*

Tamara H. Cooper
University of Adelaide
CSIRO Human Nutrition
Adelaide SA 5005, Australia
e-mail: *tamara.cooper@csiro.au*

Alex D. Corwin
GE Global Research
1 Research Circle
Niskayuna, NY 12309, USA
e-mail: *corwin@ge.com*

Maarten P. de Boer
Carnegie Mellon University
Department of Mechanical Engineering
5000 Forbes Avenue
Pittsburgh, PA 15213, USA
e-mail: *mpdebo@andrew.cmu.edu*

Dietrich Dehlinger
Lawrence Livermore National Laboratory
Engineering
Livermore, CA 94551, USA
e-mail: *dehlinger1@llnl.gov*

Frank W. DelRio
National Institute of Standards and Technology
100 Bureau Drive, Stop 8520
Gaithersburg, MD 20899-8520, USA
e-mail: *frank.delrio@nist.gov*

Michel Despont
IBM Zurich Research Laboratory
Micro- and Nanofabrication
8803 Rüschlikon, Switzerland
e-mail: *dpt@zurich.ibm.com*

Lixin Dong
Michigan State University
Electrical and Computer Engineering
2120 Engineering Building
East Lansing, MI 48824-1226, USA
e-mail: *ldong@egr.msu.edu*

Gene Dresselhaus
Massachusetts Institute of Technology
Francis Bitter Magnet Laboratory
Cambridge, MA 02139, USA
e-mail: *gene@mgm.mit.edu*

Mildred S. Dresselhaus
Massachusetts Institute of Technology
Department of Electrical Engineering
and Computer Science
Department of Physics
Cambridge, MA, USA
e-mail: *millie@mgm.mit.edu*

Urs T. Dürig
IBM Zurich Research Laboratory
Micro-/Nanofabrication
8803 Rüschlikon, Switzerland
e-mail: *drg@zurich.ibm.com*

Andreas Ebner
Johannes Kepler University Linz
Institute for Biophysics
Altenberger Str. 69
4040 Linz, Austria
e-mail: *andreas.ebner@jku.at*

Evangelos Eleftheriou
IBM Zurich Research Laboratory
8803 Rüschlikon, Switzerland
e-mail: *ele@zurich.ibm.com*

Emmanuel Flahaut
Université Paul Sabatier
CIRIMAT, Centre Interuniversitaire de Recherche
et d'Ingénierie des Matériaux, UMR 5085 CNRS
118 Route de Narbonne
31062 Toulouse, France
e-mail: *flahaut@chimie.ups-tlse.fr*

Anatol Fritsch
University of Leipzig
Institute of Experimental Physics I
Division of Soft Matter Physics
Linnéstr. 5
04103 Leipzig, Germany
e-mail: *anatol.fritsch@uni-leipzig.de*

Harald Fuchs
Universität Münster
Physikalisches Institut
Münster, Germany
e-mail: *fuchsh@uni-muenster.de*

Christoph Gerber
University of Basel
Institute of Physics
National Competence Center for Research
in Nanoscale Science (NCCR) Basel
Klingelbergstr. 82
4056 Basel, Switzerland
e-mail: *christoph.gerber@unibas.ch*

Franz J. Giessibl
Universität Regensburg
Institute of Experimental and Applied Physics
Universitätsstr. 31
93053 Regensburg, Germany
e-mail: *franz.giessibl@physik.uni-regensburg.de*

Enrico Gnecco
University of Basel
National Center of Competence in Research
Department of Physics
Klingelbergstr. 82
4056 Basel, Switzerland
e-mail: *enrico.gnecco@unibas.ch*

Stanislav N. Gorb
Max Planck Institut für Metallforschung
Evolutionary Biomaterials Group
Heisenbergstr. 3
70569 Stuttgart, Germany
e-mail: *s.gorb@mf.mpg.de*

Hermann Gruber
University of Linz
Institute of Biophysics
Altenberger Str. 69
4040 Linz, Austria
e-mail: *hermann.gruber@jku.at*

Jason Hafner
Rice University
Department of Physics and Astronomy
Houston, TX 77251, USA
e-mail: *hafner@rice.edu*

Judith A. Harrison
U.S. Naval Academy
Chemistry Department
572 Holloway Road
Annapolis, MD 21402-5026, USA
e-mail: *jah@usna.edu*

Martin Hegner
CRANN – The Naughton Institute
Trinity College, University of Dublin
School of Physics
Dublin, 2, Ireland
e-mail: *martin.hegner@tcd.ie*

Thomas Helbling
ETH Zurich
Micro and Nanosystems
Department of Mechanical
and Process Engineering
8092 Zurich, Switzerland
e-mail: *thomas.helbling@micro.mavt.ethz.ch*

Michael J. Heller
University of California San Diego
Department of Bioengineering
Dept. of Electrical and Computer Engineering
La Jolla, CA, USA
e-mail: *mjheller@ucsd.edu*

Seong-Jun Heo
Lam Research Corp.
4650 Cushing Parkway
Fremont, CA 94538, USA
e-mail: *seongjun.heo@lamrc.com*

Christofer Hierold
ETH Zurich
Micro and Nanosystems
Department of Mechanical
and Process Engineering
8092 Zurich, Switzerland
e-mail: *christofer.hierold@micro.mavt.ethz.ch*

Peter Hinterdorfer
University of Linz
Institute for Biophysics
Altenberger Str. 69
4040 Linz, Austria
e-mail: *peter.hinterdorfer@jku.at*

Dalibor Hodko
Nanogen, Inc.
10498 Pacific Center Court
San Diego, CA 92121, USA
e-mail: *dhodko@nanogen.com*

Hendrik Hölscher
Forschungszentrum Karlsruhe
Institute of Microstructure Technology
Linnéstr. 5
76021 Karlsruhe, Germany
e-mail: *hendrik.hoelscher@imt.fzk.de*

Hirotaka Hosoi
Hokkaido University
Creative Research Initiative Sousei
Kita 21, Nishi 10, Kita-ku
Sapporo, Japan
e-mail: *hosoi@cris.hokudai.ac.jp*

Katrin Hübner
Staatliche Fachoberschule Neu-Ulm
89231 Neu-Ulm, Germany
e-mail: *katrin.huebner1@web.de*

Douglas L. Irving
North Carolina State University
Materials Science and Engineering
Raleigh, NC 27695-7907, USA
e-mail: *doug_irving@ncsu.edu*

Jacob N. Israelachvili
University of California
Department of Chemical Engineering
and Materials Department
Santa Barbara, CA 93106-5080, USA
e-mail: *jacob@engineering.ucsb.edu*

Guangyao Jia
University of California, Irvine
Department of Mechanical
and Aerospace Engineering
Irvine, CA, USA
e-mail: *gjia@uci.edu*

Sungho Jin
University of California, San Diego
Department of Mechanical
and Aerospace Engineering
9500 Gilman Drive
La Jolla, CA 92093-0411, USA
e-mail: *jin@ucsd.edu*

Anne Jourdain
Interuniversity Microelectronics Center (IMEC)
Leuven, Belgium
e-mail: *jourdain@imec.be*

Yong Chae Jung
Samsung Electronics C., Ltd.
Senior Engineer Process Development Team
San #16 Banwol-Dong, Hwasung-City
Gyeonggi-Do 445-701, Korea
e-mail: *yc423.jung@samsung.com*

Harold Kahn
Case Western Reserve University
Department of Materials Science and Engineering
Cleveland, OH , USA
e-mail: *kahn@cwru.edu*

Roger Kamm
Massachusetts Institute of Technology
Department of Biological Engineering
77 Massachusetts Avenue
Cambridge, MA 02139, USA
e-mail: *rdkamm@mit.edu*

Ruti Kapon
Weizmann Institute of Science
Department of Biological Chemistry
Rehovot 76100, Israel
e-mail: *ruti.kapon@weizmann.ac.il*

Josef Käs
University of Leipzig
Institute of Experimental Physics I
Division of Soft Matter Physics
Linnéstr. 5
04103 Leipzig, Germany
e-mail: *jkaes@physik.uni-leipzig.de*

Horacio Kido
University of California at Irvine
Mechanical and Aerospace Engineering
Irvine, CA, USA
e-mail: *hkido@uci.edu*

Tobias Kießling
University of Leipzig
Institute of Experimental Physics I
Division of Soft Matter Physics
Linnéstr. 5
04103 Leipzig, Germany
e-mail: *Tobias.Kiessling@uni-leipzig.de*

Jitae Kim
University of California at Irvine
Department of Mechanical
and Aerospace Engineering
Irvine, CA, USA
e-mail: *jitaekim@uci.edu*

Jongbaeg Kim
Yonsei University
School of Mechanical Engineering
1st Engineering Bldg.
Seoul, 120-749, South Korea
e-mail: *kimjb@yonsei.ac.kr*

Nahui Kim
Samsung Advanced Institute of Technology
Research and Development
Seoul, South Korea
e-mail: *nahui.kim@samsung.com*

Kerstin Koch
Rhine-Waal University of Applied Science
Department of Life Science, Biology
and Nanobiotechnology
Landwehr 4
47533 Kleve, Germany
e-mail: *kerstin.koch@hochschule.rhein-waal.de*

Jing Kong
Massachusetts Institute of Technology
Department of Electrical Engineering
and Computer Science
Cambridge, MA, USA
e-mail: *jingkong@mit.edu*

Tobias Kraus
Leibniz-Institut für Neue Materialien gGmbH
Campus D2 2
66123 Saarbrücken, Germany
e-mail: *tobias.kraus@inm-gmbh.de*

Anders Kristensen
Technical University of Denmark
DTU Nanotech
2800 Kongens Lyngby, Denmark
e-mail: *anders.kristensen@nanotech.dtu.dk*

Ratnesh Lal
University of Chicago
Center for Nanomedicine
5841 S Maryland Av
Chicago, IL 60637, USA
e-mail: *rlal@uchicago.edu*

Jan Lammerding
Harvard Medical School
Brigham and Women's Hospital
65 Landsdowne St
Cambridge, MA 02139, USA
e-mail: *jlammerding@rics.bwh.harvard.edu*

Hans Peter Lang
University of Basel
Institute of Physics, National Competence Center
for Research in Nanoscale Science (NCCR) Basel
Klingelbergstr. 82
4056 Basel, Switzerland
e-mail: *hans-peter.lang@unibas.ch*

Carmen LaTorre
Owens Corning Science and Technology
Roofing and Asphalt
2790 Columbus Road
Granville, OH 43023, USA
e-mail: *carmen.latorre@owenscorning.com*

Christophe Laurent
Université Paul Sabatier
CIRIMAT UMR 5085 CNRS
118 Route de Narbonne
31062 Toulouse, France
e-mail: *laurent@chimie.ups-tlse.fr*

Abraham P. Lee
University of California Irvine
Department of Biomedical Engineering
Department of Mechanical
and Aerospace Engineering
Irvine, CA 92697, USA
e-mail: *aplee@uci.edu*

Stephen C. Lee
Ohio State University
Biomedical Engineering Center
Columbus, OH 43210, USA
e-mail: *lee@bme.ohio-state.edu*

Wayne R. Leifert
Adelaide Business Centre
CSIRO Human Nutrition
Adelaide SA 5000, Australia
e-mail: *wayne.leifert@csiro.au*

Liwei Lin
UC Berkeley
Mechanical Engineering Department
5126 Etcheverry
Berkeley, CA 94720-1740, USA
e-mail: *lwlin@me.berkeley.edu*

Yu-Ming Lin
IBM T.J. Watson Research Center
Nanometer Scale Science & Technology
1101 Kitchawan Road
Yorktown Heigths, NY 10598, USA
e-mail: *yming@us.ibm.com*

Marc J. Madou
University of California Irvine
Department of Mechanical and Aerospace
and Biomedical Engineering
Irvine, CA, USA
e-mail: *mmadou@uci.edu*

Othmar Marti
Ulm University
Institute of Experimental Physics
Albert-Einstein-Allee 11
89069 Ulm, Germany
e-mail: *othmar.marti@uni-ulm.de*

Jack Martin
66 Summer Street
Foxborough, MA 02035, USA
e-mail: *jack.martin@alumni.tufts.edu*

Shinji Matsui
University of Hyogo
Laboratory of Advanced Science
and Technology for Industry
Hyogo, Japan
e-mail: *matsui@lasti.u-hyogo.ac.jp*

Mehran Mehregany
Case Western Reserve University
Department of Electrical Engineering
and Computer Science
Cleveland, OH 44106, USA
e-mail: *mxm31@cwru.edu*

Etienne Menard
Semprius, Inc.
4915 Prospectus Dr.
Durham, NC 27713, USA
e-mail: *etienne.menard@semprius.com*

Ernst Meyer
University of Basel
Institute of Physics
Basel, Switzerland
e-mail: *ernst.meyer@unibas.ch*

Robert Modliński
Baolab Microsystems
Terrassa 08220, Spain
e-mail: *rmodlinski@gmx.com*

Mohammad Mofrad
University of California, Berkeley
Department of Bioengineering
Berkeley, CA 94720, USA
e-mail: *mofrad@berkeley.edu*

Marc Monthioux
CEMES – UPR A-8011 CNRS
Carbones et Matériaux Carbonés,
Carbons and Carbon-Containing Materials
29 Rue Jeanne Marvig
31055 Toulouse 4, France
e-mail: *monthiou@cemes.fr*

Markus Morgenstern
RWTH Aachen University
II. Institute of Physics B and JARA-FIT
52056 Aachen, Germany
e-mail: *mmorgens@physik.rwth-aachen.de*

Seizo Morita
Osaka University
Department of Electronic Engineering
Suita-City
Osaka, Japan
e-mail: *smorita@ele.eng.osaka-u.ac.jp*

Koichi Mukasa
Hokkaido University
Nanoelectronics Laboratory
Sapporo, Japan
e-mail: *mukasa@nano.eng.hokudai.ac.jp*

Bradley J. Nelson
Swiss Federal Institute of Technology (ETH)
Institute of Robotics and Intelligent Systems
8092 Zurich, Switzerland
e-mail: *bnelson@ethz.ch*

Michael Nosonovsky
University of Wisconsin-Milwaukee
Department of Mechanical Engineering
3200 N. Cramer St.
Milwaukee, WI 53211, USA
e-mail: *nosonovs@uwm.edu*

Hiroshi Onishi
Kanagawa Academy of Science and Technology
Surface Chemistry Laboratory
Kanagawa, Japan
e-mail: *oni@net.ksp.or.jp*

Alain Peigney
Centre Inter-universitaire de Recherche
sur l'Industrialisation des Matériaux (CIRIMAT)
Toulouse 4, France
e-mail: *peigney@chimie.ups-tlse.fr*

Oliver Pfeiffer
Individual Computing GmbH
Ingelsteinweg 2d
4143 Dornach, Switzerland
e-mail: *oliver.pfeiffer@gmail.com*

Haralampos Pozidis
IBM Zurich Research Laboratory
Storage Technologies
Rüschlikon, Switzerland
e-mail: *hap@zurich.ibm.com*

Robert Puers
Katholieke Universiteit Leuven
ESAT/MICAS
Leuven, Belgium
e-mail: *bob.puers@esat.kuleuven.ac.be*

Calvin F. Quate
Stanford University
Edward L. Ginzton Laboratory
450 Via Palou
Stanford, CA 94305-4088, USA
e-mail: *quate@stanford.edu*

Oded Rabin
University of Maryland
Department of Materials Science and Engineering
College Park, MD, USA
e-mail: *oded@umd.edu*

Françisco M. Raymo
University of Miami
Department of Chemistry
1301 Memorial Drive
Coral Gables, FL 33146-0431, USA
e-mail: *fraymo@miami.edu*

Manitra Razafinimanana
University of Toulouse III (Paul Sabatier)
Centre de Physique des Plasmas
et leurs Applications (CPPAT)
Toulouse, France
e-mail: *razafinimanana@cpat.ups-tlse.fr*

Ziv Reich
Weizmann Institute of Science Ha'Nesi Ha'Rishon
Department of Biological Chemistry
Rehovot 76100, Israel
e-mail: *ziv.reich@weizmann.ac.il*

John A. Rogers
University of Illinois
Department of Materials Science and Engineering
Urbana, IL, USA
e-mail: *jrogers@uiuc.edu*

Cosmin Roman
ETH Zurich
Micro and Nanosystems Department of Mechanical and Process Engineering
8092 Zurich, Switzerland
e-mail: *cosmin.roman@micro.mavt.ethz.ch*

Marina Ruths
University of Massachusetts Lowell
Department of Chemistry
1 University Avenue
Lowell, MA 01854, USA
e-mail: *marina_ruths@uml.edu*

Ozgur Sahin
The Rowland Institute at Harvard
100 Edwin H. Land Blvd
Cambridge, MA 02142, USA
e-mail: *sahin@rowland.harvard.edu*

Akira Sasahara
Japan Advanced Institute
of Science and Technology
School of Materials Science
1-1 Asahidai
923-1292 Nomi, Japan
e-mail: *sasahara@jaist.ac.jp*

Helmut Schift
Paul Scherrer Institute
Laboratory for Micro- and Nanotechnology
5232 Villigen PSI, Switzerland
e-mail: *helmut.schift@psi.ch*

André Schirmeisen
University of Münster
Institute of Physics
Wilhelm-Klemm-Str. 10
48149 Münster, Germany
e-mail: *schirmeisen@uni-muenster.de*

Christian Schulze
Beiersdorf AG
Research & Development
Unnastr. 48
20245 Hamburg, Germany
e-mail: *christian.schulze@beiersdorf.com;*
christian.schulze@uni-leipzig.de

Alexander Schwarz
University of Hamburg
Institute of Applied Physics
Jungiusstr. 11
20355 Hamburg, Germany
e-mail: *aschwarz@physnet.uni-hamburg.de*

Udo D. Schwarz
Yale University
Department of Mechanical Engineering
15 Prospect Street
New Haven, CT 06520-8284, USA
e-mail: *udo.schwarz@yale.edu*

Philippe Serp
Ecole Nationale Supérieure d'Ingénieurs
en Arts Chimiques et Technologiques
Laboratoire de Chimie de Coordination (LCC)
118 Route de Narbonne
31077 Toulouse, France
e-mail: *philippe.serp@ensiacet.fr*

Huamei (Mary) Shang
GE Healthcare
4855 W. Electric Ave.
Milwaukee, WI 53219, USA
e-mail: *huamei.shang@ge.com*

Susan B. Sinnott
University of Florida
Department of Materials Science and Engineering
154 Rhines Hall
Gainesville, FL 32611-6400, USA
e-mail: *ssinn@mse.ufl.edu*

Anisoara Socoliuc
SPECS Zurich GmbH
Technoparkstr. 1
8005 Zurich, Switzerland
e-mail: *socoliuc@nanonis.com*

Olav Solgaard
Stanford University
E.L. Ginzton Laboratory
450 Via Palou
Stanford, CA 94305-4088, USA
e-mail: *solgaard@stanford.edu*

Dan Strehle
University of Leipzig
Institute of Experimental Physics I
Division of Soft Matter Physics
Linnéstr. 5
04103 Leipzig, Germany
e-mail: *dan.strehle@uni-leipzig.de*

Carsten Stüber
University of Leipzig
Institute of Experimental Physics I
Division of Soft Matter Physics
Linnéstr. 5
04103 Leipzig, Germany
e-mail: *stueber@rz.uni-leipzig.de*

Yu-Chuan Su
ESS 210
Department of Engineering and System Science 101
Kuang-Fu Road
Hsinchu, 30013, Taiwan
e-mail: *ycsu@ess.nthu.edu.tw*

Kazuhisa Sueoka
Graduate School of Information Science
and Technology
Hokkaido University
Nanoelectronics Laboratory
Kita-14, Nishi-9, Kita-ku
060-0814 Sapporo, Japan
e-mail: *sueoka@nano.isthokudai.ac.jp*

Yasuhiro Sugawara
Osaka University
Department of Applied Physics
Yamada-Oka 2-1, Suita
565-0871 Osaka, Japan
e-mail: *sugawara@ap.eng.osaka-u.ac.jp*

Benjamin Sullivan
TearLab Corp.
11025 Roselle Street
San Diego, CA 92121, USA
e-mail: *bdsulliv@TearLab.com*

Paul Swanson
Nexogen, Inc.
Engineering
8360 C Camino Santa Fe
San Diego, CA 92121, USA
e-mail: *pswanson@nexogentech.com*

Yung-Chieh Tan
Washington University School of Medicine
Department of Medicine
Division of Dermatology
660 S. Euclid Ave.
St. Louis, MO 63110, USA
e-mail: *ytanster@gmail.com*

Shia-Yen Teh
University of California at Irvine
Biomedical Engineering Department
3120 Natural Sciences II
Irvine, CA 92697-2715, USA
e-mail: *steh@uci.edu*

W. Merlijn van Spengen
Leiden University
Kamerlingh Onnes Laboratory
Niels Bohrweg 2
Leiden, CA 2333, The Netherlands
e-mail: *spengen@physics.leidenuniv.nl*

Peter Vettiger
University of Neuchâtel
SAMLAB
Jaquet-Droz 1
2002 Neuchâtel, Switzerland
e-mail: *peter.vettiger@unine.ch*

Franziska Wetzel
University of Leipzig
Institute of Experimental Physics I
Division of Soft Matter Physics
Linnéstr. 5
04103 Leipzig, Germany
e-mail: *franziska.wetzel@uni-leipzig.de*

Heiko Wolf
IBM Research GmbH
Zurich Research Laboratory
Säumerstr. 4
8803 Rüschlikon, Switzerland
e-mail: *hwo@zurich.ibm.com*

Darrin J. Young
Case Western Reserve University
Department of EECS, Glennan 510
10900 Euclid Avenue
Cleveland, OH 44106, USA
e-mail: *djy@po.cwru.edu*

Babak Ziaie
Purdue University
Birck Nanotechnology Center
1205 W. State St.
West Lafayette, IN 47907-2035, USA
e-mail: *bziaie@purdue.edu*

Christian A. Zorman
Case Western Reserve University
Department of Electrical Engineering
and Computer Science
10900 Euclid Avenue
Cleveland, OH 44106, USA
e-mail: *caz@case.edu*

Jim V. Zoval
Saddleback College
Department of Math and Science
28000 Marguerite Parkway
Mission Viejo, CA 92692, USA
e-mail: *jzoval@saddleback.edu*

目　录

缩略语

Part C 扫描探针显微镜

21. 扫描探针显微镜——工作原理、检测方法和探测 ········· 3
　21.1 扫描隧道显微镜 ········· 5
　21.2 原子力显微镜 ········· 9
　21.3 原子力显微镜检测方法与分析 ········· 25
　参考文献 ········· 42

22. 扫描显微镜的普通探针和特殊探针 ········· 49
　22.1 原子力显微镜 ········· 50
　22.2 扫描隧道显微镜 ········· 60
　参考文献 ········· 61

23. 非接触式原子力显微镜及相关问题 ········· 65
　23.1 原子力显微镜 ········· 66
　23.2 半导体中的应用 ········· 71
　23.3 绝缘体中的应用 ········· 77
　23.4 分子中的应用 ········· 84
　参考文献 ········· 88

24. 低温扫描探针显微镜 ········· 93
　24.1 显微镜低温操作 ········· 94
　24.2 检测方法 ········· 96
　24.3 扫描隧道显微镜与光谱学 ········· 99
　24.4 扫描力显微镜与光谱学 ········· 118
　参考文献 ········· 130

25. 动态力显微镜中的高谐波和时变力 ········· 141
　25.1 轻敲模式原子力显微镜的探针样品的相互作用力建模 ········· 142
　25.2 增强时变力下的悬臂梁响应 ········· 144
　25.3 应用实例 ········· 150
　25.4 高谐波小振幅力显微镜 ········· 154
　参考文献 ········· 158

26. 原子力显微镜的动态模型 ... 161
26.1 单原子键的测量机制 ... 162
26.2 谐波振荡器：原子力显微镜的动态模型系统 ... 166
26.3 动态原子力显微镜的工作模型 ... 167
26.4 Q-控制 ... 180
26.5 动态原子力显微镜的耗散方法测量 ... 184
26.6 结论 ... 188
参考文献 ... 188

27. 分子识别力学显微镜：从分子键到复杂的能量形貌 ... 193
27.1 尖端化学配位 ... 194
27.2 探针表面受体的固定 ... 196
27.3 单分子识别的力学探测 ... 197
27.4 分子识别力谱的原则 ... 199
27.5 力谱识别：从孤立分子到生物膜 ... 201
27.6 图像识别 ... 209
27.7 结束语 ... 211
参考文献 ... 211

Contents

List of Abbreviations

Part C Scanning-Probe Microscopy

**21 Scanning Probe Microscopy –
Principle of Operation, Instrumentation, and Probes**
Bharat Bhushan, Othmar Marti 573
21.1 Scanning Tunneling Microscope 575
21.2 Atomic Force Microscope 579
21.3 AFM Instrumentation and Analyses 595
References 612

22 General and Special Probes in Scanning Microscopies
Jason Hafner, Edin (I-Chen) Chen, Ratnesh Lal, Sungho Jin 619
22.1 Atomic Force Microscopy 620
22.2 Scanning Tunneling Microscopy 630
References 631

23 Noncontact Atomic Force Microscopy and Related Topics
*Franz J. Giessibl, Yasuhiro Sugawara, Seizo Morita, Hirotaka Hosoi,
Kazuhisa Sueoka, Koichi Mukasa, Akira Sasahara, Hiroshi Onishi* 635
23.1 Atomic Force Microscopy (AFM) 636
23.2 Applications to Semiconductors 641
23.3 Applications to Insulators 647
23.4 Applications to Molecules 654
References 658

24 Low-Temperature Scanning Probe Microscopy
Markus Morgenstern, Alexander Schwarz, Udo D. Schwarz 663
24.1 Microscope Operation at Low Temperatures 664
24.2 Instrumentation 666
24.3 Scanning Tunneling Microscopy and Spectroscopy 669
24.4 Scanning Force Microscopy and Spectroscopy 688
References 700

**25 Higher Harmonics and Time-Varying Forces
in Dynamic Force Microscopy**
Ozgur Sahin, Calvin F. Quate, Olav Solgaard, Franz J. Giessibl 711
25.1 Modeling of Tip–Sample Interaction Forces in Tapping-Mode AFM ... 712
25.2 Enhancing the Cantilever Response to Time-Varying Forces 714

25.3	Application Examples	720
25.4	Higher-Harmonic Force Microscopy with Small Amplitudes	724
References		728

26 Dynamic Modes of Atomic Force Microscopy
André Schirmeisen, Boris Anczykowski, Hendrik Hölscher, Harald Fuchs 731

26.1	Motivation – Measurement of a Single Atomic Bond	732
26.2	Harmonic Oscillator: a Model System for Dynamic AFM	736
26.3	Dynamic AFM Operational Modes	737
26.4	Q-Control	750
26.5	Dissipation Processes Measured with Dynamic AFM	754
26.6	Conclusions	758
References		758

27 Molecular Recognition Force Microscopy: From Molecular Bonds to Complex Energy Landscapes
Peter Hinterdorfer, Andreas Ebner, Hermann Gruber, Ruti Kapon, Ziv Reich 763

27.1	Ligand Tip Chemistry	764
27.2	Immobilization of Receptors onto Probe Surfaces	766
27.3	Single-Molecule Recognition Force Detection	767
27.4	Principles of Molecular Recognition Force Spectroscopy	769
27.5	Recognition Force Spectroscopy: From Isolated Molecules to Biological Membranes	771
27.6	Recognition Imaging	779
27.7	Concluding Remarks	781
References		781

List of Abbreviations

μCP	microcontact printing
1-D	one-dimensional
18-MEA	18-methyl eicosanoic acid
2-D	two-dimensional
2-DEG	two-dimensional electron gas
3-APTES	3-aminopropyltriethoxysilane
3-D	three-dimensional

A

a-BSA	anti-bovine serum albumin
a-C	amorphous carbon
A/D	analog-to-digital
AA	amino acid
AAM	anodized alumina membrane
ABP	actin binding protein
AC	alternating-current
AC	amorphous carbon
ACF	autocorrelation function
ADC	analog-to-digital converter
ADXL	analog devices accelerometer
AFAM	atomic force acoustic microscopy
AFM	atomic force microscope
AFM	atomic force microscopy
AKD	alkylketene dimer
ALD	atomic layer deposition
AM	amplitude modulation
AMU	atomic mass unit
AOD	acoustooptical deflector
AOM	acoustooptical modulator
AP	alkaline phosphatase
APB	actin binding protein
APCVD	atmospheric-pressure chemical vapor deposition
APDMES	aminopropyldimethylethoxysilane
APTES	aminopropyltriethoxysilane
ASIC	application-specific integrated circuit
ASR	analyte-specific reagent
ATP	adenosine triphosphate

B

BAP	barometric absolute pressure
BAPDMA	behenyl amidopropyl dimethylamine glutamate
bcc	body-centered cubic
BCH	brucite-type cobalt hydroxide
BCS	Bardeen–Cooper–Schrieffer
BD	blu-ray disc
BDCS	biphenyldimethylchlorosilane
BE	boundary element
BFP	biomembrane force probe
BGA	ball grid array
BHF	buffered HF
BHPET	1,1'-(3,6,9,12,15-pentaoxapentadecane-1,15-diyl)bis(3-hydroxyethyl-1H-imidazolium-1-yl) di[bis(trifluoromethanesulfonyl)imide]
BHPT	1,1'-(pentane-1,5-diyl)bis(3-hydroxyethyl-1H-imidazolium-1-yl) di[bis(trifluoromethanesulfonyl)imide]
BiCMOS	bipolar CMOS
bioMEMS	biomedical microelectromechanical system
bioNEMS	biomedical nanoelectromechanical system
BMIM	1-butyl-3-methylimidazolium
BP	bit pitch
BPAG1	bullous pemphigoid antigen 1
BPT	biphenyl-4-thiol
BPTC	cross-linked BPT
BSA	bovine serum albumin
BST	barium strontium titanate
BTMAC	behentrimonium chloride

C

CA	constant amplitude
CA	contact angle
CAD	computer-aided design
CAH	contact angle hysteresis
cAMP	cyclic adenosine monophosphate
CAS	Crk-associated substrate
CBA	cantilever beam array
CBD	chemical bath deposition
CCD	charge-coupled device
CCVD	catalytic chemical vapor deposition
CD	compact disc
CD	critical dimension
CDR	complementarity determining region
CDW	charge density wave
CE	capillary electrophoresis
CE	constant excitation
CEW	continuous electrowetting
CG	controlled geometry
CHO	Chinese hamster ovary
CIC	cantilever in cantilever
CMC	cell membrane complex
CMC	critical micelle concentration
CMOS	complementary metal–oxide–semiconductor
CMP	chemical mechanical polishing

CNF	carbon nanofiber		DOS	density of states
CNFET	carbon nanotube field-effect transistor		DP	decylphosphonate
CNT	carbon nanotube		DPN	dip-pen nanolithography
COC	cyclic olefin copolymer		DRAM	dynamic random-access memory
COF	chip-on-flex		DRIE	deep reactive ion etching
COF	coefficient of friction		ds	double-stranded
COG	cost of goods		DSC	differential scanning calorimetry
CoO	cost of ownership		DSP	digital signal processor
COS	CV-1 in origin with SV40		DTR	discrete track recording
CP	circularly permuted		DTSSP	3,3'-dithio-bis(sulfosuccinimidylproprionate)
CPU	central processing unit			
CRP	C-reactive protein		DUV	deep-ultraviolet
CSK	cytoskeleton		DVD	digital versatile disc
CSM	continuous stiffness measurement		DWNT	double-walled CNT
CTE	coefficient of thermal expansion			
Cu-TBBP	Cu-tetra-3,5 di-tertiary-butyl-phenyl porphyrin		**E**	
CVD	chemical vapor deposition		EAM	embedded atom method
			EB	electron beam
D			EBD	electron beam deposition
			EBID	electron-beam-induced deposition
DBR	distributed Bragg reflector		EBL	electron-beam lithography
DC-PECVD	direct-current plasma-enhanced CVD		ECM	extracellular matrix
DC	direct-current		ECR-CVD	electron cyclotron resonance chemical vapor deposition
DDT	dichlorodiphenyltrichloroethane			
DEP	dielectrophoresis		ED	electron diffraction
DFB	distributed feedback		EDC	1-ethyl-3-(3-diamethylaminopropyl) carbodiimide
DFM	dynamic force microscopy			
DFS	dynamic force spectroscopy		EDL	electrostatic double layer
DGU	density gradient ultracentrifugation		EDP	ethylene diamine pyrochatechol
DI	FESPdigital instrument force modulation etched Si probe		EDTA	ethylenediamine tetraacetic acid
			EDX	energy-dispersive x-ray
DI	TESPdigital instrument tapping mode etched Si probe		EELS	electron energy loss spectra
			EFM	electric field gradient microscopy
DI	digital instrument		EFM	electrostatic force microscopy
DI	deionized		EHD	elastohydrodynamic
DIMP	diisopropylmethylphosphonate		EO	electroosmosis
DIP	dual inline packaging		EOF	electroosmotic flow
DIPS	industrial postpackaging		EOS	electrical overstress
DLC	diamondlike carbon		EPA	Environmental Protection Agency
DLP	digital light processing		EPB	electrical parking brake
DLVO	Derjaguin–Landau–Verwey–Overbeek		ESD	electrostatic discharge
DMD	deformable mirror display		ESEM	environmental scanning electron microscope
DMD	digital mirror device			
DMDM	1,3-dimethylol-5,5-dimethyl		EU	European Union
DMMP	dimethylmethylphosphonate		EUV	extreme ultraviolet
DMSO	dimethyl sulfoxide		EW	electrowetting
DMT	Derjaguin–Muller–Toporov		EWOD	electrowetting on dielectric
DNA	deoxyribonucleic acid			
DNT	2,4-dinitrotoluene		**F**	
DOD	Department of Defense			
DOE	Department of Energy		F-actin	filamentous actin
DOE	diffractive optical element		FA	focal adhesion
DOF	degree of freedom		FAA	formaldehyde–acetic acid–ethanol
DOPC	1,2-dioleoyl-sn-glycero-3-phosphocholine		FACS	fluorescence-activated cell sorting

FAK	focal adhesion kinase		HDT	hexadecanethiol
FBS	fetal bovine serum		HDTV	high-definition television
FC	flip-chip		HEK	human embryonic kidney 293
FCA	filtered cathodic arc		HEL	hot embossing lithography
fcc	face-centered cubic		HEXSIL	hexagonal honeycomb polysilicon
FCP	force calibration plot		HF	hydrofluoric
FCS	fluorescence correlation spectroscopy		HMDS	hexamethyldisilazane
FD	finite difference		HNA	hydrofluoric-nitric-acetic
FDA	Food and Drug Administration		HOMO	highest occupied molecular orbital
FE	finite element		HOP	highly oriented pyrolytic
FEM	finite element method		HOPG	highly oriented pyrolytic graphite
FEM	finite element modeling		HOT	holographic optical tweezer
FESEM	field emission SEM		HP	hot-pressing
FESP	force modulation etched Si probe		HPI	hexagonally packed intermediate
FET	field-effect transistor		HRTEM	high-resolution transmission electron microscope
FFM	friction force microscope		HSA	human serum albumin
FFM	friction force microscopy		HtBDC	hexa-*tert*-butyl-decacylene
FIB-CVD	focused ion beam chemical vapor deposition		HTCS	high-temperature superconductivity
FIB	focused ion beam		HTS	high throughput screening
FIM	field ion microscope		HUVEC	human umbilical venous endothelial cell
FIP	feline coronavirus			
FKT	Frenkel–Kontorova–Tomlinson		**I**	
FM	frequency modulation			
FMEA	failure-mode effect analysis		IBD	ion beam deposition
FP6	Sixth Framework Program		IC	integrated circuit
FP	fluorescence polarization		ICA	independent component analysis
FPR	*N*-formyl peptide receptor		ICAM-1	intercellular adhesion molecules 1
FS	force spectroscopy		ICAM-2	intercellular adhesion molecules 2
FTIR	Fourier-transform infrared		ICT	information and communication technology
FV	force–volume		IDA	interdigitated array
			IF	intermediate filament
G			IF	intermediate-frequency
			IFN	interferon
GABA	γ-aminobutyric acid		IgG	immunoglobulin G
GDP	guanosine diphosphate		IKVAV	isoleucine–lysine–valine–alanine–valine
GF	gauge factor		IL	ionic liquid
GFP	green fluorescent protein		IMAC	immobilized metal ion affinity chromatography
GMR	giant magnetoresistive		IMEC	Interuniversity MicroElectronics Center
GOD	glucose oxidase		IR	infrared
GPCR	G-protein coupled receptor		ISE	indentation size effect
GPS	global positioning system		ITO	indium tin oxide
GSED	gaseous secondary-electron detector		ITRS	International Technology Roadmap for Semiconductors
GTP	guanosine triphosphate		IWGN	Interagency Working Group on Nanoscience, Engineering, and Technology
GW	Greenwood and Williamson			
			J	
H				
			JC	jump-to-contact
HAR	high aspect ratio		JFIL	jet-and-flash imprint lithography
HARMEMS	high-aspect-ratio MEMS		JKR	Johnson–Kendall–Roberts
HARPSS	high-aspect-ratio combined poly- and single-crystal silicon			
HBM	human body model			
hcp	hexagonal close-packed			
HDD	hard-disk drive			

K

KASH	Klarsicht, ANC-1, Syne Homology
KPFM	Kelvin probe force microscopy

L

LA	lauric acid
LAR	low aspect ratio
LB	Langmuir–Blodgett
LBL	layer-by-layer
LCC	leadless chip carrier
LCD	liquid-crystal display
LCoS	liquid crystal on silicon
LCP	liquid-crystal polymer
LDL	low-density lipoprotein
LDOS	local density of states
LED	light-emitting diode
LFA-1	leukocyte function-associated antigen-1
LFM	lateral force microscope
LFM	lateral force microscopy
LIGA	Lithographie Galvanoformung Abformung
LJ	Lennard-Jones
LMD	laser microdissection
LMPC	laser microdissection and pressure catapulting
LN	liquid-nitrogen
LoD	limit-of-detection
LOR	lift-off resist
LPC	laser pressure catapulting
LPCVD	low-pressure chemical vapor deposition
LSC	laser scanning cytometry
LSN	low-stress silicon nitride
LT-SFM	low-temperature scanning force microscope
LT-SPM	low-temperature scanning probe microscopy
LT-STM	low-temperature scanning tunneling microscope
LT	low-temperature
LTM	laser tracking microrheology
LTO	low-temperature oxide
LTRS	laser tweezers Raman spectroscopy
LUMO	lowest unoccupied molecular orbital
LVDT	linear variable differential transformer

M

MALDI	matrix assisted laser desorption ionization
MAP	manifold absolute pressure
MAPK	mitogen-activated protein kinase
MAPL	molecular assembly patterning by lift-off
MBE	molecular-beam epitaxy
MC	microcantilever
MC	microcapillary
MCM	multi-chip module
MD	molecular dynamics
ME	metal-evaporated
MEMS	microelectromechanical system
MExFM	magnetic exchange force microscopy
MFM	magnetic field microscopy
MFM	magnetic force microscope
MFM	magnetic force microscopy
MHD	magnetohydrodynamic
MIM	metal–insulator–metal
MIMIC	micromolding in capillaries
MLE	maximum likelihood estimator
MOCVD	metalorganic chemical vapor deposition
MOEMS	microoptoelectromechanical system
MOS	metal–oxide–semiconductor
MOSFET	metal–oxide–semiconductor field-effect transistor
MP	metal particle
MPTMS	mercaptopropyltrimethoxysilane
MRFM	magnetic resonance force microscopy
MRFM	molecular recognition force microscopy
MRI	magnetic resonance imaging
MRP	molecular recognition phase
MscL	mechanosensitive channel of large conductance
MST	microsystem technology
MT	microtubule
mTAS	micro total analysis system
MTTF	mean time to failure
MUMP	multiuser MEMS process
MVD	molecular vapor deposition
MWCNT	multiwall carbon nanotube
MWNT	multiwall nanotube
MYD/BHW	Muller–Yushchenko–Derjaguin/Burgess–Hughes–White

N

NA	numerical aperture
NADIS	nanoscale dispensing
NASA	National Aeronautics and Space Administration
NC-AFM	noncontact atomic force microscopy
NEMS	nanoelectromechanical system
NGL	next-generation lithography
NHS	N-hydroxysuccinimidyl
NIH	National Institute of Health
NIL	nanoimprint lithography
NIST	National Institute of Standards and Technology
NMP	no-moving-part
NMR	nuclear magnetic resonance
NMR	nuclear mass resonance
NNI	National Nanotechnology Initiative

NOEMS	nanooptoelectromechanical system
NP	nanoparticle
NP	nanoprobe
NSF	National Science Foundation
NSOM	near-field scanning optical microscopy
NSTC	National Science and Technology Council
NTA	nitrilotriacetate
nTP	nanotransfer printing

O

ODA	octadecylamine
ODDMS	n-octadecyldimethyl(dimethylamino)silane
ODMS	n-octyldimethyl(dimethylamino)silane
ODP	octadecylphosphonate
ODTS	octadecyltrichlorosilane
OLED	organic light-emitting device
OM	optical microscope
OMVPE	organometallic vapor-phase epitaxy
OS	optical stretcher
OT	optical tweezers
OTRS	optical tweezers Raman spectroscopy
OTS	octadecyltrichlorosilane
oxLDL	oxidized low-density lipoprotein

P

P–V	peak-to-valley
PAA	poly(acrylic acid)
PAA	porous anodic alumina
PAH	poly(allylamine hydrochloride)
PAPP	p-aminophenyl phosphate
Pax	paxillin
PBC	periodic boundary condition
PBS	phosphate-buffered saline
PC	polycarbonate
PCB	printed circuit board
PCL	polycaprolactone
PCR	polymerase chain reaction
PDA	personal digital assistant
PDMS	polydimethylsiloxane
PDP	2-pyridyldithiopropionyl
PDP	pyridyldithiopropionate
PE	polyethylene
PECVD	plasma-enhanced chemical vapor deposition
PEEK	polyetheretherketone
PEG	polyethylene glycol
PEI	polyethyleneimine
PEN	polyethylene naphthalate
PES	photoemission spectroscopy
PES	position error signal
PET	poly(ethyleneterephthalate)
PETN	pentaerythritol tetranitrate
PFDA	perfluorodecanoic acid
PFDP	perfluorodecylphosphonate
PFDTES	perfluorodecyltriethoxysilane
PFM	photonic force microscope
PFOS	perfluorooctanesulfonate
PFPE	perfluoropolyether
PFTS	perfluorodecyltricholorosilane
PhC	photonic crystal
PI3K	phosphatidylinositol-3-kinase
PI	polyisoprene
PID	proportional–integral–differential
PKA	protein kinase
PKC	protein kinase C
PKI	protein kinase inhibitor
PL	photolithography
PLC	phospholipase C
PLD	pulsed laser deposition
PMAA	poly(methacrylic acid)
PML	promyelocytic leukemia
PMMA	poly(methyl methacrylate)
POCT	point-of-care testing
POM	polyoxy-methylene
PP	polypropylene
PPD	p-phenylenediamine
PPMA	poly(propyl methacrylate)
PPy	polypyrrole
PS-PDMS	poly(styrene-b-dimethylsiloxane)
PS/clay	polystyrene/nanoclay composite
PS	polystyrene
PSA	prostate-specific antigen
PSD	position-sensitive detector
PSD	position-sensitive diode
PSD	power-spectral density
PSG	phosphosilicate glass
PSGL-1	P-selectin glycoprotein ligand-1
PTFE	polytetrafluoroethylene
PUA	polyurethane acrylate
PUR	polyurethane
PVA	polyvinyl alcohol
PVD	physical vapor deposition
PVDC	polyvinylidene chloride
PVDF	polyvinyledene fluoride
PVS	polyvinylsiloxane
PWR	plasmon-waveguide resonance
PZT	lead zirconate titanate

Q

QB	quantum box
QCM	quartz crystal microbalance
QFN	quad flat no-lead
QPD	quadrant photodiode
QWR	quantum wire

R

RBC	red blood cell
RCA	Radio Corporation of America
RF	radiofrequency
RFID	radiofrequency identification
RGD	arginine–glycine–aspartic
RH	relative humidity
RHEED	reflection high-energy electron diffraction
RICM	reflection interference contrast microscopy
RIE	reactive-ion etching
RKKY	Ruderman–Kittel–Kasuya–Yoshida
RMS	root mean square
RNA	ribonucleic acid
ROS	reactive oxygen species
RPC	reverse phase column
RPM	revolutions per minute
RSA	random sequential adsorption
RT	room temperature
RTP	rapid thermal processing

S

SAE	specific adhesion energy
SAM	scanning acoustic microscopy
SAM	self-assembled monolayer
SARS-CoV	syndrome associated coronavirus
SATI	self-assembly, transfer, and integration
SATP	(S-acetylthio)propionate
SAW	surface acoustic wave
SB	Schottky barrier
SCFv	single-chain fragment variable
SCM	scanning capacitance microscopy
SCPM	scanning chemical potential microscopy
SCREAM	single-crystal reactive etching and metallization
SDA	scratch drive actuator
SEcM	scanning electrochemical microscopy
SEFM	scanning electrostatic force microscopy
SEM	scanning electron microscope
SEM	scanning electron microscopy
SFA	surface forces apparatus
SFAM	scanning force acoustic microscopy
SFD	shear flow detachment
SFIL	step and flash imprint lithography
SFM	scanning force microscope
SFM	scanning force microscopy
SGS	small-gap semiconducting
SICM	scanning ion conductance microscopy
SIM	scanning ion microscopy
SIP	single inline package
SKPM	scanning Kelvin probe microscopy
SL	soft lithography
SLIGA	sacrificial LIGA
SLL	sacrificial layer lithography
SLM	spatial light modulator
SMA	shape memory alloy
SMM	scanning magnetic microscopy
SNOM	scanning near field optical microscopy
SNP	single nucleotide polymorphisms
SNR	signal-to-noise ratio
SOG	spin-on-glass
SOI	silicon-on-insulator
SOIC	small outline integrated circuit
SoS	silicon-on-sapphire
SP-STM	spin-polarized STM
SPM	scanning probe microscope
SPM	scanning probe microscopy
SPR	surface plasmon resonance
sPROM	structurally programmable microfluidic system
SPS	spark plasma sintering
SRAM	static random access memory
SRC	sampling rate converter
SSIL	step-and-stamp imprint lithography
SSRM	scanning spreading resistance microscopy
STED	stimulated emission depletion
SThM	scanning thermal microscope
STM	scanning tunneling microscope
STM	scanning tunneling microscopy
STORM	statistical optical reconstruction microscopy
STP	standard temperature and pressure
STS	scanning tunneling spectroscopy
SUN	Sad1p/UNC-84
SWCNT	single-wall carbon nanotube
SWCNT	single-walled carbon nanotube
SWNT	single wall nanotube
SWNT	single-wall nanotube

T

TA	tilt angle
TASA	template-assisted self-assembly
TCM	tetracysteine motif
TCNQ	tetracyanoquinodimethane
TCP	tricresyl phosphate
TEM	transmission electron microscope
TEM	transmission electron microscopy
TESP	tapping mode etched silicon probe
TGA	thermogravimetric analysis
TI	Texas Instruments
TIRF	total internal reflection fluorescence
TIRM	total internal reflection microscopy
TLP	transmission-line pulse
TM	tapping mode
TMAH	tetramethyl ammonium hydroxide
TMR	tetramethylrhodamine
TMS	tetramethylsilane

TMS	trimethylsilyl			
TNT	trinitrotoluene			
TP	track pitch			
TPE-FCCS	two-photon excitation fluorescence cross-correlation spectroscopy			
TPI	threads per inch			
TPMS	tire pressure monitoring system			
TR	torsional resonance			
TREC	topography and recognition			
TRIM	transport of ions in matter			
TSDC	thermally stimulated depolarization current			
TTF	tetrathiafulvalene			
TV	television			

V

VBS	vinculin binding site
VCO	voltage-controlled oscillator
VCSEL	vertical-cavity surface-emitting laser
vdW	van der Waals
VHH	variable heavy–heavy
VLSI	very large-scale integration
VOC	volatile organic compound
VPE	vapor-phase epitaxy
VSC	vehicle stability control

U

UAA	unnatural AA
UHV	ultrahigh vacuum
ULSI	ultralarge-scale integration
UML	unified modeling language
UNCD	ultrananocrystalline diamond
UV	ultraviolet
UVA	ultraviolet A

X

XPS	x-ray photon spectroscopy
XRD	x-ray powder diffraction

Y

YFP	yellow fluorescent protein

Z

Z-DOL	perfluoropolyether

Part C Scanning-Probe Microscopy

**21 Scanning Probe Microscopy –
Principle of Operation, Instrumentation,
and Probes**
Bharat Bhushan, Columbus, USA
Othmar Marti, Ulm, Germany

**22 General and Special Probes
in Scanning Microscopies**
Jason Hafner, Houston, USA
Edin (I-Chen) Chen, Chung-Li, Taiwan
Ratnesh Lal, Chicago, USA
Sungho Jin, La Jolla, USA

**23 Noncontact Atomic Force Microscopy
and Related Topics**
Franz J. Giessibl, Regensburg, Germany
Yasuhiro Sugawara, Osaka, Japan
Seizo Morita, Osaka, Japan
Hirotaka Hosoi, Sapporo, Japan
Kazuhisa Sueoka, Sapporo, Japan
Koichi Mukasa, Sapporo, Japan
Akira Sasahara, Nomi, Japan
Hiroshi Onishi, Kanagawa, Japan

24 Low-Temperature Scanning Probe Microscopy
Markus Morgenstern, Aachen, Germany
Alexander Schwarz, Hamburg, Germany
Udo D. Schwarz, New Haven, USA

**25 Higher Harmonics and Time-Varying Forces
in Dynamic Force Microscopy**
Ozgur Sahin, Cambridge, USA
Calvin F. Quate, Stanford, USA
Olav Solgaard, Stanford, USA
Franz J. Giessibl, Regensburg, Germany

26 Dynamic Modes of Atomic Force Microscopy
André Schirmeisen, Münster, Germany
Boris Anczykowski, Münster, Germany
Hendrik Hölscher, Karlsruhe, Germany
Harald Fuchs, Münster, Germany

**27 Molecular Recognition Force Microscopy:
From Molecular Bonds to Complex Energy
Landscapes**
Peter Hinterdorfer, Linz, Austria
Andreas Ebner, Linz, Austria
Hermann Gruber, Linz, Austria
Ruti Kapon, Rehovot, Israel
Ziv Reich, Rehovot, Israel

21. Scanning Probe Microscopy – Principle of Operation, Instrumentation, and Probes

Bharat Bhushan, Othmar Marti

Since the introduction of the STM in 1981 and the AFM in 1985, many variations of probe-based microscopies, referred to as SPMs, have been developed. While the pure imaging capabilities of SPM techniques initially dominated applications of these methods, the physics of probe–sample interactions and quantitative analyses of tribological, electronic, magnetic, biological, and chemical surfaces using SPMs have become of increasing interest in recent years. SPMs are often associated with nanoscale science and technology, since they allow investigation and manipulation of surfaces down to the atomic scale. As our understanding of the underlying interaction mechanisms has grown, SPMs have increasingly found application in many fields beyond basic research fields. In addition, various derivatives of all these methods have been developed for special applications, some of them intended for areas other than microscopy.

This chapter presents an overview of STM and AFM and various probes (tips) used in these instruments, followed by details on AFM instrumentation and analyses.

21.1	Scanning Tunneling Microscope	575
21.1.1	The STM Design of Binnig et al.	575
21.1.2	Commercial STMs	576
21.1.3	STM Probe Construction	578
21.2	Atomic Force Microscope	579
21.2.1	The AFM Design of Binnig et al.	581
21.2.2	Commercial AFMs	581
21.2.3	AFM Probe Construction	587
21.2.4	Friction Measurement Methods	591
21.2.5	Normal Force and Friction Force Calibrations of Cantilever Beams	594
21.3	AFM Instrumentation and Analyses	595
21.3.1	The Mechanics of Cantilevers	596
21.3.2	Instrumentation and Analyses of Detection Systems for Cantilever Deflections	598
21.3.3	Combinations for 3-D Force Measurements	606
21.3.4	Scanning and Control Systems	607
References		612

The scanning tunneling microscope (STM), developed by *Binnig* and his colleagues in 1981 at the IBM Zurich Research Laboratory in Rüschlikon (Switzerland), was the first instrument capable of directly obtaining three-dimensional (3-D) images of solid surfaces with atomic resolution [21.1]. Binnig and Rohrer received a Nobel Prize in Physics in 1986 for their discovery. STMs can only be used to study surfaces which are electrically conductive to some degree. Based on their design of the STM, in 1985, *Binnig* et al. developed an atomic force microscope (AFM) to measure ultrasmall forces (less than 1 μN) between the AFM tip surface and the sample surface [21.2] (also see [21.3]). AFMs can be used to measure any engineering surface, whether it is electrically conductive or insulating. The AFM has become a popular surface profiler for topographic and normal force measurements on the micro- to nanoscale [21.4]. AFMs modified in order to measure both normal and lateral forces are called lateral force microscopes (LFMs) or friction force microscopes (FFMs) [21.5–11]. FFMs have been further modified to measure lateral forces in two orthogonal directions [21.12–16]. A number of researchers have modified and improved the original AFM and FFM designs, and have used these improved systems to measure the adhesion and friction of solid and liquid surfaces on micro- and nanoscales [21.4, 17–30]. AFMs have been used to study scratching and wear, and

Table 21.1 Comparison of various conventional microscopes with SPMs

	Optical	SEM/TEM	Confocal	SPM
Magnification	10^3	10^7	10^4	10^9
Instrument price (US$)	$10 k	$250 k	$30 k	$100 k
Technology age	200 y	40 y	20 y	20 y
Applications	Ubiquitous	Science and technology	New and unfolding	Cutting edge
Market 1993	$800 M	$400 M	$80 M	$100 M
Growth rate	10%	10%	30%	70%

to measure elastic/plastic mechanical properties (such as indentation hardness and the modulus of elasticity) [21.4, 10, 11, 21, 23, 26–29, 31–36]. AFMs have been used to manipulate individual atoms of xenon [21.37], molecules [21.38], silicon surfaces [21.39] and polymer surfaces [21.40]. STMs have been used to create nanofeatures via localized heating or by inducing chemical reactions under the STM tip [21.41–43] and through nanomachining [21.44]. AFMs have also been used for nanofabrication [21.4, 10, 45–47] and nanomachining [21.48].

STMs and AFMs are used at extreme magnifications ranging from 10^3 to 10^9 in the x-, y- and z-directions in order to image macro to atomic dimensions with high resolution and for spectroscopy. These instruments can be used in any environment, such as ambient air [21.2, 49], various gases [21.17], liquids [21.50–52], vacuum [21.1, 53], at low temperatures (lower than about 100 K) [21.54–58] and at high temperatures [21.59, 60]. Imaging in liquid allows the study of live biological samples and it also eliminates the capillary forces that are present at the tip–sample interface when imaging aqueous samples in ambient air. Low-temperature (liquid helium temperatures) imaging is useful when studying biological and organic materials and low-temperature phenomena such as superconductivity or charge-density waves. Low-temperature operation is also advantageous for high-sensitivity force mapping due to the reduced thermal vibration. They also have been used to image liquids such as liquid crystals and lubricant molecules on graphite surfaces [21.61–64]. While applications of SPM techniques initially focused on their pure imaging capabilities, research into the physics and chemistry of probe–sample interactions and SPM-based quantitative analyses of tribological, electronic, magnetic, biological, and chemical surfaces have become increasingly popular in recent years. Nanoscale science and technology is often tied to the use of SPMs since they allow investigation and manipulation of surfaces down to the atomic scale. As our understanding of the underlying interaction mechanisms has grown, SPMs and their derivatives have found applications in many fields beyond basic research fields and microscopy.

Families of instruments based on STMs and AFMs, called scanning probe microscopes (SPMs), have been developed for various applications of scientific and industrial interest. These include STM, AFM, FFM (or LFM), scanning electrostatic force microscopy (SEFM) [21.65, 66], scanning force acoustic microscopy (SFAM) (or atomic force acoustic microscopy (AFAM)) [21.21, 22, 36, 67–69], scanning magnetic microscopy (SMM) (or magnetic force microscopy (MFM)) [21.70–73], scanning near-field optical microscopy (SNOM) [21.74–77], scanning thermal microscopy (SThM) [21.78–80], scanning electrochemical microscopy (SEcM) [21.81], scanning Kelvin probe microscopy (SKPM) [21.82–86], scanning chemical potential microscopy (SCPM) [21.79], scanning ion conductance microscopy (SICM) [21.87, 88] and scanning capacitance microscopy (SCM) [21.82, 89–91]. When the technique is used to measure forces (as in AFM, FFM, SEFM, SFAM and SMM) it is also referred to as scanning force microscopy (SFM). Although these instruments offer atomic resolution and are ideal for basic research, they are also used for cutting-edge industrial applications which do not require atomic resolution. The commercial production of SPMs started with the STM in 1987 and the AFM in 1989 by Digital Instruments, Inc. (Santa Barbara, USA). For comparisons of SPMs with other microscopes, see Table 21.1 (Veeco Instruments, Inc., Santa Barbara, USA). Numbers of these instruments are equally divided between the US, Japan and Europe, with the following split between industry/university and government laboratories: 50/50, 70/30, and 30/70, respectively. It is clear that research and industrial applications of SPMs are expanding rapidly.

21.1 Scanning Tunneling Microscope

The principle of electron tunneling was first proposed by *Giaever* [21.93]. He envisioned that if a potential difference is applied to two metals separated by a thin insulating film, a current will flow because of the ability of electrons to penetrate a potential barrier. To be able to measure a tunneling current, the two metals must be spaced no more than 10 nm apart. *Binnig* et al. [21.1] introduced vacuum tunneling combined with lateral scanning. The vacuum provides the ideal barrier for tunneling. The lateral scanning allows one to image surfaces with exquisite resolution – laterally to less than 1 nm and vertically to less than 0.1 nm – sufficient to define the position of single atoms. The very high vertical resolution of the STM is obtained because the tunnel current varies exponentially with the distance between the two electrodes; that is, the metal tip and the scanned surface. Typically, the tunneling current decreases by a factor of 2 as the separation is increased by 0.2 nm. Very high lateral resolution depends upon sharp tips. Binnig et al. overcame two key obstacles by damping external vibrations and moving the tunneling probe in close proximity to the sample. Their instrument is called the scanning tunneling microscope (STM). Today's STMs can be used in ambient environments for atomic-scale imaging of surfaces. Excellent reviews on this subject have been presented by *Hansma* and *Tersoff* [21.92], *Sarid* and *Elings* [21.94], *Durig* et al. [21.95]; *Frommer* [21.96], *Güntherodt* and *Wiesendanger* [21.97], *Wiesendanger* and *Güntherodt* [21.98], *Bonnell* [21.99], *Marti* and *Amrein* [21.100], *Stroscio* and *Kaiser* [21.101], and *Güntherodt* et al. [21.102].

The principle of the STM is straightforward. A sharp metal tip (one electrode of the tunnel junction) is brought close enough (0.3–1 nm) to the surface to be investigated (the second electrode) to make the tunneling current measurable at a convenient operating voltage (10 mV–1 V). The tunneling current in this case varies from 0.2 to 10 nA. The tip is scanned over the surface at a distance of 0.3–1 nm, while the tunneling current between it and the surface is measured. The STM can be operated in either the constant current mode or the constant height mode (Fig. 21.1). The left-hand column of Fig. 21.1 shows the basic constant current mode of operation. A feedback network changes the height of the tip z to keep the current constant. The displacement of the tip, given by the voltage applied to the piezoelectric drive, then yields a topographic map of the surface. Alternatively, in the constant height mode,

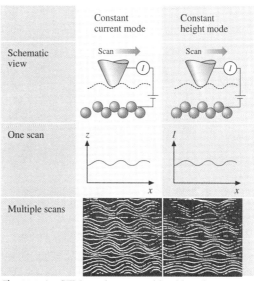

Fig. 21.1 An STM can be operated in either the constant-current or the constant-height mode. The images are of graphite in air (after [21.92])

a metal tip can be scanned across a surface at nearly constant height and constant voltage while the current is monitored, as shown in the right-hand column of Fig. 21.1. In this case, the feedback network responds just rapidly enough to keep the average current constant. The current mode is generally used for atomic-scale images; this mode is not practical for rough surfaces. A three-dimensional picture $[z(x, y)]$ of a surface consists of multiple scans $[z(x)]$ displayed laterally to each other in the y-direction. It should be noted that if different atomic species are present in a sample, the different atomic species within a sample may produce different tunneling currents for a given bias voltage. Thus the height data may not be a direct representation of the topography of the surface of the sample.

21.1.1 The STM Design of Binnig et al.

Figure 21.2 shows a schematic of an AFM designed by *Binnig* and *Rohrer* and intended for operation in ultrahigh vacuum [21.1, 103]. The metal tip was fixed to rectangular piezodrives P_x, P_y, and P_z made out of commercial piezoceramic material for scanning. The sample is mounted via either superconducting magnetic levitation or a two-stage spring system to achieve a sta-

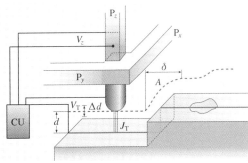

Fig. 21.2 Principle of operation of the STM, from *Binnig* and *Rohrer* [21.103]

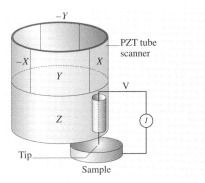

Fig. 21.3 Principle of operation of a commercial STM. A sharp tip attached to a piezoelectric tube scanner is scanned on a sample

ble gap width of about 0.02 nm. The tunnel current J_T is a sensitive function of the gap width d where $J_T \propto V_T \exp(-A\phi^{1/2}d)$. Here V_T is the bias voltage, ϕ is the average barrier height (work function) and the constant $A = 1.025\,\text{eV}^{-1/2}\,\text{Å}^{-1}$. With a work function of a few eV, J_T changes by an order of magnitude for an angstrom change in d. If the current is kept constant to within, for example, 2%, then the gap d remains constant to within 1 pm. For operation in the constant current mode, the control unit CU applies a voltage V_z to the piezo P_z such that J_T remains constant when scanning the tip with P_y and P_x over the surface. At a constant work function ϕ, $V_z(V_x, V_y)$ yields the roughness of the surface $z(x, y)$ directly, as illustrated by a surface step at A. Smearing the step, δ (lateral resolution) is on the order of $(R)^{1/2}$, where R is the radius of the curvature of the tip. Thus, a lateral resolution of about 2 nm requires tip radii on the order of 10 nm. A 1 mm diameter solid rod ground at one end at roughly 90° yields overall tip radii of only a few hundred nanometers, the presence of rather sharp microtips on the relatively dull end yields a lateral resolution of about 2 nm. In situ sharpening of the tips, achieved by gently touching the surface, brings the resolution down to the 1 nm range; by applying high fields (on the order of 10^8 V/cm) for, say, half an hour, resolutions considerably below 1 nm can be reached. Most experiments have been performed with tungsten wires either ground or etched to a typical radius of 0.1–10 μm. In some cases, in situ processing of the tips has been performed to further reduce tip radii.

21.1.2 Commercial STMs

There are a number of commercial STMs available on the market. Digital Instruments, Inc., introduced the first commercial STM, the Nanoscope I, in 1987. In the recent Nanoscope IV STM, intended for operation in ambient air, the sample is held in position while a piezoelectric crystal in the form of a cylindrical tube (referred to as a PZT tube scanner) scans the sharp metallic probe over the surface in a raster pattern while sensing and relaying the tunneling current to the control station (Fig. 21.3). The digital signal processor (DSP) calculates the tip–sample separation required by sensing the tunneling current flowing between the sample and the tip. The bias voltage applied between the sample and the tip encourages the tunneling current to flow. The DSP completes the digital feedback loop by relaying the desired voltage to the piezoelectric tube. The STM can operate in either the *constant height* or the *constant current* mode, and this can be selected using the control panel. In the constant current mode, the feedback gains are set high, the tunneling tip closely tracks the sample surface, and the variation in the tip height required to maintain constant tunneling current is measured by the change in the voltage applied to the piezo tube. In the constant height mode, the feedback gains are set low, the tip remains at a nearly constant height as it sweeps over the sample surface, and the tunneling current is imaged.

Physically, the Nanoscope STM consists of three main parts: the head, which houses the piezoelectric tube scanner which provides three-dimensional tip motion and the preamplifier circuit for the tunneling current (FET input amplifier) mounted on the top of the head; the base on which the sample is mounted; and the base support, which supports the base and head [21.4]. The base accommodates samples which are up to 10 mm by 20 mm and 10 mm thick. Scan sizes available for the STM are 0.7 µm (for atomic resolution), 12 µm, 75 µm and 125 µm square.

The scanning head controls the three-dimensional motion of the tip. The removable head consists of a piezo tube scanner, about 12.7 mm in diameter, mounted into an Invar shell, which minimizes vertical thermal drift because of the good thermal match between the piezo tube and the Invar. The piezo tube has separate electrodes for x-, y- and z-motion, which are driven by separate drive circuits. The electrode configuration (Fig. 21.3) provides x- and y-motions which are perpendicular to each other, it minimizes horizontal and vertical coupling, and it provides good sensitivity. The vertical motion of the tube is controlled by the Z-electrode, which is driven by the feedback loop. The x- and y-scanning motions are each controlled by two electrodes which are driven by voltages of the same magnitude but opposite signs. These electrodes are called $-y$, $-x$, $+y$, and $+x$. Applying complimentary voltages allows a short, stiff tube to provide a good scan range without the need for a large voltage. The motion of the tip that arises due to external vibrations is proportional to the square of the ratio of vibration frequency to the resonant frequency of the tube. Therefore, to minimize the tip vibrations, the resonant frequencies of the tube are high: about 60 kHz in the vertical direction and about 40 kHz in the horizontal direction. The tip holder is a stainless steel tube with an inner diameter of 300 µm when 250 µm diameter tips are used, which is mounted in ceramic in order to minimize the mass at the end of the tube. The tip is mounted either on the front edge of the tube (to keep the mounting mass low and the resonant frequency high) (Fig. 21.3) or the center of the tube for large-range scanners, namely 75 and 125 µm (to preserve the symmetry of the scanning). This commercial STM accepts any tip with a 250 µm diameter shaft. The piezotube requires x–y-calibration, which is carried out by imaging an appropriate calibration standard. Cleaved graphite is used for heads with small scan lengths while two-dimensional grids (a gold-plated rule) can be used for long-range heads.

The Invar base holds the sample in position, supports the head, and provides coarse x–y-motion for the sample. A sprung-steel sample clip with two thumb screws holds the sample in place. An x–y-translation stage built into the base allows the sample to be repositioned under the tip. Three precision screws arranged in a triangular pattern support the head and provide coarse and fine adjustment of the tip height. The base support consists of the base support ring and the motor housing. The stepper motor enclosed in the motor housing allows the tip to be engaged and withdrawn from the surface automatically.

Samples to be imaged with the STM must be conductive enough to allow a few nanoamperes of current to flow from the bias voltage source to the area to be scanned. In many cases, nonconductive samples can be coated with a thin layer of a conductive material to facilitate imaging. The bias voltage and the tunneling current depend on the sample. Usually they are set to a standard value for engagement and fine tuned to enhance the quality of the image. The scan size depends on the sample and the features of interest. A maximum scan rate of 122 Hz can be used. The maximum scan rate is usually related to the scan size. Scan rates above 10 Hz are used for small scans (typically 60 Hz for atomic-scale imaging with a 0.7 µm scanner). The scan rate should be lowered for large scans, especially if the sample surfaces are rough or contain large steps. Moving the tip

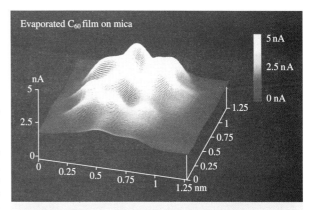

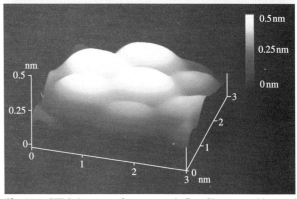

Fig. 21.4 STM images of evaporated C_{60} film on gold-coated freshly cleaved mica obtained using a mechanically sheared Pt-Ir (80/20) tip in constant height mode (after [21.104])

quickly along the sample surface at high scan rates with large scan sizes will usually lead to a tip crash. Essentially, the scan rate should be inversely proportional to the scan size (typically 2–4 Hz for a scan size of 1 μm, 0.5–1 Hz for 12 μm, and 0.2 Hz for 125 μm). The scan rate (in length/time) is equal to the scan length divided by the scan rate in Hz. For example, for a scan size of 10 μm × 10 μm scanned at 0.5 Hz, the scan rate is 10 μm/s. 256 × 256 data formats are the most common. The lateral resolution at larger scans is approximately equal to scan length divided by 256.

Figure 21.4 shows sample STM images of an evaporated C_{60} film on gold-coated freshly-cleaved mica taken at room temperature and ambient pressure [21.104]. Images were obtained with atomic resolution at two scan sizes. Next we describe some STM designs which are available for special applications.

Electrochemical STM

The electrochemical STM is used to perform and monitor the electrochemical reactions inside the STM. It includes a microscope base with an integral potentiostat, a short head with a 0.7 μm scan range and a differential preamp as well as the software required to operate the potentiostat and display the result of the electrochemical reaction.

Standalone STM

Standalone STMs are available to scan large samples. In this case, the STM rests directly on the sample. It is available from Digital Instruments in scan ranges of 12 and 75 μm. It is similar to the standard STM design except the sample base has been eliminated.

21.1.3 STM Probe Construction

The STM probe has a cantilever integrated with a sharp metal tip with a low aspect ratio (tip length/tip shank) to minimize flexural vibrations. Ideally, the tip should be atomically sharp, but in practice most tip preparation methods produce a tip with a rather ragged profile that consists of several asperities where the one closest to the surface is responsible for tunneling. STM cantilevers with sharp tips are typically fabricated from metal wires (the metal can be tungsten (W), platinum-iridium (Pt-Ir), or gold (Au)) and are sharpened by grinding, cutting with a wire cutter or razor blade, field emission/evaporation, ion milling, fracture, or electrochemical polishing/etching [21.105, 106]. The two most commonly used tips are made from either Pt-Ir (80/20) alloy or tungsten wire. Iridium is used to provide stiff-

Fig. 21.5 Schematic of a typical tungsten cantilever with a sharp tip produced by electrochemical etching

ness. The Pt-Ir tips are generally formed mechanically and are readily available. The tungsten tips are etched from tungsten wire by an electrochemical process, for example by using 1 M KOH solution with a platinum electrode in a electrochemical cell at about 30 V. In general, Pt-Ir tips provide better atomic resolution than tungsten tips, probably due to the lower reactivity of Pt. However, tungsten tips are more uniformly shaped and may perform better on samples with steeply sloped features. The tungsten wire diameter used for the cantilever is typically 250 μm, with the radius of curvature ranging from 20 to 100 nm and a cone angle ranging from 10 to 60° (Fig. 21.5). The wire can be bent in an L shape, if so required, for use in the instrument. For calculations of the normal spring constant and the natural frequency of round cantilevers, see *Sarid* and *Elings* [21.94].

High aspect ratio, controlled geometry (CG) Pt-Ir probes are commercially available to image deep trenches (Fig. 21.6). These probes are electrochemically etched from Pt-Ir (80/20) wire and are polished

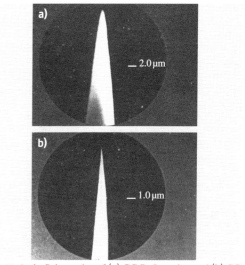

Fig. 21.6a,b Schematics of (**a**) CG Pt-Ir probe, and (**b**) CG Pt-Ir FIB milled probe

to a specific shape which is consistent from tip to tip. The probes have a full cone angle of $\approx 15°$, and a tip radius of less than 50 nm. To image very deep trenches ($> 0.25\,\mu$m) and nanofeatures, focused ion beam (FIB)-milled CG probes with extremely sharp tips (radii < 5 nm) are used. The Pt-Ir probes are coated with a nonconducting film (not shown in the figure) for electrochemistry. These probes are available from Materials Analytical Services (Raleigh, USA).

Pt alloy and W tips are very sharp and give high resolution, but are fragile and sometimes break when contacting a surface. Diamond tips have been used by *Kaneko* and *Oguchi* [21.107]. Diamond tips made conductive by boron ion implantation were found to be chip-resistant.

21.2 Atomic Force Microscope

Like the STM, the AFM relies on a scanning technique to produce very high resolution 3-D images of sample surfaces. The AFM measures ultrasmall forces (less than 1 nN) present between the AFM tip surface and a sample surface. These small forces are measured by measuring the motion of a very flexible cantilever beam with an ultrasmall mass. While STMs require the surface being measured be electrically conductive, AFMs are capable of investigating the surfaces of both conductors and insulators on an atomic scale if suitable techniques for measuring the cantilever motion are used. During the operation of a high-resolution AFM, the sample is generally scanned instead of the tip (unlike for STM) because the AFM measures the relative displacement between the cantilever surface and the reference surface and any cantilever movement from scanning would add unwanted vibrations. However, for measurements of large samples, AFMs are available where the tip is scanned and the sample is stationary. As long as the AFM is operated in the so-called contact mode, little if any vibration is introduced.

The AFM combines the principles of the STM and the stylus profiler (Fig. 21.7). In an AFM, the force between the sample and tip is used (rather than the tunneling current) to sense the proximity of the tip to the sample. The AFM can be used either in the static or the dynamic mode. In the static mode, also referred to as the repulsive or contact mode [21.2], a sharp tip at the end of the cantilever is brought into contact with the surface of the sample. During initial contact, the atoms at the end of the tip experience a very weak repulsive force due to electronic orbital overlap with the atoms in the surface of the sample. The force acting on the tip causes the cantilever to deflect, which is measured by tunneling, capacitive, or optical detectors. The deflection can be measured to within 0.02 nm, so a force as low as 0.2 nN (corresponding to a normal pressure of ≈ 200 MPa for a Si_3N_4 tip with a radius of about 50 nm against single-crystal silicon) can be detected for typical cantilever spring constant of 10 N/m. (To put these number in perspective, individual atoms and human hair are typically a fraction of a nanometer and about 75 μm in diameter, respectively, and a drop of water and an eyelash have masses of about 10 μN and 100 nN, respectively.) In the dynamic mode of operation, also referred to as attractive force imaging or noncontact imaging mode, the tip is brought into close proximity to (within a few nanometers of), but not in contact with, the sample. The cantilever is deliberately vibrated in either amplitude modulation (AM) mode [21.65] or frequency modulation (FM) mode [21.65, 94, 108, 109]. Very weak van der Waals attractive forces are present at the tip–sample interface. Although the normal pressure exerted at the interface is zero in this technique (in order to avoid any surface deformation), it is slow and difficult to use, and is rarely used outside of research environments. The surface topography is measured by laterally scanning the sample under the tip while simultaneously measuring the separation-dependent force or force gradient (derivative) between the tip and the surface (Fig. 21.7). In the contact (static) mode, the

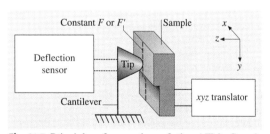

Fig. 21.7 Principle of operation of the AFM. Sample mounted on a piezoelectric scanner is scanned against a short tip and the cantilever deflection is usually measured using a laser deflection technique. The force (in contact mode) or the force gradient (in noncontact mode) is measured during scanning

interaction force between tip and sample is measured by monitoring the cantilever deflection. In the noncontact (or dynamic) mode, the force gradient is obtained by vibrating the cantilever and measuring the shift in the resonant frequency of the cantilever. To obtain topographic information, the interaction force is either recorded directly, or used as a control parameter for a feedback circuit that maintains the force or force derivative at a constant value. Using an AFM operated in the contact mode, topographic images with a vertical resolution of less than 0.1 nm (as low as 0.01 nm) and a lateral resolution of about 0.2 nm have been obtained [21.3, 50, 110–114]. Forces of 10 nN to 1 pN are measurable with a displacement sensitivity of 0.01 nm. These forces are comparable to the forces associated with chemical bonding, for example 0.1 μN for an ionic bond and 10 pN for a hydrogen bond [21.2]. For further reading, see [21.94–96, 100, 102, 115–119].

Lateral forces applied at the tip during scanning in the contact mode affect roughness measurements [21.120]. To minimize the effects of friction and other lateral forces on topography measurements in the contact mode, and to measure the topographies of soft surfaces, AFMs can be operated in the so-called tapping or force modulation mode [21.32, 121].

The STM is ideal for atomic-scale imaging. To obtain atomic resolution with the AFM, the spring constant of the cantilever should be weaker than the equivalent spring between atoms. For example, the vibration frequencies ω of atoms bound in a molecule or in a crystalline solid are typically 10^{13} Hz or higher. Combining this with an atomic mass m of $\approx 10^{-25}$ kg gives an interatomic spring constant k, given by $\omega^2 m$, of around 10 N/m [21.115]. (For comparison, the spring constant of a piece of household aluminium foil that is 4 mm long and 1 mm wide is about 1 N/m.) Therefore, a cantilever beam with a spring constant of about 1 N/m or lower is desirable. Tips must be as sharp as possible, and tip radii of 5 to 50 nm are commonly available.

Atomic resolution cannot be achieved with these tips at normal loads in the nN range. Atomic structures at these loads have been obtained from lattice imaging or by imaging the crystal's periodicity. Reported data show either perfectly ordered periodic atomic structures or defects on a larger lateral scale, but no well-defined, laterally resolved atomic-scale defects like those seen in images routinely obtained with a STM. Interatomic forces with one or several atoms in contact are 20–40 or 50–100 pN, respectively. Thus, atomic resolution with an AFM is only possible with a sharp tip on a flexible cantilever at a net repulsive force of 100 pN

or lower [21.122]. Upon increasing the force from 10 pN, *Ohnesorge* and *Binnig* [21.122] observed that monoatomic steplines were slowly wiped away and a perfectly ordered structure was left. This observation explains why mostly defect-free atomic resolution has been observed with AFM. Note that for atomic-resolution measurements, the cantilever should not be so soft as to avoid jumps. Further note that performing measurements in the noncontact imaging mode may be desirable for imaging with atomic resolution.

The key component in an AFM is the sensor used to measure the force on the tip due to its interaction with the sample. A cantilever (with a sharp tip) with an extremely low spring constant is required for high vertical and lateral resolutions at small forces (0.1 nN or lower), but a high resonant frequency is desirable (about 10 to 100 kHz) at the same time in order to minimize the sensitivity to building vibrations, which occur at around 100 Hz. This requires a spring with an extremely low vertical spring constant (typically 0.05 to 1 N/m) as well as a low mass (on the order of 1 ng). Today, the most advanced AFM cantilevers are microfabricated from silicon or silicon nitride using photolithographic techniques. Typical lateral dimensions are on the order of 100 μm, with thicknesses on the order of 1 μm. The force on the tip due to its interaction with the sample is sensed by detecting the deflection of the compliant lever with a known spring constant. This cantilever deflection (displacement smaller than 0.1 nm) has been measured by detecting a tunneling current similar to that used in the STM in the pioneering work of *Binnig* et al. [21.2] and later used by *Giessibl*

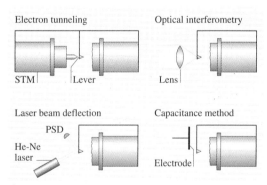

Fig. 21.8 Schematics of the four detection systems to measure cantilever deflection. In each set-up, the sample mounted on piezoelectric body is shown *on the right*, the cantilever *in the middle*, and the corresponding deflection sensor *on the left* (after [21.118])

et al. [21.56], by capacitance detection [21.123, 124], piezoresistive detection [21.125, 126], and by four optical techniques, namely (1) optical interferometry [21.5, 6, 127, 128] using optical fibers [21.57, 129] (2) optical polarization detection [21.72, 130], (3) laser diode feedback [21.131] and (4) optical (laser) beam deflection [21.7, 8, 53, 111, 112]. Schematics of the four more commonly used detection systems are shown in Fig. 21.8. The tunneling method originally used by *Binnig* et al. [21.2] in the first version of the AFM uses a second tip to monitor the deflection of the cantilever with its force sensing tip. Tunneling is rather sensitive to contaminants and the interaction between the tunneling tip and the rear side of the cantilever can become comparable to the interaction between the tip and sample. Tunneling is rarely used and is mentioned mainly for historical reasons. *Giessibl* et al. [21.56] have used it for a low-temperature AFM/STM design. In contrast to tunneling, other deflection sensors are placed far from the cantilever, at distances of micrometers to tens of millimeters. The optical techniques are believed to be more sensitive, reliable and easily implemented detection methods than the others [21.94, 118]. The optical beam deflection method has the largest working distance, is insensitive to distance changes and is capable of measuring angular changes (friction forces); therefore, it is the most commonly used in commercial SPMs.

Almost all SPMs use piezo translators to scan the sample, or alternatively to scan the tip. An electric field applied across a piezoelectric material causes a change in the crystal structure, with expansion in some directions and contraction in others. A net change in volume also occurs [21.132]. The first STM used a piezo tripod for scanning [21.1]. The piezo tripod is one way to generate three-dimensional movement of a tip attached at its center. However, the tripod needs to be fairly large (≈ 50 mm) to get a suitable range. Its size and asymmetric shape makes it susceptible to thermal drift. Tube scanners are widely used in AFMs [21.133]. These provide ample scanning range with a small size. Electronic control systems for AFMs are based on either analog or digital feedback. Digital feedback circuits are better suited for ultralow noise operation.

Images from the AFMs need to be processed. An ideal AFM is a noise-free device that images a sample with perfect tips of known shape and has a perfectly linear scanning piezo. In reality, scanning devices are affected by distortions and these distortions must be corrected for. The distortions can be linear and nonlinear. Linear distortions mainly result from imperfections in the machining of the piezo translators, causing crosstalk between the Z-piezo to the x- and y-piezos, and vice versa. Nonlinear distortions mainly result from the presence of a hysteresis loop in piezoelectric ceramics. They may also occur if the scan frequency approaches the upper frequency limit of the x- and y-drive amplifiers or the upper frequency limit of the feedback loop (z-component). In addition, electronic noise may be present in the system. The noise is removed by digital filtering in real space [21.134] or in the spatial frequency domain (Fourier space) [21.135].

Processed data consists of many tens of thousand of points per plane (or data set). The outputs from the first STM and AFM images were recorded on an x–y-chart recorder, with the z-value plotted against the tip position in the fast scan direction. Chart recorders have slow responses, so computers are used to display the data these days. The data are displayed as wire mesh displays or grayscale displays (with at least 64 shades of gray).

21.2.1 The AFM Design of Binnig et al.

In the first AFM design developed by *Binnig* et al. [21.2], AFM images were obtained by measuring the force exerted on a sharp tip created by its proximity to the surface of a sample mounted on a 3-D piezoelectric scanner. The tunneling current between the STM tip and the backside of the cantilever beam to which the tip was attached was measured to obtain the normal force. This force was kept at a constant level with a feedback mechanism. The STM tip was also mounted on a piezoelectric element to maintain the tunneling current at a constant level.

21.2.2 Commercial AFMs

A review of early designs of AFMs has been presented by *Bhushan* [21.4]. There are a number of commercial AFMs available on the market. Major manufacturers of AFMs for use in ambient environments are: Digital Instruments, Inc., Topometrix Corp. and other subsidiaries of Veeco Instruments, Inc., Molecular Imaging Corp. (Phoenix, USA), Quesant Instrument Corp. (Agoura Hills, USA), Nanoscience Instruments, Inc. (Phoenix, USA), Seiko Instruments (Chiba, Japan); and Olympus (Tokyo, Japan). AFM/STMs for use in UHV environments are manufactured by Omicron Vakuumphysik GmbH (Taunusstein, Germany).

We describe here two commercial AFMs – small-sample and large-sample AFMs – for operation in the contact mode, produced by Digital Instruments, Inc.,

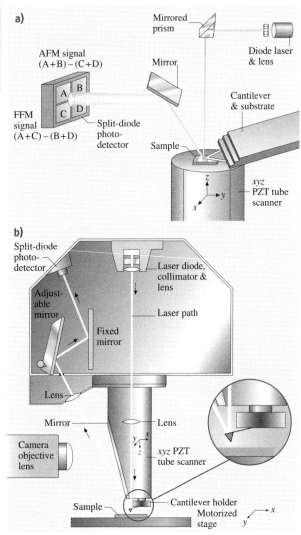

Fig. 21.9a,b Principles of operation of (**a**) a commercial small-sample AFM/FFM, and (**b**) a large-sample AFM/FFM

the cantilever deflection error signal. The AFM operates in both *constant height* and *constant force* modes. The DSP always adjusts the distance between the sample and the tip according to the cantilever deflection error signal, but if the feedback gains are low the piezo remains at an almost *constant height* and the cantilever deflection data is collected. With high gains, the piezo height changes to keep the cantilever deflection nearly constant (so the force is constant), and the change in piezo height is collected by the system.

In the operation of a commercial small-sample AFM (as shown in Fig. 21.9a), the sample (which is generally no larger than $10\,\text{mm} \times 10\,\text{mm}$) is mounted on a PZT tube scanner, which consists of separate electrodes used to precisely scan the sample in the x–y-plane in a raster pattern and to move the sample in the vertical (z-) direction. A sharp tip at the free end of a flexible cantilever is brought into contact with the sample. Features on the sample surface cause the cantilever to deflect in the vertical and lateral directions as the sample moves under the tip. A laser beam from a diode laser (5 mW max. peak output at 670 nm) is directed by a prism onto the back of a cantilever near its free end, tilted downward at about $10°$ with respect to the horizontal plane. The reflected beam from the vertex of the cantilever is directed through a mirror onto a quad photodetector (split photodetector with four quadrants) (commonly called a position-sensitive detector or PSD, produced by Silicon Detector Corp., Camarillo, USA). The difference in signal between the top and bottom photodiodes provides the AFM signal, which is a sensitive measure of the cantilever vertical deflection. The topographic features of the sample cause the tip to deflect in the vertical

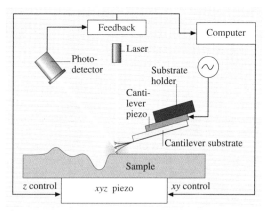

Fig. 21.10 Schematic of tapping mode used for surface roughness measurements

with scanning lengths ranging from about $0.7\,\mu\text{m}$ (for atomic resolution) to about $125\,\mu\text{m}$ [21.9, 111, 114, 136]. The original design of these AFMs comes from *Meyer* and *Amer* [21.53]. Basically, the AFM scans the sample in a raster pattern while outputting the cantilever deflection error signal to the control station. The cantilever deflection (or the force) is measured using a laser deflection technique (Fig. 21.9). The DSP in the workstation controls the z-position of the piezo based on

direction as the sample is scanned under the tip. This tip deflection will change the direction of the reflected laser beam, changing the intensity difference between the top and bottom sets of photodetectors (AFM signal). In a mode of operation called the height mode, used for topographic imaging or for any other operation in which the normal force applied is to be kept constant, a feedback circuit is used to modulate the voltage applied to the PZT scanner in order to adjust the height of the PZT, so that the cantilever vertical deflection (given by the intensity difference between the top and bottom detector) will remain constant during scanning. The PZT height variation is thus a direct measure of the surface roughness of the sample.

In a large-sample AFM, force sensors based on optical deflection methods or scanning units are mounted on the microscope head (Fig. 21.9b). Because of the unwanted vibrations caused by cantilever movement, the lateral resolution of this design is somewhat poorer than the design in Fig. 21.9a in which the sample is scanned instead of the cantilever beam. The advantage of the large-sample AFM is that large samples can be easily measured.

Most AFMs can be used for topography measurements in the so-called tapping mode (intermittent contact mode), in what is also referred to as dynamic force microscopy. In the tapping mode, during the surface scan, the cantilever/tip assembly is sinusoidally vibrated by a piezo mounted above it, and the oscillating tip slightly taps the surface at the resonant frequency of the cantilever (70–400 kHz) with a constant (20–100 nm) amplitude of vertical oscillation, and a feedback loop keeps the average normal force constant (Fig. 21.10). The oscillating amplitude is kept large enough that the tip does not get stuck to the sample due to adhesive attraction. The tapping mode is used in topography measurements to minimize the effects of friction and other lateral forces to measure the topography of soft surfaces.

Topographic measurements can be made at any scanning angle. At first glance, the scanning angle may not appear to be an important parameter. However, the friction force between the tip and the sample will affect the topographic measurements in a parallel scan (scanning along the long axis of the cantilever). This means that a perpendicular scan may be more desirable. Generally, one picks a scanning angle which gives the same topographic data in both directions; this angle may be slightly different to that for the perpendicular scan.

The left-hand and right-hand quadrants of the photodetector are used to measure the friction force applied at the tip surface during sliding. In the so-called friction mode, the sample is scanned back and forth in a direction orthogonal to the long axis of the cantilever beam. Friction force between the sample and the tip will twist the cantilever. As a result, the laser beam will be deflected out of the plane defined by the incident beam and the beam is reflected vertically from an untwisted cantilever. This produces a difference in laser beam intensity between the beams received by the left-hand and right-hand sets of quadrants of the photodetector. The intensity difference between the two sets of detectors (FFM signal) is directly related to the degree of twisting and hence to the magnitude of the friction force. This method provides three-dimensional maps of the friction force. One problem associated with this method is that any misalignment between the laser beam and the photodetector axis introduces errors into the measurement. However, by following the procedures developed by *Ruan* and *Bhushan* [21.136], in which the average FFM signal for the sample scanned in two opposite directions is subtracted from the friction profiles of each of the two scans, the misalignment effect can be eliminated. By following the friction force calibration procedures developed by *Ruan* and *Bhushan* [21.136], voltages corresponding to friction forces can be converted to force units. The coefficient of friction is obtained from the slope of the friction force data measured as a function of the normal load, which typically ranges from 10 to 150 nN. This approach eliminates any contributions from adhesive forces [21.10]. To calculate the coefficient of friction based on a single point measurement, the friction force should be divided by the sum of the normal load applied and the intrinsic adhesive force. Furthermore, it should be pointed out that the coefficient of friction is not independent of load for single-asperity contact. This is discussed in more detail later.

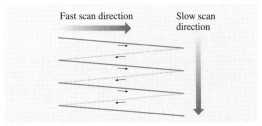

Fig. 21.11 Schematic of triangular pattern trajectory of the AFM tip as the sample is scanned in two dimensions. During imaging, data are only recorded during scans along the *solid scan lines*

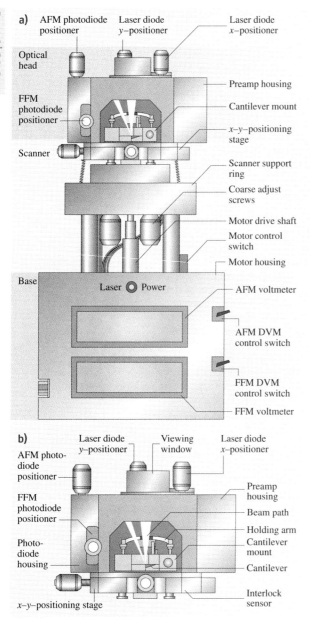
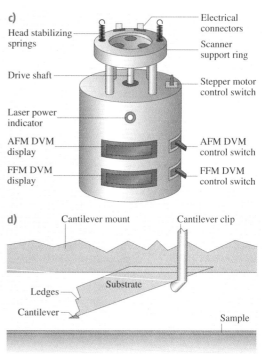

Fig. 21.12a–d Schematics of a commercial AFM/FFM made by Digital Instruments, Inc. (a) Front view, (b) optical head, (c) base, and (d) cantilever substrate mounted on cantilever mount (not to scale)

scan rates of less than 0.5 to 122 Hz are typically used. Higher scan rates are used for smaller scan lengths. For example, the scan rates in the fast and slow scan directions for an area of 10 μm × 10 μm scanned at 0.5 Hz are 10 μm/s and 20 nm/s, respectively.

We now describe the construction of a small-sample AFM in more detail. It consists of three main parts: the optical head which senses the cantilever deflection; a PZT tube scanner which controls the scanning motion of the sample mounted on one of its ends; and the base, which supports the scanner and head and includes circuits for the deflection signal (Fig. 21.12a). The AFM connects directly to a control system. The optical head consists of a laser diode stage, a photodiode stage preamp board, the cantilever mount and its holding arm, and the deflected beam reflecting mirror, which reflects the deflected beam toward the photodiode (Fig. 21.12b). The laser diode stage is a tilt stage used to adjust the position of the laser beam relative to the cantilever. It consists of the laser diode, collimator, focusing lens, base-

The tip is scanned in such a way that its trajectory on the sample forms a triangular pattern (Fig. 21.11). Scanning speeds in the fast and slow scan directions depend on the scan area and scan frequency. Scan sizes ranging from less than 1 nm × 1 nm to 125 μm × 125 μm and

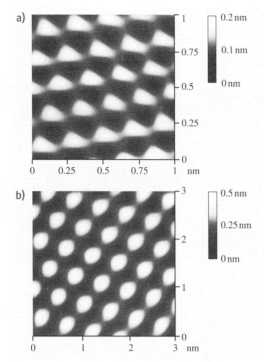

Fig. 21.13a,b Typical AFM images of freshly-cleaved (**a**) highly oriented pyrolytic graphite and (**b**) mica surfaces taken using a square pyramidal Si_3N_4 tip

ner fits into the scanner support ring mounted on the base of the microscope (Fig. 21.12c). The stepper motor is controlled manually with the switch on the upper surface of the base and automatically by the computer during the tip–engage and tip–withdraw processes.

The scan sizes available for these instruments are $0.7\,\mu m$, $12\,\mu m$ and $125\,\mu m$. The scan rate must be decreased as the scan size is increased. A maximum scan rate of 122 Hz can be used. Scan rates of about 60 Hz should be used for small scan lengths ($0.7\,\mu m$). Scan rates of 0.5 to 2.5 Hz should be used for large scans on samples with tall features. High scan rates help reduce drift, but they can only be used on flat samples with small scan sizes. The scan rate or the scanning speed (length/time) in the fast scan direction is equal to twice the scan length multiplied by the scan rate in Hz, and in the slow direction it is equal to the scan length multiplied by the scan rate in Hz divided by number of data points in the transverse direction. For example, for a scan size of $10\,\mu m \times 10\,\mu m$ scanned at 0.5 Hz, the scan rates in the fast and slow scan directions are $10\,\mu m/s$ and $20\,nm/s$, respectively. Normally 256×256 data points are taken for each image. The lateral resolution at larger scans is approximately equal to the scan length divided by 256. The piezo tube requires x–y-calibration, which is carried out by imaging an appropriate calibration standard. Cleaved graphite is used for small scan heads, while two-dimensional grids (a gold-plated rule) can be used for long-range heads.

plate, and the x- and y-laser diode positioners. The positioners are used to place the laser spot on the end of the cantilever. The photodiode stage is an adjustable stage used to position the photodiode elements relative to the reflected laser beam. It consists of the split photodiode, the base plate, and the photodiode positioners. The deflected beam reflecting mirror is mounted on the upper left in the interior of the head. The cantilever mount is a metal (for operation in air) or glass (for operation in water) block which holds the cantilever firmly at the proper angle (Fig. 21.12d). Next, the tube scanner consists of an Invar cylinder holding a single tube made of piezoelectric crystal which imparts the necessary three-dimensional motion to the sample. Mounted on top of the tube is a magnetic cap on which the steel sample puck is placed. The tube is rigidly held at one end with the sample mounted on the other end of the tube. The scanner also contains three fine-pitched screws which form the mount for the optical head. The optical head rests on the tips of the screws, which are used to adjust the position of the head relative to the sample. The scan-

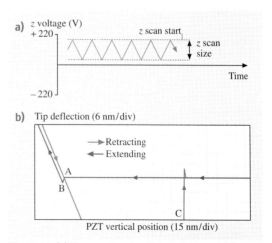

Fig. 21.14 (**a**) Force calibration Z waveform, and (**b**) a typical force–distance curve for a tip in contact with a sample. Contact occurs at point B; tip breaks free of adhesive forces at point C as the sample moves away from the tip

Examples of AFM images of freshly cleaved highly oriented pyrolytic (HOP) graphite and mica surfaces are shown in Fig. 21.13 [21.50, 110, 114]. Images with near-atomic resolution are obtained.

The force calibration mode is used to study interactions between the cantilever and the sample surface. In the force calibration mode, the x- and y-voltages applied to the piezo tube are held at zero and a sawtooth voltage is applied to the z-electrode of the piezo tube (Fig. 21.14a). At the start of the force measurement the cantilever is in its rest position. By changing the applied voltage, the sample can be moved up and down relative to the stationary cantilever tip. As the piezo moves the sample up and down, the cantilever deflection signal from the photodiode is monitored. The force–distance curve, a plot of the cantilever tip deflection signal as a function of the voltage applied to the piezo tube, is obtained. Figure 21.14b shows the typical features of a force–distance curve. The arrowheads indicate the direction of piezo travel. As the piezo extends, it approaches the tip, which is in mid-air at this point and hence shows no deflection. This is indicated by the flat portion of the curve. As the tip approaches the sample to within a few nanometers (point A), an attractive force kicks in between the atoms of the tip surface and the atoms of the surface of the sample. The tip is pulled towards the sample and contact occurs at point B on the graph. From this point on, the tip is in contact with the surface, and as the piezo extends further, the tip gets deflected further. This is represented by the sloped portion of the curve. As the piezo retracts, the tip moves beyond the zero deflection (flat) line due to attractive forces (van der Waals forces and long-range meniscus forces), into the adhesive regime. At point C in the graph, the tip snaps free of the adhesive forces, and is again in free air. The horizontal distance between points B and C along the retrace line gives the distance moved by the tip in the adhesive regime. Multiplying this distance by the stiffness of the cantilever gives the adhesive force. Incidentally, the horizontal shift between the loading and unloading curves results from the hysteresis in the PZT tube [21.4].

Multimode Capabilities

The multimode AFM can be used for topography measurements in the contact mode and tapping mode, described earlier, and for measurements of lateral (friction) force, electric force gradients and magnetic force gradients.

The multimode AFM, when used with a grounded conducting tip, can be used to measure electric field gradients by oscillating the tip near its resonant frequency. When the lever encounters a force gradient from the electric field, the effective spring constant of the cantilever is altered, changing its resonant frequency. Depending on which side of the resonance curve is chosen, the oscillation amplitude of the cantilever increases or decreases due to the shift in the resonant frequency. By recording the amplitude of the cantilever, an image revealing the strength of the electric field gradient is obtained.

In the magnetic force microscope (MFM), used with a magnetically coated tip, static cantilever deflection is detected when a magnetic field exerts a force on the tip, and MFM images of magnetic materials can be obtained. MFM sensitivity can be enhanced by oscillating the cantilever near its resonant frequency. When the tip encounters a magnetic force gradient, the effective spring constant (and hence the resonant frequency) is shifted. By driving the cantilever above or below the resonant frequency, the oscillation amplitude varies as the resonance shifts. An image of the magnetic field gradient is obtained by recording the oscillation amplitude as the tip is scanned over the sample.

Topographic information is separated from the electric field gradient and magnetic field images using the so-called lift mode. In lift mode, measurements are taken in two passes over each scan line. In the first pass, topographical information is recorded in the standard tapping mode, where the oscillating cantilever lightly taps the surface. In the second pass, the tip is lifted to a user-selected separation (typically 20–200 nm) between the tip and local surface topography. By using stored topographical data instead of standard feedback, the tip–sample separation can be kept constant. In this way, the cantilever amplitude can be used to measure electric field force gradients or relatively weak but long-range magnetic forces without being influenced by topographic features. Two passes are made for every scan line, producing separate topographic and magnetic force images.

Electrochemical AFM

This option allows one to perform electrochemical reactions on the AFM. The technique involves a potentiostat, a fluid cell with a transparent cantilever holder and electrodes, and the software required to operate the potentiostat and display the results of the electrochemical reaction.

21.2.3 AFM Probe Construction

Various probes (cantilevers and tips) are used for AFM studies. The cantilever stylus used in the AFM should meet the following criteria: (1) low normal spring constant (stiffness); (2) high resonant frequency; (3) high cantilever quality factor Q; (4) high lateral spring constant (stiffness); (5) short cantilever length; (6) incorporation of components (such as mirror) for deflection sensing; and (7) a sharp protruding tip [21.137]. In order to register a measurable deflection with small forces, the cantilever must flex with a relatively low force (on the order of few nN), requiring vertical spring constants of 10^{-2} to 10^2 N/m for atomic resolution in the contact profiling mode. The data rate or imaging rate in the AFM is limited by the mechanical resonant frequency of the cantilever. To achieve a large imaging bandwidth, the AFM cantilever should have a resonant frequency of more than about 10 kHz (30–100 kHz is preferable), which makes the cantilever the least sensitive part of the system. Fast imaging rates are not just a matter of convenience, since the effects of thermal drifts are more pronounced with slow scanning speeds. The combined requirements of a low spring constant and a high resonant frequency are met by reducing the mass of the cantilever. The quality factor Q ($= \omega_R/(c/m)$, where ω_R is the resonant frequency of the damped oscillator, c is the damping constant and m is the mass of the oscillator) should have a high value for some applications. For example, resonance curve detection is a sensitive modulation technique for measuring small force gradients in noncontact imaging. Increasing the Q increases the sensitivity of the measurements. Mechanical Q values of 100–1000 are typical. In contact modes, the Q value is of less importance. A high lateral cantilever spring constant is desirable in order to reduce the effect of lateral forces in the AFM, as frictional forces can cause appreciable lateral bending of the cantilever. Lateral bending results in erroneous topography measurements. For friction measurements, cantilevers with reduced lateral rigidity are preferred. A sharp protruding tip must be present at the end of the cantilever to provide a well-defined interaction with the sample over a small area. The tip radius should be much smaller than the radii of the corrugations in the sample in order for these to be measured accurately. The lateral spring constant depends critically on the tip length. Additionally, the tip should be centered at the free end.

In the past, cantilevers have been cut by hand from thin metal foils or formed from fine wires. Tips for these cantilevers were prepared by attaching diamond fragments to the ends of the cantilevers by hand, or in the case of wire cantilevers, electrochemically etching the wire to a sharp point. Several cantilever geometries for wire cantilevers have been used. The simplest geometry is the L-shaped cantilever, which is usually made by bending a wire at a 90° angle. Other geometries include single-V and double-V geometries, with a sharp tip attached at the apex of the V, and double-X configuration with a sharp tip attached at the intersection [21.31, 138]. These cantilevers can be constructed with high vertical spring constants. For example, a double-cross cantilever with an effective spring constant of 250 N/m was used by *Burnham* and *Colton* [21.31]. The small size and low mass needed in the AFM make hand fabrication of the cantilever a difficult process with poor reproducibility. Conventional microfabrication techniques are ideal for constructing planar thin-film structures which have submicron lateral dimensions. The triangular (V-shaped) cantilevers have improved (higher) lateral spring constants in comparison to rectangular cantilevers. In terms of spring constants, the triangular cantilevers are approximately equivalent to two rectangular cantilevers placed in parallel [21.137]. Although the macroscopic radius of a photolithographically patterned corner is seldom much less than about 50 nm, microscopic asperities on the etched surface provide tips with near-atomic dimensions.

Cantilevers have been used from a whole range of materials. Cantilevers made of Si_3N_4, Si, and dia-

Table 21.2 Relevant properties of materials used for cantilevers

Property	Young's modulus (E) (GPa)	Density (ρg) (kg/m³)	Microhardness (GPa)	Speed of sound ($\sqrt{E/\rho}$) (m/s)
Diamond	900–1050	3515	78.4–102	17 000
Si_3N_4	310	3180	19.6	9900
Si	130–188	2330	9–10	8200
W	350	19 310	3.2	4250
Ir	530	–	≈ 3	5300

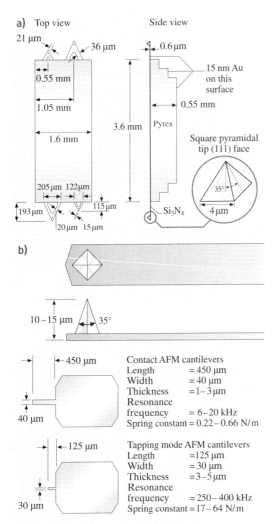

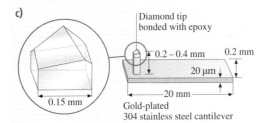

Fig. 21.15a–c Schematics of (**a**) triangular cantilever beam with square-pyramidal tips made of PECVD Si_3N_4, (**b**) rectangular cantilever beams with square-pyramidal tips made of etched single-crystal silicon, and (**c**) rectangular cantilever stainless steel beam with three-sided pyramidal natural diamond tip ◀

mond are the most common. The Young's modulus and the density are the material parameters that determine the resonant frequency, aside from the geometry. Table 21.2 shows the relevant properties and the speed of sound, indicative of the resonant frequency for a given shape. Hardness is an important indicator of the durability of the cantilever, and is also listed in the table. Materials used for STM cantilevers are also included.

Silicon nitride cantilevers are less expensive than those made of other materials. They are very rugged and well suited to imaging in almost all environments. They are especially compatible with organic and biological materials. Microfabricated triangular silicon nitride beams with integrated square pyramidal tips made using plasma-enhanced chemical vapor deposition (PECVD) are the most common [21.137]. Four cantilevers, marketed by Digital Instruments, with different sizes and spring constants located on cantilever substrate made of boron silicate glass (Pyrex), are shown in Figs. 21.15a and 21.16. The two pairs of

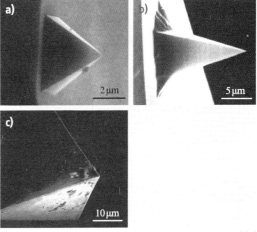

Fig. 21.16a–c SEM micrographs of a square-pyramidal PECVD Si_3N_4 tip (**a**), a square-pyramidal etched single-crystal silicon tip (**b**), and a three-sided pyramidal natural diamond tip (**c**)

Table 21.3 Measured vertical spring constants and natural frequencies of triangular (V-shaped) cantilevers made of PECVD Si_3N_4 (data provided by Digital Instruments, Inc.)

Cantilever dimension	Spring constant (k_z) (N/m)	Natural frequency (ω_0) (kHz)
115 μm long, narrow leg	0.38	40
115 μm long, wide leg	0.58	40
193 μm long, narrow leg	0.06	13–22
193 μm long, wide leg	0.12	13–22

Table 21.4 Vertical (k_z), lateral (k_y), and torsional (k_{yT}) spring constants of rectangular cantilevers made of Si (IBM) and PECVD Si_3N_4 (source: Veeco Instruments, Inc.)

Dimensions/stiffness	Si cantilever	Si_3N_4 cantilever
Length L (μm)	100	100
Width b (μm)	10	20
Thickness h (μm)	1	0.6
Tip length ℓ (μm)	5	3
k_z (N/m)	0.4	0.15
k_y (N/m)	40	175
k_{yT} (N/m)	120	116
ω_0 (kHz)	≈ 90	≈ 65

Note: $k_z = Ebh^3/(4L^3)$, $k_y = Eb^3h/(4\ell^3)$, $k_{yT} = Gbh^3/(3L\ell^2)$, and $\omega_0 = [k_z/(m_c + 0.24bhL\rho)]^{1/2}$, where E is Young's modulus, G is the modulus of rigidity [$= E/2(1+\nu)$, ν is Poisson's ratio], ρ is the mass density of the cantilever, and m_c is the concentrated mass of the tip (≈ 4 ng) [21.94]. For Si, $E = 130$ GPa, $\rho g = 2300$ kg/m^3, and $\nu = 0.3$. For Si_3N_4, $E = 150$ GPa, $\rho g = 3100$ kg/m^3, and $\nu = 0.3$

cantilevers on each substrate measure about 115 and 193 μm from the substrate to the apex of the triangular cantilever, with base widths of 122 and 205 μm, respectively. The cantilever legs, which are of the same thickness (0.6 μm) in all the cantilevers, are available in wide and narrow forms. Only one cantilever is selected and used from each substrate. The calculated spring constants and measured natural frequencies for each of the configurations are listed in Table 21.3. The most commonly used cantilever beam is the 115 μm long, wide-legged cantilever (vertical spring constant = 0.58 N/m). Cantilevers with smaller spring constants should be used on softer samples. The pyramidal tip is highly symmetric, and the end has a radius of about 20–50 nm. The side walls of the tip have a slope of 35° and the lengths of the edges of the tip at the cantilever base are about 4 μm.

An alternative to silicon nitride cantilevers with integrated tips are microfabricated single-crystal silicon cantilevers with integrated tips. Si tips are sharper than Si_3N_4 tips because they are formed directly by anisotropic etching of single-crystal Si, rather than through the use of an etch pit as a mask for the deposited material [21.139]. Etched single-crystal n-type silicon rectangular cantilevers with square pyramidal tips of radii < 10 nm for contact and tapping mode (tapping-mode etched silicon probe or TESP) AFMs are commercially available from Digital Instruments and Nanosensors GmbH, Aidlingen, Germany (Figs. 21.15b and 21.16). Spring constants and resonant frequencies are also presented in the Fig. 21.15b.

Commercial triangular Si_3N_4 cantilevers have a typical width : thickness ratio of 10 to 30, which results in spring constants that are 100 to 1000 times stiffer in the lateral direction than in the normal direction. Therefore, these cantilevers are not well suited for torsion. For friction measurements, the torsional spring constant should be minimized in order to be sensitive to the lateral force. Rather long cantilevers with small thicknesses and large tip lengths are most suitable. Rectangular beams have smaller torsional spring constants than the triangular (V-shaped) cantilevers. Table 21.4 lists the spring constants (with the full length of the beam used) in three directions for typical rectangular beams. We note that the lateral and torsional spring constants are about two orders of magnitude larger than the normal spring constants. A cantilever beam required for the tapping mode is quite stiff and may not be sensitive enough for friction measurements. *Meyer* et al. [21.140] used a specially designed rectangular silicon cantilever with length = 200 μm, width = 21 μm, thickness = 0.4 μm, tip length = 12.5 μm and shear modulus = 50 GPa, giving a normal spring constant of 0.007 N/m and a torsional spring constant of 0.72 N/m, which gives a lateral force sensitivity of 10 pN and an angle of resolution of 10^{-7} rad. Using this particular geometry, the sensitivity to lateral forces can be improved by about a factor of 100 compared with commercial V-shaped Si_3N_4 or the rectangular Si or Si_3N_4 cantilevers used by *Meyer* and *Amer* [21.8], with torsional spring constants of ≈ 100 N/m. *Ruan* and *Bhushan* [21.136] and *Bhushan* and *Ruan* [21.9] used 115 μm long, wide-legged V-shaped cantilevers made of Si_3N_4 for friction measurements.

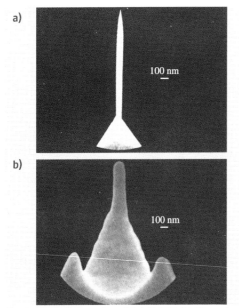

Fig. 21.17a,b Schematics of (**a**) HART Si_3N_4 probe, and (**b**) an FIB-milled Si_3N_4 probe

For scratching, wear and indentation studies, single-crystal natural diamond tips ground to the shape of a three-sided pyramid with an apex angle of either 60° or 80° and a point sharpened to a radius of about 100 nm are commonly used [21.4, 10] (Figs. 21.15c and 21.16). The tips are bonded with conductive epoxy to a gold-plated 304 stainless steel spring sheet (length = 20 mm, width = 0.2 mm, thickness = 20 to 60 μm) which acts as a cantilever. The free length of the spring is varied in order to change the beam stiffness. The normal spring constant of the beam ranges from about 5 to 600 N/m for a 20 μm thick beam. The tips are produced by R-DEC Co., Tsukuba, Japan.

High aspect ratio tips are used to image within trenches. Examples of two probes used are shown in Fig. 21.17. These high aspect ratio tip (HART) probes are produced from conventional Si_3N_4 pyramidal probes. Through a combination of focused ion beam (FIB) and high-resolution scanning electron microscopy (SEM) techniques, a thin filament is grown at the apex of the pyramid. The probe filament is $\approx$ 1 μm long and 0.1 μm in diameter. It tapers to an extremely sharp point (with a radius that is better than the resolutions of most SEMs). The long thin shape and sharp radius make it ideal for imaging within *vias* of microstructures and trenches (> 0.25 μm). This is, however, unsuitable for

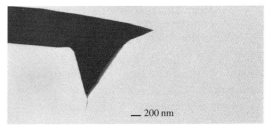

Fig. 21.18 SEM micrograph of a multiwall carbon nanotube (MWNT) tip physically attached to a single-crystal silicon, square-pyramidal tip (courtesy of Piezomax Technologies, Inc.)

imaging structures at the atomic level, since probe flexing can create image artefacts. A FIB-milled probe is used for atomic-scale imaging, which is relatively stiff yet allows for closely spaced topography. These probes start out as conventional Si_3N_4 pyramidal probes, but the pyramid is FIB-milled until a small cone shape is formed which has a high aspect ratio and is 0.2–0.3 μm in length. The milled probes permit nanostructure resolution without sacrificing rigidity. These types of probes are manufactured by various manufacturers including Materials Analytical Services.

Carbon nanotube tips with small diameters and high aspect ratios are used for high-resolution imaging of surfaces and of deep trenches, in the tapping mode or the noncontact mode. Single-wall carbon nanotubes (SWNTs) are microscopic graphitic cylinders that are 0.7 to 3 nm in diameter and up to many microns in length. Larger structures called multiwall carbon nanotubes (MWNTs) consist of nested, concentrically arranged SWNTs and have diameters of 3 to 50 nm. MWNT carbon nanotube AFM tips are produced by manual assembly [21.141], chemical vapor deposition (CVD) synthesis, and a hybrid fabrication process [21.142]. Figure 21.18 shows a TEM micrograph of a carbon nanotube tip, ProbeMax, commercially produced by mechanical assembly by Piezomax Technologies, Inc. (Middleton, USA). To fabricate these tips, MWNTs are produced using a carbon arc and they are physically attached to the single-crystal silicon, square-pyramidal tips in the SEM, using a manipulator and the SEM stage to independently control the nanotubes and the tip. When the nanotube is first attached to the tip, it is usually too long to image with. It is shortened by placing it in an AFM and applying voltage between the tip and the sample. Nanotube tips are also commercially produced by CVD synthesis by NanoDevices (Santa Barbara, USA).

21.2.4 Friction Measurement Methods

The two methods for performing friction measurements that are based on the work by *Ruan* and *Bhushan* [21.136] are now described in more detail (also see [21.8]). The scanning angle is defined as the angle relative to the *y*-axis in Fig. 21.19a. This is also the long axis of the cantilever. The zero-degree scanning angle corresponds to the sample scan in the *y*-direction, and the 90° scanning angle corresponds to the sample scan perpendicular to this axis in the *x*–*y*-plane (along *x*-axis). If both the *y*- and −*y*-directions are scanned, we call this a *parallel scan*. Similarly, a *perpendicular scan* means that both the *x*- and −*x*-directions are scanned. The direction of sample travel for each of these two methods is illustrated in Fig. 21.19b.

Using method 1 (*height* mode with parallel scans) in addition to topographic imaging, it is also possible to measure friction force when the sample scanning direction is parallel to the *y*-direction (parallel scan). If there was no friction force between the tip and the moving sample, the topographic feature would be the only factor that would cause the cantilever to be deflected vertically. However, friction force does exist on all surfaces that are in contact where one of the surfaces is moving relative to the other. The friction force between the sample and the tip will also cause the cantilever to be deflected. We assume that the normal force between the sample and the tip is W_0 when the sample is stationary (W_0 is typically 10 to 200 nN), and the friction force between the sample and the tip is W_f as the sample is scanned by the tip. The direction of the friction force (W_f) is reversed as the scanning direction of the sample is reversed from the positive (y) to the negative ($-y$) direction ($W_{f(y)} = -W_{f(-y)}$).

When the vertical cantilever deflection is set at a constant level, it is the total force (normal force and friction force) applied to the cantilever that keeps the

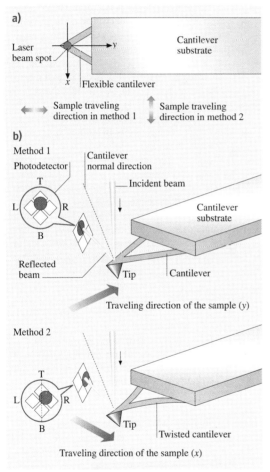

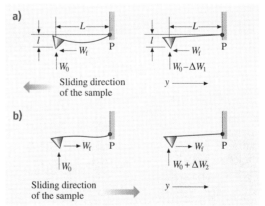

Fig. 21.19 (a) Schematic defining the *x*- and *y*-directions relative to the cantilever, and showing the direction of sample travel in two different measurement methods discussed in the text. (b) Schematic of deformation of the tip and cantilever shown as a result of sliding in the *x*- and *y*-directions. A twist is introduced to the cantilever if the scanning is performed in the *x*-direction ((b), *lower part*) (after [21.136])

Fig. 21.20 (a) Schematic showing an additional bending of the cantilever due to friction force when the sample is scanned in the *y*- or −*y*-directions (*left*). (b) This effect can be canceled out by adjusting the piezo height using a feedback circuit (*right*) (after [21.136])

cantilever deflection at this level. Since the friction force is directed in the opposite direction to the direction of travel of the sample, the normal force will have to be adjusted accordingly when the sample reverses its traveling direction, so that the total deflection of the cantilever will remain the same. We can calculate the difference in the normal force between the two directions of travel for a given friction force W_f. First, since the deflection is constant, the total moment applied to the cantilever is constant. If we take the reference point to be the point where the cantilever joins the cantilever holder (substrate), point P in Fig. 21.20, we have the following relationship

$$(W_0 - \Delta W_1)L + W_f \ell$$
$$= (W_0 + \Delta W_2)L - W_f \ell \quad (21.1)$$

or

$$(\Delta W_1 + \Delta W_2)L = 2W_f \ell . \quad (21.2)$$

Thus

$$W_f = (\Delta W_1 + \Delta W_2)L/(2\ell) , \quad (21.3)$$

where ΔW_1 and ΔW_2 are the absolute values of the changes in normal force when the sample is traveling in the $-y$- and y-directions, respectively, as shown in Fig. 21.20; L is the length of the cantilever; ℓ is the vertical distance between the end of the tip and point P. The coefficient of friction (μ) between the tip and the sample is then given as

$$\mu = \frac{W_f}{W_0} = \left(\frac{(\Delta W_1 + \Delta W_2)}{W_0}\right)\left(\frac{L}{2\ell}\right) . \quad (21.4)$$

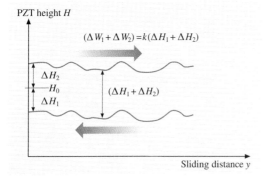

Fig. 21.21 Schematic illustration of the height difference for the piezoelectric tube scanner as the sample is scanned in the y- and $-y$-directions

There are adhesive and interatomic attractive forces between the cantilever tip and the sample at all times. The adhesive force can be due to water from the capillary condensation and other contaminants present at the surface, which form meniscus bridges [21.4, 143, 144] and the interatomic attractive force includes van der Waals attractions [21.18]. If these forces (and the effect of indentation too, which is usually small for rigid samples) can be neglected, the normal force W_0 is then equal to the initial cantilever deflection H_0 multiplied by the spring constant of the cantilever. ($\Delta W_1 + \Delta W_2$) can be derived by multiplying the same spring constant by the change in height of the piezo tube between the two traveling directions (y- and $-y$-directions) of the sample. This height difference is denoted as ($\Delta H_1 + \Delta H_2$), shown schematically in Fig. 21.21. Thus, (21.4) can be rewritten as

$$\mu = \frac{W_f}{W_0} = \left(\frac{(\Delta H_1 + \Delta H_2)}{H_0}\right)\left(\frac{L}{2\ell}\right) . \quad (21.5)$$

Since the vertical position of the piezo tube is affected by the topographic profile of the sample surface in addition to the friction force being applied at the tip, this difference must be found point-by-point at the same location on the sample surface, as shown in Fig. 21.21. Subtraction of point-by-point measurements may introduce errors, particularly for rough samples. We will come back to this point later. In addition, precise measurements of L and ℓ (which should include the cantilever angle) are also required.

If the adhesive force between the tip and the sample is large enough that it cannot be neglected, it should be included in the calculation. However, determinations of this force can involve large uncertainties, which is introduced into (21.5). An alternative approach is to make the measurements at different normal loads and to use $\Delta(H_0)$ and $\Delta(\Delta H_1 + \Delta H_2)$ in (21.5). Another comment on (21.5) is that, since only the ratio between $(\Delta H_1 + \Delta H_2)$ and H_0 enters this equation, the vertical position of the piezo tube H_0 and the difference in position $(\Delta H_1 + \Delta H_2)$ can be in volts as long as the vertical travel of the piezo tube and the voltage applied to have a linear relationship. However, if there is a large nonlinearity between the piezo tube traveling distance and the applied voltage, this nonlinearity must be included in the calculation.

It should also be pointed out that (21.4) and (21.5) are derived under the assumption that the friction force W_f is the same for the two scanning directions of the sample. This is an approximation, since the normal force is slightly different for the two scans and the

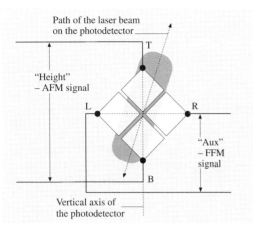

Fig. 21.22 The trajectory of the laser beam on the photodetectors as the cantilever is vertically deflected (with no torsional motion) with respect to the laser beam for a misaligned photodetector. For a change of normal force (vertical deflection of the cantilever), the laser beam is projected to a different position on the detector. Due to a misalignment, the projected trajectory of the laser beam on the detector is not parallel with the detector vertical axis (the line T–B) (after [21.136])

friction may be direction-dependent. However, this difference is much smaller than W_0 itself. We can ignore the second-order correction.

Method 2 (*aux* mode with perpendicular scan) of measuring friction was suggested by *Meyer* and *Amer* [21.8]. The sample is scanned perpendicular to the long axis of the cantilever beam (along the x- or $-x$-direction in Fig. 21.19a) and the outputs from the two horizontal quadrants of the photodiode detector are measured. In this arrangement, as the sample moves under the tip, the friction force will cause the cantilever to twist. Therefore, the light intensity between the left and right (L and R in Fig. 21.19b, right) detectors will be different. The differential signal between the left and right detectors is denoted the FFM signal $[(L-R)/(L+R)]$. This signal can be related to the degree of twisting, and hence to the magnitude of friction force. Again, because possible errors in measurements of the normal force due to the presence of adhesive force at the tip–sample interface, the slope of the friction data (FFM signal versus normal load) needs to be measured for an accurate value of the coefficient of friction.

While friction force contributes to the FFM signal, friction force may not be the only contributing factor in commercial FFM instruments (for example,

NanoScope IV). One can see this if we simply engage the cantilever tip with the sample. The left and right detectors can be balanced beforehand by adjusting the positions of the detectors so that the intensity difference between these two detectors is zero (FFM signal is zero). Once the tip is engaged with the sample, this signal is no longer zero, even if the sample is not moving in the x–y-plane with no friction force applied. This would be a detrimental effect. It has to be understood and eliminated from the data acquisition before any quantitative measurement of friction force is made.

One of the reasons for this observation is as follows. The detectors may not have been properly aligned with respect to the laser beam. To be precise, the vertical axis of the detector assembly (the line joining T–B in Fig. 21.22) is not in the plane defined by the incident laser beam and the beam reflected from the untwisted cantilever (we call this plane the *beam plane*). When the cantilever vertical deflection changes due to a change in the normal force applied (without the sample being scanned in the x–y-plane), the laser beam will be reflected up and down and form a projected trajectory on the detector. (Note that this trajectory is in the defined beam plane.) If this trajectory is not coincident with the vertical axis of the detector, the laser beam will not evenly bisect the left and right quadrants of the detectors, even under the condition of no torsional motion of the cantilever (Fig. 21.22). Thus, when the laser beam is reflected up and down due a change in the normal force, the intensity difference between the left and right detectors will also change. In other words, the FFM signal will change as the normal force applied to the tip is changed, even if the tip is not experiencing any friction force. This (FFM) signal is unrelated to friction force or to the actual twisting of the cantilever. We will call this part of the FFM signal FFM_F, and the part which is truly related to friction force FFM_T.

The FFM_F signal can be eliminated. One way of doing this is as follows. First the sample is scanned in both the x- and the $-x$-directions and the FFM signals for scans in each direction are recorded. Since the friction force reverses its direction of action when the scanning direction is reversed from the x- to the $-x$-direction, the FFM_T signal will change signs as the scanning direction of the sample is reversed ($FFM_T(x) = -FFM_T(-x)$). Hence the FFM_T signal will be canceled out if we take the sum of the FFM signals for the two scans. The average value of the two scans will be related to FFM_F due to the misalignment,

$$FFM(x) + FFM(-x) = 2FFM_F.\qquad(21.6)$$

This value can therefore be subtracted from the original FFM signals of each of these two scans to obtain the true FFM signal (FFM_T). Or, alternatively, by taking the difference of the two FFM signals, one gets the FFM_T value directly

$$FFM(x) - FFM(-x) = FFM_T(x) - FFM_T(-x)$$
$$= 2 FFM_T(x). \qquad (21.7)$$

Ruan and *Bhushan* [21.136] have shown that the error signal (FFM_F) can be very large compared to the friction signal FFM_T, so correction is required.

Now we compare the two methods. The method of using the *height* mode and parallel scanning (method 1) is very simple to use. Technically, this method can provide 3-D friction profiles and the corresponding topographic profiles. However, there are some problems with this method. Under most circumstances, the piezo scanner displays hysteresis when the traveling direction of the sample is reversed. Therefore, the measured surface topographic profiles will be shifted relative to each other along the y-axis for the two opposite (y and $-y$) scans. This would make it difficult to measure the local difference in height of the piezo tube for the two scans. However, the average difference in height between the two scans and hence the average friction can still be measured. The measurement of average friction can serve as an internal means of friction force calibration. Method 2 is a more desirable approach. The subtraction of the FFM_F signal from FFM for the two scans does not introduce any error into local friction force data. An ideal approach when using this method would be to add the average values of the two profiles in order to get the error component (FFM_F) and then subtract this component from either profile to get true friction profiles in either directions. By performing measurements at various loads, we can get the average value of the coefficient of friction which then can be used to convert the friction profile to the coefficient of friction profile. Thus, any directionality and local variations in friction can be easily measured. In this method, since topography data are not affected by friction, accurate topography data can be measured simultaneously with friction data and a better localized relationship between the two can be established.

21.2.5 Normal Force and Friction Force Calibrations of Cantilever Beams

Based on *Ruan* and *Bhushan* [21.136], we now discuss normal force and friction force calibrations. In order to calculate the absolute values of normal and friction forces in Newtons using the measured AFM and FFM_T voltage signals, it is necessary to first have an accurate value of the spring constant of the cantilever (k_c). The spring constant can be calculated using the geometry and the physical properties of the cantilever material [21.8, 94, 137]. However, the properties of the PECVD Si_3N_4 (used to fabricate cantilevers) can be different from those of the bulk material. For example, using ultrasonics, we found the Young's modulus of the cantilever beam to be about 238 ± 18 GPa, which is less than that of bulk Si_3N_4 (310 GPa). Furthermore, the thickness of the beam is nonuniform and difficult to measure precisely. Since the stiffness of a beam goes as the cube of thickness, minor errors in precise measurements of thickness can introduce substantial stiffness errors. Thus one should measure the spring constant of the cantilever experimentally. *Cleveland* et al. [21.145] measured normal spring constants by measuring resonant frequencies of beams.

For normal spring constant measurement, *Ruan* and *Bhushan* [21.136] used a stainless steel spring sheet of known stiffness (width = 1.35 mm, thickness = 15 µm, free hanging length = 5.2 mm). One end of the spring was attached to the sample holder and the other end was made to contact with the cantilever tip during the measurement (Fig. 21.23). They measured the piezo travel for a given cantilever deflection. For a rigid sample (such as diamond), the piezo travel Z_t (measured from the point where the tip touches the sample) should equal

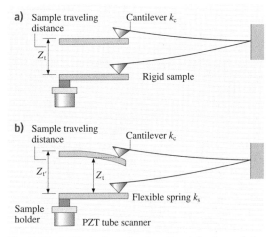

Fig. 21.23a,b Illustration showing the deflection of the cantilever as it is pushed by (**a**) a rigid sample, (**b**) a flexible spring sheet (after [21.136])

the cantilever deflection. To maintain the cantilever deflection at the same level using a flexible spring sheet, the new piezo travel $Z_{t'}$ would need to be different from Z_t. The difference between $Z_{t'}$ and Z_t corresponds to the deflection of the spring sheet. If the spring constant of the spring sheet is k_s, the spring constant of the cantilever k_c can be calculated by

$$(Z_{t'} - Z_t)k_s = Z_t k_c$$

or

$$k_c = k_s(Z_{t'} - Z_t)/Z_t \,. \tag{21.8}$$

The spring constant of the spring sheet (k_s) used in this study is calculated to be 1.54 N/m. For the wide-legged cantilever used in our study (length = 115 μm, base width = 122 μm, leg width = 21 μm and thickness = 0.6 μm), k_c was measured to be 0.40 N/m instead of the 0.58 N/m reported by its manufacturer – Digital Instruments, Inc. To relate the photodiode detector output to the cantilever deflection in nanometers, they used the same rigid sample to push against the AFM tip. Since the cantilever vertical deflection equals the sample traveling distance measured from the point where the tip touches the sample for a rigid sample, the photodiode output observed as the tip is pushed by the sample can be converted directly to the cantilever deflection. For these measurements, they found the conversion factor to be 20 nm/V.

The normal force applied to the tip can be calculated by multiplying the cantilever vertical deflection by the cantilever spring constant for samples that have very small adhesion with the tip. If the adhesive force between the sample and the tip is large, it should be included in the normal force calculation. This is particularly important in atomic-scale force measurements, because the typical normal force that is measured in this region is in the range of a few hundreds of nN to a few mN. The adhesive force could be comparable to the applied force.

The conversion of friction signal (from FFM_T) to friction force is not as straightforward. For example, one can calculate the degree of twisting for a given friction force using the geometry and the physical properties of the cantilever [21.53, 144]. One would need information about the detector such as its quantum efficiency, laser power, gain and so on in order to be able convert the signal into the degree of twisting. Generally speaking, this procedure can not be accomplished without having some detailed information about the instrument. This information is not usually provided by the manufacturer. Even if this information is readily available, errors may still occur when using this approach because there will always be variations as a result of the instrumental set-up. For example, it has been noticed that the measured FFM_T signal varies for the same sample when different AFM microscopes from the same manufacturer are used. This means that one can not calibrate the instrument experimentally using this calculation. O'Shea et al. [21.144] did perform a calibration procedure in which the torsional signal was measured as the sample was displaced a known distance laterally while ensuring that the tip did not slide over the surface. However, it is difficult to verify that tip sliding does not occur.

A new method of calibration is therefore required. There is a simpler, more direct way of doing this. The first method described above (method 1) of measuring friction can provide an absolute value of the coefficient of friction directly. It can therefore be used as an internal calibration technique for data obtained using method 2. Or, for a polished sample, which introduces the least error into friction measurements taken using method 1, method 1 can be used to calibrate the friction force for method 2. Then this calibration can be used for measurements taken using method 2. In method 1, the length of the cantilever required can be measured using an optical microscope; the length of the tip can be measured using a scanning electron microscope. The relative angle between the cantilever and the horizontal sample surface can be measured directly. This enables the coefficient of friction to be measured with few unknown parameters. The friction force can then be calculated by multiplying the coefficient of friction by the normal load. The FFM_T signal obtained using method 2 is then converted into the friction force. For their instrument, they found the conversion to be 8.6 nN/V.

21.3 AFM Instrumentation and Analyses

The performance of AFMs and the quality of AFM images greatly depend on the instrument available and the probes (cantilever and tips) in use. This section describes the mechanics of cantilevers, instrumentation and analysis of force detection systems for cantilever deflections, and scanning and control systems.

21.3.1 The Mechanics of Cantilevers

Stiffness and Resonances of Lumped Mass Systems

All of the building blocks of an AFM, including the body of the microscope itself and the force-measuring cantilevers, are mechanical resonators. These resonances can be excited either by the surroundings or by the rapid movement of the tip or the sample. To avoid problems due to building- or air-induced oscillations, it is of paramount importance to optimize the design of the AFM for high resonant frequencies. This usually means decreasing the size of the microscope [21.146]. By using cube-like or sphere-like structures for the microscope, one can considerably increase the lowest eigenfrequency. The fundamental natural frequency ω_0 of any spring is given by

$$\omega_0 = \frac{1}{2\pi} \sqrt{\frac{k}{m_{\text{eff}}}}, \tag{21.9}$$

where k is the spring constant (stiffness) in the normal direction and m_{eff} is the effective mass. The spring constant k of a cantilever beam with uniform cross section (Fig. 21.24) is given by [21.147]

$$k = \frac{3EI}{L^3}, \tag{21.10}$$

where E is the Young's modulus of the material, L is the length of the beam and I is the moment of inertia of the cross section. For a rectangular cross section with a width b (perpendicular to the deflection) and a height h one obtains the following expression for I

$$I = \frac{bh^3}{12}. \tag{21.11}$$

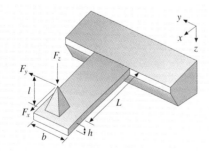

Fig. 21.24 A typical AFM cantilever with length L, width b, and height h. The height of the tip is ℓ. The material is characterized by the Young's modulus E, the shear modulus G and the mass density ρ. Normal (F_z), axial (F_x) and lateral (F_y) forces exist at the end of the tip

Combining (21.9)–(21.11), we get an expression for ω_0

$$\omega_0 = \sqrt{\frac{Ebh^3}{4L^3 m_{\text{eff}}}}. \tag{21.12}$$

The effective mass can be calculated using Raleigh's method. The general formula using Raleigh's method for the kinetic energy T of a bar is

$$T = \frac{1}{2} \int_0^L \frac{m}{L} \left(\frac{\partial z(x)}{\partial t} \right)^2 dx. \tag{21.13}$$

For the case of a uniform beam with a constant cross section and length L, one obtains for the deflection $z(x) = z_{\max} \left[1 - (3x/2L) + (x^3/2L^3) \right]$. Inserting $z_{\max}$ into (21.13) and solving the integral gives

$$T = \frac{1}{2} \int_0^L \frac{m}{L} \left[\frac{\partial z_{\max}(x)}{\partial t} \left(1 - \frac{3x}{2L} \right) + \left(\frac{x^3}{L^3} \right) \right]^2 dx$$

$$= \frac{1}{2} m_{\text{eff}} (z_{\max} t)^2,$$

which gives

$$m_{\text{eff}} = \frac{9}{20} m. \tag{21.14}$$

Substituting (21.14) into (21.12) and noting that $m = \rho L b h$, where ρ is the mass density, one obtains the following expression

$$\omega_0 = \left(\frac{\sqrt{5}}{3} \sqrt{\frac{E}{\rho}} \right) \frac{h}{L^2}. \tag{21.15}$$

It is evident from (21.15) that one way to increase the natural frequency is to choose a material with a high ratio E/ρ; see Table 21.2 for typical values of $\sqrt{E/\rho}$ for various commonly used materials. Another way to increase the lowest eigenfrequency is also evident in (21.15). By optimizing the ratio h/L^2, one can increase the resonant frequency. However, it does not help to make the length of the structure smaller than the width or height. Their roles will just be interchanged. Hence the optimum structure is a cube. This leads to the design rule that long, thin structures like sheet metal should be avoided. For a given resonant frequency, the quality factor Q should be as low as possible. This means that an inelastic medium such as rubber should be in contact with the structure in order to convert kinetic energy into heat.

Stiffness and Resonances of Cantilevers

Cantilevers are mechanical devices specially shaped to measure tiny forces. The analysis given in the previous section is applicable. However, to better understand the intricacies of force detection systems, we will discuss the example of a cantilever beam with uniform cross section (Fig. 21.24). The bending of a beam due to a normal load on the beam is governed by the Euler equation [21.147]

$$M = EI(x) \frac{d^2 z}{dx^2}, \qquad (21.16)$$

where M is the bending moment acting on the beam cross section. $I(x)$ is the moment of inertia of the cross section with respect to the neutral axis, defined by

$$I(x) = \int_z \int_y z^2 \, dy \, dz. \qquad (21.17)$$

For a normal force F_z acting at the tip,

$$M(x) = (L - x) F_z \qquad (21.18)$$

since the moment must vanish at the endpoint of the cantilever. Integrating (21.16) for a normal force F_z acting at the tip and observing that EI is a constant for beams with a uniform cross section, one gets

$$z(x) = \frac{L^3}{6EI} \left(\frac{x}{L}\right)^2 \left(3 - \frac{x}{L}\right) F_z. \qquad (21.19)$$

The slope of the beam is

$$z'(x) = \frac{Lx}{2EI} \left(2 - \frac{x}{L}\right) F_z. \qquad (21.20)$$

From (21.19) and (21.20), at the end of the cantilever (for $x = L$), for a rectangular beam, and by using an expression for I in (21.11), one gets

$$z(L) = \frac{4}{Eb} \left(\frac{L}{h}\right)^3 F_z, \qquad (21.21)$$

$$z'(L) = \frac{3}{2} \left(\frac{z}{L}\right). \qquad (21.22)$$

Now, the stiffness in the normal (z) direction k_z is

$$k_z = \frac{F_z}{z(L)} = \frac{Eb}{4} \left(\frac{h}{L}\right)^3. \qquad (21.23)$$

and the change in angular orientation of the end of cantilever beam is

$$\Delta\alpha = \frac{3}{2} \frac{z}{L} = \frac{6}{Ebh} \left(\frac{L}{h}\right)^2 F_z. \qquad (21.24)$$

Now we ask what will, to a first-order approximation, happen if we apply a lateral force F_y to the end of the tip (Fig. 21.24). The cantilever will bend sideways and it will twist. The stiffness in the lateral (y) direction k_y can be calculated with (21.23) by exchanging b and h

$$k_y = \frac{Eh}{4} \left(\frac{b}{L}\right)^3. \qquad (21.25)$$

Therefore, the bending stiffness in the lateral direction is larger than the stiffness for bending in the normal direction by $(b/h)^2$. The twisting or torsion on the other hand is more complicated to handle. For a wide, thin cantilever ($b \gg h$) we obtain torsional stiffness along y-axis k_{yT}

$$k_{yT} = \frac{Gbh^3}{3L\ell^2}, \qquad (21.26)$$

where G is the modulus of rigidity ($= E/2(1+\nu)$; ν is Poisson's ratio). The ratio of the torsional stiffness to the lateral bending stiffness is

$$\frac{k_{yT}}{k_y} = \frac{1}{2} \left(\frac{\ell b}{hL}\right)^2, \qquad (21.27)$$

where we assume $\nu = 0.333$. We see that thin, wide cantilevers with long tips favor torsion while cantilevers with square cross sections and short tips favor bending. Finally, we calculate the ratio between the torsional stiffness and the normal bending stiffness,

$$\frac{k_{yT}}{k_z} = 2 \left(\frac{L}{\ell}\right)^2. \qquad (21.28)$$

Equations (21.26) to (21.28) hold in the case where the cantilever tip is exactly in the middle axis of the cantilever. Triangular cantilevers and cantilevers with tips which are not on the middle axis can be dealt with by finite element methods.

The third possible deflection mode is the one from the force on the end of the tip along the cantilever axis, F_x (Fig. 21.24). The bending moment at the free end of the cantilever is equal to $F_x \ell$. This leads to the following modification of (21.18) for forces F_z and F_x

$$M(x) = (L - x) F_z + F_x \ell. \qquad (21.29)$$

Integration of (21.16) now leads to

$$z(x) = \frac{1}{2EI} \left[Lx^2 \left(1 - \frac{x}{3L}\right) F_z + \ell x^2 F_x \right] \qquad (21.30)$$

and

$$z'(x) = \frac{1}{EI} \left[\frac{Lx}{2} \left(2 - \frac{x}{L}\right) F_z + \ell x F_x \right]. \qquad (21.31)$$

Evaluating (21.30) and (21.31) at the end of the cantilever, we get the deflection and the tilt

$$z(L) = \frac{L^2}{EI} \left(\frac{L}{3} F_z - \frac{\ell}{2} F_x \right),$$

$$z'(L) = \frac{L}{EI} \left(\frac{L}{2} F_z + \ell F_x \right). \qquad (21.32)$$

From these equations, one gets

$$F_z = \frac{12EI}{L^3}\left[z(L) - \frac{Lz'(L)}{2}\right],$$

$$F_x = \frac{2EI}{\ell L^2}\left[2Lz'(L) - 3z(L)\right]. \quad (21.33)$$

A second class of interesting properties of cantilevers is their resonance behavior. For cantilever beams, one can calculate the resonant frequencies [21.147, 148]

$$\omega_n^{\text{free}} = \frac{\lambda_n^2}{2\sqrt{3}}\frac{h}{L^2}\sqrt{\frac{E}{\rho}} \quad (21.34)$$

with $\lambda_0 = (0.596864\ldots)\pi$, $\lambda_1 = (1.494175\ldots)\pi$, $\lambda_n \to (n+1/2)\pi$. The subscript n represents the order of the frequency, such as the fundamental, the second mode, and the nth mode.

A similar equation to (21.34) holds for cantilevers in rigid contact with the surface. Since there is an additional restriction on the movement of the cantilever, namely the location of its endpoint, the resonant frequency increases. Only the terms of λ_n change to [21.148]

$$\lambda_0' = (1.2498763\ldots)\pi, \quad \lambda_1' = (2.2499997\ldots)\pi,$$
$$\lambda_n' \to (n+1/4)\pi. \quad (21.35)$$

The ratio of the fundamental resonant frequency during contact to the fundamental resonant frequency when not in contact is 4.3851.

For the torsional mode we can calculate the resonant frequencies as

$$\omega_0^{\text{tors}} = 2\pi\frac{h}{Lb}\sqrt{\frac{G}{\rho}}. \quad (21.36)$$

For cantilevers in rigid contact with the surface, we obtain the following expression for the fundamental resonant frequency [21.148]

$$\omega_0^{\text{tors, contact}} = \frac{\omega_0^{\text{tors}}}{\sqrt{1+3(2L/b)^2}}. \quad (21.37)$$

The amplitude of the thermally induced vibration can be calculated from the resonant frequency using

$$\Delta z_{\text{therm}} = \sqrt{\frac{k_B T}{k}}, \quad (21.38)$$

where k_B is Boltzmann's constant and T is the absolute temperature. Since AFM cantilevers are resonant structures, sometimes with rather high Q values, the thermal noise is not as evenly distributed as (21.38) suggests. The spectral noise density below the peak of the response curve is [21.148]

$$z_0 = \sqrt{\frac{4k_B T}{k\omega_0 Q}} \quad (\text{in m}/\sqrt{\text{Hz}}), \quad (21.39)$$

where Q is the quality factor of the cantilever, described earlier.

21.3.2 Instrumentation and Analyses of Detection Systems for Cantilever Deflections

A summary of selected detection systems was provided in Fig. 21.8. Here we discuss the pros and cons of various systems in detail.

Optical Interferometer Detection Systems
Soon after the first papers on the AFM [21.2] appeared, which used a tunneling sensor, an instrument based on an interferometer was published [21.149]. The sensitivity of the interferometer depends on the wavelength of the light employed in the apparatus. Figure 21.25 shows the principle of such an interferometric design. The light incident from the left is focused by a lens onto the cantilever. The reflected light is collimated by the same lens and interferes with the light reflected at the flat. To separate the reflected light from the incident light, a $\lambda/4$ plate converts the linearly polarized incident light into circularly polarized light. The reflected light is made linearly polarized again by the $\lambda/4$-plate, but with a polarization orthogonal to that of the incident light. The polarizing beam splitter then deflects the reflected light to the photodiode.

Homodyne Interferometer. To improve the signal-to-noise ratio of the interferometer, the cantilever is driven by a piezo near its resonant frequency. The amplitude Δz of the cantilever as a function of driving

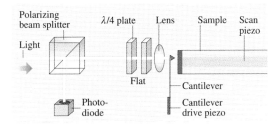

Fig. 21.25 Principle of an interferometric AFM. The light from the laser light source is polarized by the polarizing beam splitter and focused onto the back of the cantilever. The light passes twice through a quarter-wave plate and is hence orthogonally polarized to the incident light. The second arm of the interferometer is formed by the flat. The interference pattern is modulated by the oscillating cantilever

frequency Ω is

$$\Delta z(\Omega) = \Delta z_0 \frac{\Omega_0^2}{\sqrt{(\Omega^2 - \Omega_0^2)^2 + \frac{\Omega^2 \Omega_0^2}{Q^2}}}, \quad (21.40)$$

where Δz_0 is the constant drive amplitude and Ω_0 the resonant frequency of the cantilever. The resonant frequency of the cantilever is given by the effective potential

$$\Omega_0 = \sqrt{\left(k + \frac{\partial^2 U}{\partial z^2}\right) \frac{1}{m_{\text{eff}}}}, \quad (21.41)$$

where U is the interaction potential between the tip and the sample. Equation (21.41) shows that an attractive potential decreases Ω_0. The change in Ω_0 in turn results in a change in Δz (21.40). The movement of the cantilever changes the path difference in the interferometer. The light reflected from the cantilever with amplitude $A_{\ell,0}$ and the reference light with amplitude $A_{r,0}$ interfere on the detector. The detected intensity $I(t) = [A_\ell(t) + A_r(t)]^2$ consists of two constant terms and a fluctuating term

$$2A_\ell(t) A_r(t)$$
$$= A_{\ell,0} A_{r,0} \sin\left[\omega t + \frac{4\pi\delta}{\lambda} + \frac{4\pi\Delta z}{\lambda} \sin(\Omega t)\right] \sin(\omega t). \quad (21.42)$$

Here ω is the frequency of the light, λ is the wavelength of the light, δ is the path difference in the interferometer, and Δz is the instantaneous amplitude of the cantilever, given according to (21.40) and (21.41) as a function of Ω, k, and U. The time average of (21.42) then becomes

$$\langle 2A_\ell(t) A_r(t)\rangle_T \propto \cos\left[\frac{4\pi\delta}{\lambda} + \frac{4\pi\Delta z}{\lambda}\sin(\Omega t)\right]$$
$$\approx \cos\left(\frac{4\pi\delta}{\lambda}\right) - \sin\left[\frac{4\pi\Delta z}{\lambda}\sin(\Omega t)\right]$$
$$\approx \cos\left(\frac{4\pi\delta}{\lambda}\right) - \frac{4\pi\Delta z}{\lambda}\sin(\Omega t). \quad (21.43)$$

Here all small quantities have been omitted and functions with small arguments have been linearized. The amplitude of Δz can be recovered with a lock-in technique. However, (21.43) shows that the measured amplitude is also a function of the path difference δ in the interferometer. Hence, this path difference δ must be very stable. The best sensitivity is obtained when $\sin(4\delta/\lambda) \approx 0$.

Heterodyne Interferometer. This influence is not present in the heterodyne detection scheme shown in Fig. 21.26. Light incident from the left with a frequency ω is split into a reference path (upper path in Fig. 21.26) and a measurement path. Light in the measurement path is shifted in frequency to $\omega_1 = \omega + \Delta\omega$ and focused onto the cantilever. The cantilever oscillates at the frequency Ω, as in the homodyne detection scheme. The reflected light $A_\ell(t)$ is collimated by the same lens and interferes on the photodiode with the reference light $A_r(t)$. The fluctuating term of the intensity is given by

$$2A_\ell(t) A_r(t)$$
$$= A_{\ell,0} A_{r,0} \sin\left[(\omega + \Delta\omega)t + \frac{4\pi\delta}{\lambda} + \frac{4\pi\Delta z}{\lambda}\sin(\Omega t)\right]\sin(\omega t), \quad (21.44)$$

where the variables are defined as in (21.42). Setting the path difference $\sin(4\pi\delta/\lambda) \approx 0$ and taking the time average, omitting small quantities and linearizing functions with small arguments, we get

$$\langle 2A_\ell(t) A_r(t)\rangle_T$$
$$\propto \cos\left[\Delta\omega t + \frac{4\pi\delta}{\lambda} + \frac{4\pi\Delta z}{\lambda}\sin(\Omega t)\right]$$
$$= \cos\left(\Delta\omega t + \frac{4\pi\delta}{\lambda}\right)\cos\left[\frac{4\pi\Delta z}{\lambda}\sin(\Omega t)\right]$$
$$-\sin\left(\Delta\omega t + \frac{4\pi\delta}{\lambda}\right)\sin\left[\frac{4\pi\Delta z}{\lambda}\sin(\Omega t)\right]$$

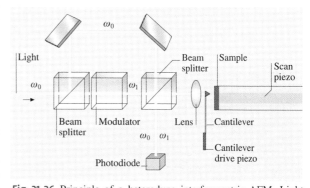

Fig. 21.26 Principle of a heterodyne interferometric AFM. Light with frequency ω_0 is split into a reference path (upper path) and a measurement path. The light in the measurement path is frequency shifted to ω_1 by an acousto-optical modulator (or an electro-optical modulator). The light reflected from the oscillating cantilever interferes with the reference beam on the detector

$$\approx \cos\left(\frac{4\pi\delta}{\lambda}\right) - \sin\left[\frac{4\pi\Delta z}{\lambda}\sin(\Omega t)\right]$$

$$\approx \cos\left(\Delta\omega t + \frac{4\pi\delta}{\lambda}\right)\left[1 - \frac{8\pi^2\Delta z^2}{\lambda^2}\sin(\Omega t)\right]$$

$$- \frac{4\pi\Delta z}{\lambda}\sin\left(\Delta\omega t + \frac{4\pi\delta}{\lambda}\right)\sin(\Omega t)$$

$$= \cos\left(\Delta\omega t + \frac{4\pi\delta}{\lambda}\right) - \frac{8\pi^2\Delta z^2}{\lambda^2}\cos\left(\Delta\omega t + \frac{4\pi\delta}{\lambda}\right)$$

$$\times \sin(\Omega t) - \frac{4\pi\Delta z}{\lambda}\sin\left(\Delta\omega t + \frac{4\pi\delta}{\lambda}\right)\sin(\Omega t)$$

$$= \cos\left(\Delta\omega t + \frac{4\pi\delta}{\lambda}\right) - \frac{4\pi^2\Delta z^2}{\lambda^2}\cos\left(\Delta\omega t + \frac{4\pi\delta}{\lambda}\right)$$

$$+ \frac{4\pi^2\Delta z^2}{\lambda^2}\cos\left(\Delta\omega t + \frac{4\pi\delta}{\lambda}\right)\cos(2\Omega t)$$

$$- \frac{4\pi\Delta z}{\lambda}\sin\left(\Delta\omega t + \frac{4\pi\delta}{\lambda}\right)\sin(\Omega t)$$

$$= \cos\left(\Delta\omega t + \frac{4\pi\delta}{\lambda}\right)\left(1 - \frac{4\pi^2\Delta z^2}{\lambda^2}\right)$$

$$+ \frac{2\pi^2\Delta z^2}{\lambda^2}\left\{\cos\left[(\Delta\omega + 2\Omega)t + \frac{4\pi\delta}{\lambda}\right]\right.$$

$$\left. + \cos\left[(\Delta\omega - 2\Omega)t + \frac{4\pi\delta}{\lambda}\right]\right\}$$

$$+ \frac{2\pi\Delta z}{\lambda}\left\{\cos\left[(\Delta\omega + \Omega)t + \frac{4\pi\delta}{\lambda}\right]\right.$$

$$\left. + \cos\left[(\Delta\omega - \Omega)t + \frac{4\pi\delta}{\lambda}\right]\right\}. \quad (21.45)$$

Multiplying electronically the components oscillating at $\Delta\omega$ and $\Delta\omega + \Omega$ and rejecting any product except the one oscillating at Ω we obtain

$$A = \frac{2\Delta z}{\lambda}\left(1 - \frac{4\pi^2\Delta z^2}{\lambda^2}\right)\cos\left[(\Delta\omega + 2\Omega)t + \frac{4\pi\delta}{\lambda}\right]$$

$$\times \cos\left(\Delta\omega t + \frac{4\pi\delta}{\lambda}\right)$$

$$= \frac{\Delta z}{\lambda}\left(1 - \frac{4\pi^2\Delta z^2}{\lambda^2}\right)\left\{\cos\left[(2\Delta\omega + \Omega)t + \frac{8\pi\delta}{\lambda}\right]\right.$$

$$\left. + \cos(\Omega t)\right\}$$

$$\approx \frac{\pi\Delta z}{\lambda}\cos(\Omega t). \quad (21.46)$$

Unlike in the homodyne detection scheme, the recovered signal is independent from the path difference δ of the interferometer. Furthermore, a lock-in amplifier with the reference set $\sin(\Delta\omega t)$ can measure the path difference δ independent of the cantilever oscillation. If necessary, a feedback circuit can keep $\delta = 0$.

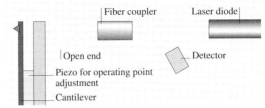

Fig. 21.27 A typical set-up for a fiber-optic interferometer readout

Fiber-Optical Interferometer. The fiber-optical interferometer [21.129] is one of the simplest interferometers to build and use. Its principle is sketched in Fig. 21.27. The light of a laser is fed into an optical fiber. Laser diodes with integrated fiber pigtails are convenient light sources. The light is split in a fiber-optic beam splitter into two fibers. One fiber is terminated by index-matching oil to avoid any reflections back into the fiber. The end of the other fiber is brought close to the cantilever in the AFM. The emerging light is partially reflected back into the fiber by the cantilever. Most of the light, however, is lost. This is not a big problem since only 4% of the light is reflected at the end of the fiber, at the glass–air interface. The two reflected light waves interfere with each other. The product is guided back into the fiber coupler and again split into two parts. One half is analyzed by the photodiode. The other half is fed back into the laser. Communications grade laser diodes are sufficiently resistant to feedback to be operated in this environment. They have, however, a bad coherence length, which in this case does not matter, since the optical path difference is in any case no larger than 5 μm. Again the end of the fiber has to be positioned on a piezo drive to set the distance between the fiber and the cantilever to $\lambda(n + 1/4)$.

Nomarski-Interferometer. Another way to minimize the optical path difference is to use the Nomarski interferometer [21.130]. Figure 21.28 shows a schematic of the microscope. The light from a laser is focused on the cantilever by lens. A birefringent crystal (for instance calcite) between the cantilever and the lens, which has its optical axis 45° off the polarization direction of the light, splits the light beam into two paths, offset by a distance given by the length of the crystal. Birefringent crystals have varying indices of refraction. In calcite, one crystal axis has a lower index than the other two. This means that certain light rays will propagate at different speeds through the crystal than others. By choosing the correct polarization, one can

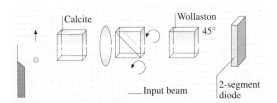

Fig. 21.28 Principle of Nomarski AFM. The circularly polarized input beam is deflected to the left by a nonpolarizing beam splitter. The light is focused onto a cantilever. The calcite crystal between the lens and the cantilever splits the circular polarized light into two spatially separated beams with orthogonal polarizations. The two light beams reflected from the lever are superimposed by the calcite crystal and collected by the lens. The resulting beam is again circularly polarized. A Wollaston prism produces two interfering beams with a $\pi/2$ phase shift between them. The minimal path difference accounts for the excellent stability of this microscope

select the ordinary ray or the extraordinary ray or one can get any mixture of the two rays. A detailed description of birefringence can be found in textbooks (e.g., [21.150]). A calcite crystal deflects the extraordinary ray at an angle of $6°$ within the crystal. Any separation can be set by choosing a suitable length for the calcite crystal.

The focus of one light ray is positioned near the free end of the cantilever while the other is placed close to the clamped end. Both arms of the interferometer pass through the same space, except for the distance between the calcite crystal and the lever. The closer the calcite crystal is placed to the lever, the less influence disturbances like air currents have.

Sarid [21.116] has given values for the sensitivities of different interferometeric detection systems. Table 21.5 presents a summary of his results.

Optical Lever

The most common cantilever deflection detection system is the optical lever [21.53, 111]. This method, depicted in Fig. 21.29, employs the same technique as light beam deflection galvanometers. A fairly well collimated light beam is reflected off a mirror and projected to a receiving target. Any change in the angular position of the mirror will change the position where the light ray hits the target. Galvanometers use optical path lengths of several meters and scales projected onto the target wall are also used to monitor changes in position.

In an AFM using the optical lever method, a photodiode segmented into two (or four) closely spaced devices detects the orientation of the end of the cantilever. Initially, the light ray is set to hit the photodiodes

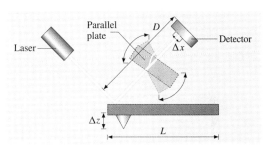

Fig. 21.29 Set-up for an optical lever detection microscope

Table 21.5 Noise in interferometers. F is the finesse of the cavity in the homodyne interferometer, P_i the incident power, P_d is the power on the detector, η is the sensitivity of the photodetector and RIN is the relative intensity noise of the laser. P_R and P_S are the power in the reference and sample beam in the heterodyne interferometer. P is the power in the Nomarski interferometer, $\delta\theta$ is the phase difference between the reference and the probe beam in the Nomarski interferometer. B is the bandwidth, e is the electron charge, λ is the wavelength of the laser, k the cantilever stiffness, ω_0 is the resonant frequency of the cantilever, Q is the quality factor of the cantilever, T is the temperature, and δi is the variation in current i

	Homodyne interferometer, fiber-optic interferometer	Heterodyne interferometer	Nomarski interferometer
Laser noise $\langle \delta i^2 \rangle_L$	$\frac{1}{4}\eta^2 F^2 P_i^2$ RIN	$\eta^2\left(P_R^2 + P_S^2\right)$ RIN	$\frac{1}{16}\eta^2 P^2 \delta\theta$
Thermal noise $\langle \delta i^2 \rangle_T$	$\frac{16\pi^2}{\lambda^2}\eta^2 F^2 P_i^2 \frac{4k_B TBQ}{\omega_0 k}$	$\frac{4\pi^2}{\lambda^2}\eta^2 P_d^2 \frac{4k_B TBQ}{\omega_0 k}$	$\frac{\pi^2}{\lambda^2}\eta^2 P^2 \frac{4k_B TBQ}{\omega_0 k}$
Shot noise $\langle \delta i^2 \rangle_S$	$4e\eta P_d B$	$2e\eta (P_R + P_S) B$	$\frac{1}{2} e\eta PB$

in the middle of the two subdiodes. Any deflection of the cantilever will cause an imbalance of the number of photons reaching the two halves. Hence the electrical currents in the photodiodes will be unbalanced too. The difference signal is further amplified and is the input signal to the feedback loop. Unlike the interferometric AFMs, where a modulation technique is often necessary to get a sufficient signal-to-noise ratio, most AFMs employing the optical lever method are operated in a static mode. AFMs based on the optical lever method are universally used. It is the simplest method for constructing an optical readout and it can be confined in volumes that are smaller than 5 cm in side length.

The optical lever detection system is a simple yet elegant way to detect normal and lateral force signals simultaneously [21.7, 8, 53, 111]. It has the additional advantage that it is a remote detection system.

Implementations. Light from a laser diode or from a super luminescent diode is focused on the end of the cantilever. The reflected light is directed onto a quadrant diode that measures the direction of the light beam. A Gaussian light beam far from its waist is characterized by an opening angle β. The deflection of the light beam by the cantilever surface tilted by an angle α is 2α. The intensity on the detector then shifts to the side by the product of 2α and the separation between the detector and the cantilever. The readout electronics calculates the difference in the photocurrents. The photocurrents, in turn, are proportional to the intensity incident on the diode.

The output signal is hence proportional to the change in intensity on the segments

$$I_{\text{sig}} \propto 4\frac{\alpha}{\beta} I_{\text{tot}} . \tag{21.47}$$

For the sake of simplicity, we assume that the light beam is of uniform intensity with its cross section increasing in proportion to the distance between the cantilever and the quadrant detector. The movement of the center of the light beam is then given by

$$\Delta x_{\text{Det}} = \Delta z \frac{D}{L} . \tag{21.48}$$

The photocurrent generated in a photodiode is proportional to the number of incoming photons hitting it. If the light beam contains a total number of N_0 photons, then the change in difference current becomes

$$\Delta (I_{\text{R}} - I_{\text{L}}) = \Delta I = \text{const } \Delta z \, D \, N_0 . \tag{21.49}$$

Combining (21.48) and (21.49), one obtains that the difference current ΔI is independent of the separation of the quadrant detector and the cantilever. This relation is true if the light spot is smaller than the quadrant detector. If it is greater, the difference current ΔI becomes smaller with increasing distance. In reality, the light beam has a Gaussian intensity profile. For small movements Δx (compared to the diameter of the light spot at the quadrant detector), (21.49) still holds. Larger movements Δx, however, will introduce a nonlinear response. If the AFM is operated in a constant force mode, only small movements Δx of the light spot will occur. The feedback loop will cancel out all other movements.

The scanning of a sample with an AFM can twist the microfabricated cantilevers because of lateral forces [21.5, 7, 8] and affect the images [21.120]. When the tip is subjected to lateral forces, it will twist the cantilever and the light beam reflected from the end of the cantilever will be deflected perpendicular to the ordinary deflection direction. For many investigations this influence of lateral forces is unwanted. The design of the triangular cantilevers stems from the desire to minimize the torsion effects. However, lateral forces open up a new dimension in force measurements. They allow, for instance, two materials to be distinguished because of their different friction coefficients, or adhesion energies to be determined. To measure lateral forces, the original optical lever AFM must be modified. The only modification compared with Fig. 21.29 is the use of a quadrant detector photodiode instead of a two-segment photodiode and the necessary readout electronics (Fig. 21.9a). The electronics calculates the following signals

$$\begin{aligned} U_{\text{normal force}} &= \alpha \left[\left(I_{\text{upper left}} + I_{\text{upper right}} \right) \right. \\ &\quad \left. - \left(I_{\text{lower left}} + I_{\text{lower right}} \right) \right] , \\ U_{\text{lateral force}} &= \beta \left[\left(I_{\text{upper left}} + I_{\text{lower left}} \right) \right. \\ &\quad \left. - \left(I_{\text{upper right}} + I_{\text{lower right}} \right) \right] . \end{aligned} \tag{21.50}$$

The calculation of the lateral force as a function of the deflection angle does not have a simple solution for cross sections other than circles. An approximate formula for the angle of twist for rectangular beams is [21.151]

$$\theta = \frac{M_{\text{t}} L}{\beta G b^3 h} , \tag{21.51}$$

where $M_{\text{t}} = F_y \ell$ is the external twisting moment due to lateral force F_y and β a constant determined by the value of h/b. For the equation to hold, h has to be larger than b.

Inserting the values for a typical microfabricated cantilever with integrated tips

$b = 6 \times 10^{-7}$ m,
$h = 10^{-5}$ m,
$L = 10^{-4}$ m,
$\ell = 3.3 \times 10^{-6}$ m,
$G = 5 \times 10^{10}$ Pa,
$\beta = 0.333$ (21.52)

into (21.51) we obtain the relation

$$F_y = 1.1 \times 10^{-4}\,\text{N} \times \theta\,. \tag{21.53}$$

Typical lateral forces are of the order of 10^{-10} N.

Sensitivity. The sensitivity of this set-up has been calculated in various papers [21.116, 148, 152]. Assuming a Gaussian beam, the resulting output signal as a function of the deflection angle is dispersion-like. Equation (21.47) shows that the sensitivity can be increased by increasing the intensity of the light beam I_tot or by decreasing the divergence of the laser beam. The upper bound of the intensity of the light I_tot is given by saturation effects on the photodiode. If we decrease the divergence of a laser beam we automatically increase the beam waist. If the beam waist becomes larger than the width of the cantilever we start to get diffraction. Diffraction sets a lower bound on the divergence angle. Hence one can calculate the optimal beam waist w_opt and the optimal divergence angle β [21.148, 152]

$$w_\text{opt} \approx 0.36 b\,,$$
$$\theta_\text{opt} \approx 0.89 \frac{\lambda}{b}\,. \tag{21.54}$$

The optimal sensitivity of the optical lever then becomes

$$\varepsilon\,[\text{mW/rad}] = 1.8 \frac{b}{\lambda} I_\text{tot}\,[\text{mW}]\,. \tag{21.55}$$

The angular sensitivity of the optical lever can be measured by introducing a parallel plate into the beam. Tilting the parallel plate results in a displacement of the beam, mimicking an angular deflection.

Additional noise sources can be considered. Of little importance is the quantum mechanical uncertainty of the position [21.148, 152], which is, for typical cantilevers at room temperature

$$\Delta z = \sqrt{\frac{\hbar}{2m\omega_0}} = 0.05\,\text{fm}\,, \tag{21.56}$$

where $\hbar$ is the Planck constant ($= 6.626 \times 10^{-34}$ J s). At very low temperatures and for high-frequency cantilevers this could become the dominant noise source. A second noise source is the shot noise of the light. The shot noise is related to the particle number. We can calculate the number of photons incident on the detector using

$$n = \frac{I\tau}{\hbar\omega} = \frac{I\lambda}{2\pi B\hbar c} = 1.8 \times 10^9\,\frac{I[\text{W}]}{B[\text{Hz}]}\,, \tag{21.57}$$

where I is the intensity of the light, τ the measurement time, $B = 1/\tau$ the bandwidth, and c the speed of light. The shot noise is proportional to the square root of the number of particles. Equating the shot noise signal with the signal resulting from the deflection of the cantilever one obtains

$$\Delta z_\text{shot} = 68 \frac{L}{w} \sqrt{\frac{B\,[\text{kHz}]}{I\,[\text{mW}]}}\,[\text{fm}]\,, \tag{21.58}$$

where w is the diameter of the focal spot. Typical AFM set-ups have a shot noise of 2 pm. The thermal noise can be calculated from the equipartition principle. The amplitude at the resonant frequency is

$$\Delta z_\text{therm} = 129 \sqrt{\frac{B}{k\,[\text{N/m}]\,\omega_0 Q}}\,[\text{pm}]\,. \tag{21.59}$$

A typical value is 16 pm. Upon touching the surface, the cantilever increases its resonant frequency by a factor of 4.39. This results in a new thermal noise amplitude of 3.2 pm for the cantilever in contact with the sample.

Piezoresistive Detection

Implementation. A piezoresistive cantilever is an alternative detection system which is not as widely used as the optical detection schemes [21.125, 126, 132]. This cantilever is based on the fact that the resistivities of certain materials, in particular Si, change with the applied

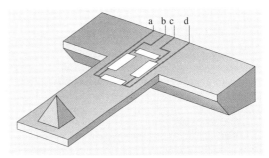

Fig. 21.30 A typical set-up for a piezoresistive readout

stress. Figure 21.30 shows a typical implementation of a piezo-resistive cantilever. Four resistances are integrated on the chip, forming a Wheatstone bridge. Two of the resistors are in unstrained parts of the cantilever, and the other two measure the bending at the point of the maximal deflection. For instance, when an AC voltage is applied between terminals a and c, one can measure the detuning of the bridge between terminals b and d. With such a connection the output signal only varies due to bending, not due to changes in the ambient temperature and thus the coefficient of the piezoresistance.

Sensitivity. The resistance change is [21.126]

$$\frac{\Delta R}{R_0} = \Pi \delta, \quad (21.60)$$

where Π is the tensor element of the piezo-resistive coefficients, δ the mechanical stress tensor element and R_0 the equilibrium resistance. For a single resistor, they separate the mechanical stress and the tensor element into longitudinal and transverse components

$$\frac{\Delta R}{R_0} = \Pi_t \delta_t + \Pi_l \delta_l. \quad (21.61)$$

The maximum values of the stress components are $\Pi_t = -64.0 \times 10^{-11}\,\mathrm{m^2/N}$ and $\Pi_l = -71.4 \times 10^{-11}\,\mathrm{m^2/N}$ for a resistor oriented along the (110) direction in silicon [21.126]. In the resistor arrangement of Fig. 21.30, two of the resistors are subject to the longitudinal piezo-resistive effect and two of them are subject to the transversal piezo-resistive effect. The sensitivity of that set-up is about four times that of a single resistor, with the advantage that temperature effects cancel to first order. The resistance change is then calculated as

$$\frac{\Delta R}{R_0} = \Pi \frac{3Eh}{2L^2} \Delta z = \Pi \frac{6L}{bh^2} F_z, \quad (21.62)$$

where $\Pi = 67.7 \times 10^{-11}\,\mathrm{m^2/N}$ is the averaged piezo-resistive coefficient. Plugging in typical values for the dimensions (Fig. 21.24) ($L = 100\,\mu\mathrm{m}$, $b = 10\,\mu\mathrm{m}$, $h = 1\,\mu\mathrm{m}$), one obtains

$$\frac{\Delta R}{R_0} = \frac{4 \times 10^{-5}}{\mathrm{nN}} F_z. \quad (21.63)$$

The sensitivity can be tailored by optimizing the dimensions of the cantilever.

Capacitance Detection

The capacitance of an arrangement of conductors depends on the geometry. Generally speaking, the capacitance increases for decreasing separations. Two

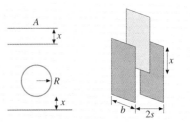

Fig. 21.31 Three possible arrangements of a capacitive readout. The *upper left* diagram shows a cross section through a parallel plate capacitor. The *lower left* diagram shows the geometry of a sphere versus a plane. The *right-hand* diagram shows the linear (but more complicated) capacitive readout

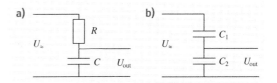

Fig. 21.32a,b Measuring the capacitance. (**a**) Low pass filter, (**b**) capacitive divider. C (*left*) and C_2 (*right*) are the capacitances under test

parallel plates form a simple capacitor (Fig. 21.31, upper left), with capacitance

$$C = \frac{\varepsilon \varepsilon_0 A}{x}, \quad (21.64)$$

where A is the area of the plates, assumed equal, and x is the separation. Alternatively one can consider a sphere versus an infinite plane (Fig. 21.31, lower left). Here the capacitance is [21.116]

$$C = 4\pi\varepsilon_0 R \sum_{n=2}^{\infty} \frac{\sinh(\alpha)}{\sinh(n\alpha)} \quad (21.65)$$

where R is the radius of the sphere, and α is defined by

$$\alpha = \ln\left(1 + \frac{z}{R} + \sqrt{\frac{z^2}{R^2} + 2\frac{z}{R}}\right). \quad (21.66)$$

One has to bear in mind that the capacitance of a parallel plate capacitor is a nonlinear function of the separation. One can circumvent this problem using a voltage divider. Figure 21.32a shows a low-pass filter. The output voltage is given by

$$U_{\text{out}} = U_\approx \frac{\frac{1}{j\omega C}}{R + \frac{1}{j\omega C}} = U_\approx \frac{1}{j\omega C R + 1}$$
$$\cong \frac{U_\approx}{j\omega C R}. \qquad (21.67)$$

Here C is given by (21.64), ω is the excitation frequency and j is the imaginary unit. The approximate relation at the end is true when $\omega C R \gg 1$. This is equivalent to the statement that C is fed by a current source, since R must be large in this set-up. Plugging (21.64) into (21.67) and neglecting the phase information, one obtains

$$U_{\text{out}} = \frac{U_\approx x}{\omega R \varepsilon \varepsilon_0 A}, \qquad (21.68)$$

which is linear in the displacement x.

Figure 21.32b shows a capacitive divider. Again the output voltage U_{out} is given by

$$U_{\text{out}} = U_\approx \frac{C_1}{C_2 + C_1} = U_\approx \frac{C_1}{\frac{\varepsilon \varepsilon_0 A}{x} + C_1}. \qquad (21.69)$$

If there is a stray capacitance C_s then (21.69) is modified as

$$U_{\text{out}} = U_\approx \frac{C_1}{\frac{\varepsilon \varepsilon_0 A}{x} + C_s + C_1}. \qquad (21.70)$$

Provided $C_s + C_1 \ll C_2$, one has a system which is linear in x. The driving voltage $U_\approx$ must be large (more than 100 V) to gave an output voltage in the range of 1 V. The linearity of the readout depends on the capacitance C_1 (Fig. 21.33).

Another idea is to keep the distance constant and to change the relative overlap of the plates (Fig. 21.31, right side). The capacitance of the moving center plate versus the stationary outer plates becomes

$$C = C_s + 2\frac{\varepsilon \varepsilon_0 b x}{s}, \qquad (21.71)$$

where the variables are defined in Fig. 21.31. The stray capacitance comprises all effects, including the capacitance of the fringe fields. When the length x is comparable to the width b of the plates, one can safely assume that the stray capacitance is constant and independent of x. The main disadvantage of this set-up is that it is not as easily incorporated into a microfabricated device as the others.

Sensitivity. The capacitance itself is not a measure of the sensitivity, but its derivative is indicative of the signals one can expect. Using the situation described in Fig. 21.31 (upper left) and in (21.64), one obtains for the parallel plate capacitor

$$\frac{dC}{dx} = -\frac{\varepsilon \varepsilon_0 A}{x^2}. \qquad (21.72)$$

Assuming a plate area A of 20 μm by 40 μm and a separation of 1 μm, one obtains a capacitance of 31 fF (neglecting stray capacitance and the capacitance of the connection leads) and a dC/dx of 3.1×10^{-8} F/m = 31 fF/μm. Hence it is of paramount importance to maximize the area between the two contacts and to minimize the distance x. The latter however is far from being trivial. One has to go to the limits of microfabrication to achieve a decent sensitivity.

If the capacitance is measured by the circuit shown in Fig. 21.32, one obtains for the sensitivity

$$\frac{dU_{\text{out}}}{U_\approx} = \frac{dx}{\omega R \varepsilon \varepsilon_0 A}. \qquad (21.73)$$

Using the same value for A as above, setting the reference frequency to 100 kHz, and selecting $R = 1\,\text{G}\Omega$, we get the relative change in the output voltage U_{out} as

$$\frac{dU_{\text{out}}}{U_\approx} = \frac{22.5 \times 10^{-6}}{\text{Å}} \times dx. \qquad (21.74)$$

A driving voltage of 45 V then translates to a sensitivity of 1 mV/Å. A problem in this set-up is the stray capacitances. They are in parallel to the original capacitance and decrease the sensitivity considerably.

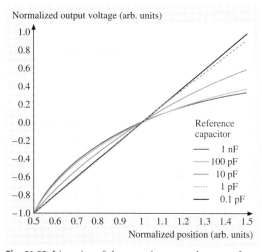

Fig. 21.33 Linearity of the capacitance readout as a function of the reference capacitor

Alternatively, one could build an oscillator with this capacitance and measure the frequency. RC-oscillators typically have an oscillation frequency of

$$f_{\text{res}} \propto \frac{1}{RC} = \frac{x}{R\varepsilon\varepsilon_0 A} \,. \tag{21.75}$$

Again the resistance R must be of the order of $1\,\text{G}\Omega$ when stray capacitances C_s are neglected. However C_s is of the order of $1\,\text{pF}$. Therefore one gets $R = 10\,\text{M}\Omega$. Using these values, the sensitivity becomes

$$\mathrm{d}f_{\text{res}} = \frac{C\,\mathrm{d}x}{R(C+C_s)^2 x} \approx \frac{0.1\,\text{Hz}}{\text{Å}}\,\mathrm{d}x \,. \tag{21.76}$$

The bad thing is that the stray capacitances have made the signal nonlinear again. The linearized set-up in Fig. 21.31 has a sensitivity of

$$\frac{\mathrm{d}C}{\mathrm{d}x} = 2\frac{\varepsilon\varepsilon_0 b}{s} \,. \tag{21.77}$$

Substituting typical values ($b = 10\,\mu\text{m}$, $s = 1\,\mu\text{m}$), one gets $\mathrm{d}C/\mathrm{d}x = 1.8 \times 10^{-10}\,\text{F/m}$. It is noteworthy that the sensitivity remains constant for scaled devices.

Implementations. Capacitance readout can be achieved in different ways [21.123, 124]. All include an alternating current or voltage with frequencies in the 100 kHz to 100 MHz range. One possibility is to build a tuned circuit with the capacitance of the cantilever determining the frequency. The resonance frequency of a high-quality Q tuned circuit is

$$\omega_0 = (LC)^{-1/2} \,, \tag{21.78}$$

where L is the inductance of the circuit. The capacitance C includes not only the sensor capacitance but also the capacitance of the leads. The precision of a frequency measurement is mainly determined by the ratio of L and C

$$Q = \left(\frac{L}{C}\right)^{1/2} \frac{1}{R} \,. \tag{21.79}$$

Here R symbolizes the losses in the circuit. The higher the quality, the more precise the frequency measurement. For instance, a frequency of 100 MHz and a capacitance of 1 pF gives an inductance of 250 μH. The quality then becomes 2.5×10^8. This value is an upper limit, since losses are usually too high.

Using a value of $\mathrm{d}C/\mathrm{d}x = 31\,\text{fF}/\mu\text{m}$, one gets $\Delta C/\text{Å} = 3.1\,\text{aF}/\text{Å}$. With a capacitance of 1 pF, one gets

$$\frac{\Delta\omega}{\omega} = \frac{1}{2}\frac{\Delta C}{C} \,,$$

$$\Delta\omega = 100\,\text{MHz} \times \frac{1}{2}\frac{3.1\,\text{aF}}{1\,\text{pF}} = 155\,\text{Hz} \,. \tag{21.80}$$

This is the frequency shift for a deflection of 1 Å. The calculation shows that this is a measurable quantity. The quality also indicates that there is no physical reason why this scheme should not work.

21.3.3 Combinations for 3-D Force Measurements

Three-dimensional force measurements are essential if one wants to know all of the details of the interaction between the tip and the cantilever. The straightforward attempt to measure three forces is complicated, since force sensors such as interferometers or capacitive sensors need a minimal detection volume, which is often too large. The second problem is that the force-sensing tip has to be held in some way. This implies that one of the three Cartesian axes is stiffer than the others.

However, by combining different sensors it is possible to achieve this goal. Straight cantilevers are employed for these measurements, because they can be handled analytically. The key observation is that the optical lever method does not determine the position of the end of the cantilever. It measures the orientation. In the previous sections, one has always made use of the fact that, for a force along one of the orthogonal symmetry directions at the end of the cantilever (normal force, lateral force, force along the cantilever beam axis), there is a one-to-one correspondence of the tilt angle and the deflection. The problem is that the force along the cantilever beam axis and the normal force create a deflection in the same direction. Hence, what is called the normal force component is actually a mixture of two forces. The deflection of the cantilever is the third quantity, which is not considered in most of the AFMs. A fiber-optic interferometer in parallel with the optical lever measures the deflection. Three measured quantities then allow the separation of the three orthonormal force directions, as is evident from (21.27) and (21.33) [21.12–16].

Alternatively, one can put the fast scanning direction along the axis of the cantilever. Forward and backward scans then exert opposite forces F_x. If the piezo movement is linearized, both force components in AFM based on optical lever detection can be determined. In this case, the normal force is simply the average of the forces in the forward and backward direction. The force F_x is the difference in the forces measured in the forward and backward directions.

21.3.4 Scanning and Control Systems

Almost all SPMs use piezo translators to scan the tip or the sample. Even the first STM [21.1, 103] and some of its predecessors [21.153, 154] used them. Other materials or set-ups for nanopositioning have been proposed, but they have not been successful [21.155, 156].

Piezo Tubes

A popular solution is tube scanners (Fig. 21.34). They are now widely used in SPMs due to their simplicity and their small size [21.133, 157]. The outer electrode is segmented into four equal sectors of 90°. Opposite sectors are driven by signals of the same magnitude, but opposite sign. This gives, through bending, two-dimensional movement on (approximately) a sphere. The inner electrode is normally driven by the z-signal. It is possible, however, to use only the outer electrodes for scanning and for the z-movement. The main drawback of applying the z-signal to the outer electrodes is that the applied voltage is the sum of both the x- or y-movements and the z-movement. Hence a larger scan size effectively reduces the available range for the z-control.

Piezo Effect

An electric field applied across a piezoelectric material causes a change in the crystal structure, with expansion in some directions and contraction in others. Also, a net volume change occurs [21.132]. Many SPMs use the transverse piezo electric effect, where the applied electric field E is perpendicular to the expansion/contraction direction.

$$\Delta L = L \left(\boldsymbol{E} \cdot \boldsymbol{n} \right) d_{31} = L \frac{V}{t} d_{31} \,, \quad (21.81)$$

where d_{31} is the transverse piezoelectric constant, V is the applied voltage, t is the thickness of the piezo slab or the distance between the electrodes where the voltage is applied, L is the free length of the piezo slab, and $\boldsymbol{n}$ is the direction of polarization. Piezo translators based on the transverse piezoelectric effect have a wide range of sensitivities, limited mainly by mechanical stability and breakdown voltage.

Scan Range

The scanning range of a piezotube is difficult to calculate [21.157–159]. The bending of the tube depends on the electric fields and the nonuniform strain induced. A finite element calculation where the piezo tube was divided into 218 identical elements was used [21.158] to calculate the deflection. On each node, the mechanical stress, the stiffness, the strain and the piezoelectric stress were calculated when a voltage was applied on one electrode. The results were found to be linear on the first iteration and higher order corrections were very small even for large electrode voltages. It was found that, to first order, the x- and z-movement of the tube could be reasonably well approximated by assuming that the piezo tube is a segment of a torus. Using this model, one obtains

$$\mathrm{d}x = (V_+ - V_-) |d_{31}| \frac{L^2}{2td} \,, \quad (21.82)$$

$$\mathrm{d}z = (V_+ + V_- - 2V_z) |d_{31}| \frac{L}{2t} \,, \quad (21.83)$$

where $|d_{31}|$ is the coefficient of the transversal piezoelectric effect, L is the tube's free length, t is the tube's wall thickness, d is the tube's diameter, V_+ is the voltage on the positive outer electrode, while V_- is the voltage of the opposite quadrant negative electrode and V_z is the voltage of the inner electrode.

The cantilever or sample mounted on the piezotube has an additional lateral movement because the point of measurement is not in the endplane of the piezotube. The additional lateral displacement of the end of the tip is $\ell \sin \varphi \approx \ell \varphi$, where ℓ is the tip length and φ is the deflection angle of the end surface. Assuming that the sample or cantilever is always perpendicular to the end of the walls of the tube, and calculating with the torus model, one gets for the angle

$$\varphi = \frac{L}{R} = \frac{2\mathrm{d}x}{L} \,, \quad (21.84)$$

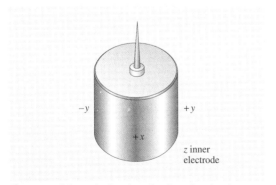

Fig. 21.34 Schematic drawing of a piezoelectric tube scanner. The piezo ceramic is molded into a tube form. The outer electrode is separated into four segments and connected to the scanning voltage. The z-voltage is applied to the inner electrode

where R is the radius of curvature of the piezo tube. Using the result of (21.84), one obtains for the additional x-movement

$$dx_{add} = \ell\varphi = \frac{2\,dx\ell}{L}$$
$$= (V_+ - V_-)|d_{31}|\frac{\ell L}{td} \quad (21.85)$$

and for the additional z-movement due to the x-movement

$$dz_{add} = \ell - \ell\cos\varphi = \frac{\ell\varphi^2}{2} = \frac{2\ell(dx)^2}{L^2}$$
$$= (V_+ - V_-)^2|d_{31}|^2\frac{\ell L^2}{2t^2d^2}. \quad (21.86)$$

Carr [21.158] assumed for his finite element calculations that the top of the tube was completely free to move and, as a consequence, the top surface was distorted, leading to a deflection angle that was about half that of the geometrical model. Depending on the attachment of the sample or the cantilever, this distortion may be smaller, leading to a deflection angle in-between that of the geometrical model and the one from the finite element calculation.

Nonlinearities and Creep

Piezo materials with a high conversion ratio (a large d_{31} or small electrode separations with large scanning ranges) are hampered by substantial hysteresis resulting in a deviation from linearity by more than 10%. The sensitivity of the piezo ceramic material (mechanical displacement divided by driving voltage) decreases with reduced scanning range, whereas the hysteresis is reduced. Careful selection of the material used for the piezo scanners, the design of the scanners, and of the operating conditions is necessary to obtain optimum performance.

Passive Linearization: Calculation. The analysis of images affected by piezo nonlinearities [21.160–163] shows that the dominant term is

$$x = AV + BV^2, \quad (21.87)$$

where x is the excursion of the piezo, V is the applied voltage and A and B are two coefficients describing the sensitivity of the material. Equation (21.87) holds for scanning from $V = 0$ to large V. For the reverse direction, the equation becomes

$$x = \tilde{A}V - \tilde{B}(V - V_{max})^2, \quad (21.88)$$

where $\tilde{A}$ and $\tilde{B}$ are the coefficients for the back scan and V_{max} is the applied voltage at the turning point. Both equations demonstrate that the true x-travel is small at the beginning of the scan and becomes larger towards the end. Therefore, images are stretched at the beginning and compressed at the end.

Similar equations hold for the slow scan direction. The coefficients, however, are different. The combined action causes a greatly distorted image. This distortion can be calculated. The data acquisition systems record the signal as a function of V. However the data is measured as a function of x. Therefore we have to distribute the x-values evenly across the image. This can be done by inverting an approximation of (21.87). First we write

$$x = AV\left(1 - \frac{B}{A}V\right). \quad (21.89)$$

For $B \ll A$ we can approximate

$$V = \frac{x}{A}. \quad (21.90)$$

We now substitute (21.90) into the nonlinear term of (21.89). This gives

$$x = AV\left(1 + \frac{Bx}{A^2}\right),$$
$$V = \frac{x}{A}\frac{1}{(1 + Bx/A^2)} \approx \frac{x}{A}\left(1 - \frac{Bx}{A^2}\right). \quad (21.91)$$

Hence an equation of the type

$$x_{true} = x(\alpha - \beta x/x_{max})$$
$$\text{with} \quad 1 = \alpha - \beta \quad (21.92)$$

takes out the distortion of an image. α and β are dependent on the scan range, the scan speed and on the scan history, and have to be determined with exactly the same settings as for the measurement. x_{max} is the maximal scanning range. The condition for α and β guarantees that the image is transformed onto itself.

Similar equations to the empirical one shown above (21.92) can be derived by analyzing the movements of domain walls in piezo ceramics.

Passive Linearization: Measuring the Position. An alternative strategy is to measure the positions of the piezo translators. Several possibilities exist.

1. The interferometers described above can be used to measure the elongation of the piezo elongation. The fiber-optic interferometer is especially easy to implement. The coherence length of the laser only limits the measurement range. However, the signal is of a periodic nature. Hence direct use of the signal in a feedback circuit for the position is not

possible. However, as a measurement tool and, especially, as a calibration tool, the interferometer is without competition. The wavelength of the light, for instance that in a He-Ne laser, is so well defined that the precision of the other components determines the error of the calibration or measurement.

2. The movement of the light spot on the quadrant detector can be used to measure the position of a piezo [21.164]. The output current changes by $0.5\,\text{A/cm} \times P(\text{W})/R(\text{cm})$. Typical values ($P = 1\,\text{mW}$, $R = 0.001\,\text{cm}$) give $0.5\,\text{A/cm}$. The noise limit is typically $0.15\,\text{nm} \times \sqrt{\Delta f(\text{Hz})/H(\text{W/cm}^2)}$. Again this means that the laser beam above would have a 0.1 nm noise limitation for a bandwidth of 21 Hz. The advantage of this method is that, in principle, one can linearize two axes with only one detector.

3. A knife-edge blocking part of a light beam incident on a photodiode can be used to measure the position of the piezo. This technique, commonly used in optical shear force detection [21.75, 165], has a sensitivity of better than 0.1 nm.

4. The capacitive detection [21.166, 167] of the cantilever deflection can be applied to the measurement of the piezo elongation. Equations (21.64) to (21.79) apply to the problem. This technique is used in some commercial instruments. The difficulties lie in the avoidance of fringe effects at the borders of the two plates. While conceptually simple, one needs the latest technology in surface preparation to get a decent linearity. The electronic circuits used for the readout are often proprietary.

5. Linear variable differential transformers (LVDT) are a convenient way to measure positions down to 1 nm. They can be used together with a solid state joint set-up, as often used for large scan range stages. Unlike capacitive detection, there are few difficulties in implementation. The sensors and the detection circuits LVDTs are available commercially.

6. A popular measurement technique is the use of strain gauges. They are especially sensitive when mounted on a solid state joint where the curvature is maximal. The resolution depends mainly on the induced curvature. A precision of 1 nm is attainable. The signals are low – a Wheatstone bridge is needed for the readout.

Active Linearization. Active linearization is done with feedback systems. Sensors need to be monotonic. Hence all of the systems described above, with the exception of the interferometers, are suitable. The most common solutions include the strain gauge approach, capacitance measurement or the LVDT, which are all electronic solutions. Optical detection systems have the disadvantage that the intensity enters into the calibration.

Alternative Scanning Systems

The first STMs were based on piezo tripods [21.1]. The piezo tripod (Fig. 21.35) is an intuitive way to generate the three-dimensional movement of a tip attached to its center. However, to get a suitable stability and scanning range, the tripod needs to be fairly large (about 50 mm). Some instruments use piezo stacks instead of monolithic piezoactuators. They are arranged in a tripod. Piezo stacks are thin layers of piezoactive materials glued together to form a device with up to $200\,\mu\text{m}$ of actuation range. Preloading with a suitable metal casing reduces the nonlinearity.

If one tries to construct a homebuilt scanning system, the use of linearized scanning tables is recommended. They are built around solid state joints and actuated by piezo stacks. The joints guarantee that the movement is parallel with little deviation from the predefined scanning plane. Due to the construction it is easy to add measurement devices such as capacitive sensors, LVDTs or strain gauges, which are essential for a closed loop linearization. Two-dimensional tables can be bought from several manufacturers. They have linearities of better than 0.1% and a noise level of 10^{-4} to 10^{-5} for the maximal scanning range.

Control Systems

Basics. The electronics and software play an important role in the optimal performance of an SPM. Control electronics and software are supplied with commercial SPMs. Electronic control systems can use either analog or digital feedback. While digital feedback of-

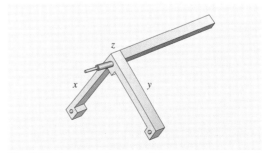

Fig. 21.35 An alternative type of piezo scanner: the tripod

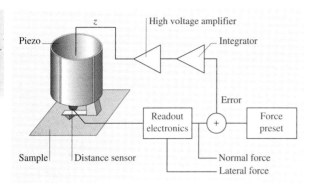

Fig. 21.36 Block schematic of the feedback control loop of an AFM

fers greater flexibility and ease of configuration, analog feedback circuits might be better suited for ultralow noise operation. We will describe here the basic set-ups for AFMs.

Figure 21.36 shows a block schematic of a typical AFM feedback loop. The signal from the force transducer is fed into the feedback loop, which consists mainly of a subtraction stage to get an error signal and an integrator. The gain of the integrator (high gain corresponds to short integration times) is set as high as possible without generating more than 1% overshoot. High gain minimizes the error margin of the current and forces the tip to follow the contours of constant density of states as well as possible. This operating mode is known as constant force mode. A high-voltage amplifier amplifies the outputs of the integrator. As AFMs using piezotubes usually require ± 150 V at the output, the output of the integrator needs to be amplified by a high-voltage amplifier.

In order to scan the sample, additional voltages at high tension are required to drive the piezo. For example, with a tube scanner, four scanning voltages are required, namely $+V_x$, $-V_x$, $+V_y$ and $-V_y$. The x- and y-scanning voltages are generated in a scan generator (analog or computer-controlled). Both voltages are input to the two respective power amplifiers. Two inverting amplifiers generate the input voltages for the other two power amplifiers. The topography of the sample surface is determined by recording the input voltage to the high-voltage amplifier for the z-channel as a function of x and y (constant force mode).

Another operating mode is the variable force mode. The gain in the feedback loop is lowered and the scanning speed increased such that the force on the cantilever is no longer constant. Here the force is recorded as a function of x and y.

Force Spectroscopy. Four modes of spectroscopic imaging are in common use with force microscopes: measuring lateral forces, $\partial F/\partial z$, $\partial F/\partial x$ spatially resolved, and measuring force versus distance curves. Lateral forces can be measured by detecting the deflection of a cantilever in a direction orthogonal to the normal direction. The optical lever deflection method does this most easily. Lateral force measurements give indications of adhesion forces between the tip and the sample.

$\partial F/\partial z$ measurements probe the local elasticity of the sample surface. In many cases the measured quantity originates from a volume of a few cubic nanometers. The $\partial F/\partial z$ or local stiffness signal is proportional to Young's modulus, as far as one can define this quantity. Local stiffness is measured by vibrating the cantilever by a small amount in the z-direction. The expected signal for very stiff samples is zero: for very soft samples one also gets, independent of the stiffness, a constant signal. This signal is again zero for the optical lever deflection and equal to the driving amplitude for interferometric measurements. The best sensitivity is obtained when the compliance of the cantilever matches the stiffness of the sample.

A third spectroscopic quantity is the lateral stiffness. It is measured by applying a small modulation in the x-direction on the cantilever. The signal is again optimal when the lateral compliance of the cantilever matches the lateral stiffness of the sample. The lateral stiffness is, in turn, related to the shear modulus of the sample.

Detailed information on the interaction of the tip and the sample can be gained by measuring force versus distance curves. The cantilevers need to have enough compliance to avoid instabilities due to the attractive forces on the sample.

Using the Control Electronics as a Two-Dimensional Measurement Tool. Usually the control electronics of an AFM is used to control the x- and y-piezo signals while several data acquisition channels record the position-dependent signals. The control electronics can be used in another way: they can be viewed as a two-dimensional function generator. What is normally the x- and y-signal can be used to control two independent variables of an experiment. The control logic of the AFM then ensures that the available parameter space is systematically probed at equally spaced points. An example is friction force curves measured along a line across a step on graphite.

Figure 21.37 shows the connections. The z-piezo is connected as usual, like the x-piezo. However, the y-output is used to command the desired input parameter. The offset of the y-channel determines the position of the tip on the sample surface, together with the x-channel.

Some Imaging Processing Methods

The visualization and interpretation of images from AFMs is intimately connected to the processing of these images. An ideal AFM is a noise-free device that images a sample with perfect tips of known shape and has perfect linear scanning piezos. In reality, AFMs are not that ideal. The scanning device in an AFM is affected by distortions. The distortions are both linear and nonlinear. Linear distortions mainly result from imperfections in the machining of the piezotranslators causing crosstalk from the z-piezo to the x- and y-piezos, and vice versa. Among the linear distortions, there are two kinds which are very important. First, scanning piezos invariably have different sensitivities along the different scan axes due to variations in the piezo material and uneven electrode areas. Second, the same reasons might cause the scanning axes to be nonorthogonal. Furthermore, the plane in which the piezoscanner moves for constant height z is hardly ever coincident with the sample plane. Hence, a linear ramp is added to the sample data. This ramp is especially bothersome when the height z is displayed as an intensity map.

The nonlinear distortions are harder to deal with. They can affect AFM data for a variety of reasons. First, piezoelectric ceramics do have a hysteresis loop, much like ferromagnetic materials. The deviations of piezoceramic materials from linearity increase with increasing amplitude of the driving voltage. The mechanical position for one voltage depends on the previously applied voltages to the piezo. Hence, to get the best positional accuracy, one should always approach a point on the sample from the same direction. Another type of nonlinear distortion of images occurs when the scan frequency approaches the upper frequency limits of the x- and y-drive amplifiers or the upper frequency limit of the feedback loop (z-component). This distortion, due to the feedback loop, can only be minimized by reducing the scan frequency. On the other hand, there is a simple way to reduce distortions due to the x- and y-piezo drive amplifiers. To keep the system as simple as possible, one normally uses a triangular waveform to drive the scanning piezos. However, triangular waves contain frequency components as multiples of the scan frequency. If the cut-off frequencies of the x- and y-drive electronics or of the feedback loop are too close to the scanning frequency (two or three times the scanning frequency), the triangular drive voltage is rounded off at the turn-

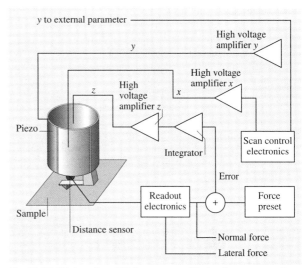

Fig. 21.37 Wiring of an AFM to measure friction force curves along a line

ing points. This rounding error causes, first, a distortion of the scan linearity and, second, through phase lags, the projection of part of the backward scan onto the forward scan. This type of distortion can be minimized by carefully selecting the scanning frequency and by using driving voltages for the x- and y-piezos with waveforms like trapezoidal waves, which are closer to a sine wave. The values measured for x-, y- or z-piezos are affected by noise. The origin of this noise can be either electronic, disturbances, or a property of the sample surface due to adsorbates. In addition to this incoherent noise, interference with main and other equipment nearby might be present. Depending on the type of noise, one can filter it in real space or in Fourier space. The most important part of image processing is to visualize the measured data. Typical AFM data sets can consist of many thousands to over a million points per plane. There may be more than one image plane present. The AFM data represents a topography in various data spaces.

Most commercial data acquisition systems implicitly use some kind of data processing. Since the original data is commonly subject to slopes on the surface, most programs use some kind of slope correction. The least disturbing way is to subtract a plane $z(x, y) = Ax + By + C$ from the data. The coefficients are determined by fitting $z(x, y)$ to the data. Another operation is to subtract a second-order function such as $z(x, y) = Ax^2 + By^2 + Cxy + Dx + Ey + F$. Again, the parameters are determined with a fit. This function is appropriate for almost planar data, where the nonlinearity of the piezos caused the distortion.

In the image processing software from Digital Instruments, up to three operations are performed on the raw data. First, a zero-order flatten is applied. The flatten operation is used to eliminate image bow in the slow scan direction (caused by a physical bow in the instrument itself), slope in the slow scan direction, and bands in the image (caused by differences in the scan height from one scan line to the next). The flattening operation takes each scan line and subtracts the average value of the height along each scan line from each point in that scan line. This brings each scan line to the same height. Next, a first-order plane fit is applied in the fast scan direction. The plane-fit operation is used to eliminate bow and slope in the fast scan direction. The plane fit operation calculates a best fit plane for the image and subtracts it from the image. This plane has a constant nonzero slope in the fast scan direction. In some cases a higher order polynomial *plane* may be required. Depending upon the quality of the raw data, the flattening operation and/or the plane fit operation may not be required at all.

References

21.1 G. Binnig, H. Rohrer, C. Gerber, E. Weibel: Surface studies by scanning tunneling microscopy, Phys. Rev. Lett. **49**, 57–61 (1982)

21.2 G. Binnig, C.F. Quate, C. Gerber: Atomic force microscope, Phys. Rev. Lett. **56**, 930–933 (1986)

21.3 G. Binnig, C. Gerber, E. Stoll, T.R. Albrecht, C.F. Quate: Atomic resolution with atomic force microscope, Europhys. Lett. **3**, 1281–1286 (1987)

21.4 B. Bhushan: *Handbook of Micro-/Nanotribology*, 2nd edn. (CRC, Boca Raton 1999)

21.5 C.M. Mate, G.M. McClelland, R. Erlandsson, S. Chiang: Atomic-scale friction of a tungsten tip on a graphite surface, Phys. Rev. Lett. **59**, 1942–1945 (1987)

21.6 R. Erlandsson, G.M. McClelland, C.M. Mate, S. Chiang: Atomic force microscopy using optical interferometry, J. Vac. Sci. Technol. A **6**, 266–270 (1988)

21.7 O. Marti, J. Colchero, J. Mlynek: Combined scanning force and friction microscopy of mica, Nanotechnology **1**, 141–144 (1990)

21.8 G. Meyer, N.M. Amer: Simultaneous measurement of lateral and normal forces with an optical-beam-deflection atomic force microscope, Appl. Phys. Lett. **57**, 2089–2091 (1990)

21.9 B. Bhushan, J. Ruan: Atomic-scale friction measurements using friction force microscopy: Part II – Application to magnetic media, ASME J. Tribol. **116**, 389–396 (1994)

21.10 B. Bhushan, V.N. Koinkar, J. Ruan: Microtribology of magnetic media, Proc. Inst. Mech. Eng. Part J **208**, 17–29 (1994)

21.11 B. Bhushan, J.N. Israelachvili, U. Landman: Nanotribology: Friction, wear, and lubrication at the atomic scale, Nature **374**, 607–616 (1995)

21.12 S. Fujisawa, M. Ohta, T. Konishi, Y. Sugawara, S. Morita: Difference between the forces measured by an optical lever deflection and by an optical interferometer in an atomic force microscope, Rev. Sci. Instrum. **65**, 644–647 (1994)

21.13 S. Fujisawa, E. Kishi, Y. Sugawara, S. Morita: Fluctuation in 2-dimensional stick-slip phenomenon observed with 2-dimensional frictional force microscope, Jpn. J. Appl. Phys. **33**, 3752–3755 (1994)

21.14 S. Grafstrom, J. Ackermann, T. Hagen, R. Neumann, O. Probst: Analysis of lateral force effects on the topography in scanning force microscopy, J. Vac. Sci. Technol. B **12**, 1559–1564 (1994)

21.15 R.M. Overney, H. Takano, M. Fujihira, W. Paulus, H. Ringsdorf: Anisotropy in friction and molecular stick-slip motion, Phys. Rev. Lett. **72**, 3546–3549 (1994)

21.16 R.J. Warmack, X.Y. Zheng, T. Thundat, D.P. Allison: Friction effects in the deflection of atomic force microscope cantilevers, Rev. Sci. Instrum. **65**, 394–399 (1994)

21.17 N.A. Burnham, D.D. Domiguez, R.L. Mowery, R.J. Colton: Probing the surface forces of monolayer films with an atomic force microscope, Phys. Rev. Lett. **64**, 1931–1934 (1990)

21.18 N.A. Burham, R.J. Colton, H.M. Pollock: Interpretation issues in force microscopy, J. Vac. Sci. Technol. A **9**, 2548–2556 (1991)

21.19 C.D. Frisbie, L.F. Rozsnyai, A. Noy, M.S. Wrighton, C.M. Lieber: Functional group imaging by chemical force microscopy, Science **265**, 2071–2074 (1994)

21.20 V.N. Koinkar, B. Bhushan: Microtribological studies of unlubricated and lubricated surfaces using atomic force/friction force microscopy, J. Vac. Sci. Technol. A **14**, 2378–2391 (1996)

21.21 V. Scherer, B. Bhushan, U. Rabe, W. Arnold: Local elasticity and lubrication measurements using atomic force and friction force microscopy at ultrasonic frequencies, IEEE Trans. Magn. **33**, 4077–4079 (1997)

21.22 V. Scherer, W. Arnold, B. Bhushan: Lateral force microscopy using acoustic friction force microscopy, Surf. Interface Anal. **27**, 578–587 (1999)

21.23 B. Bhushan, S. Sundararajan: Micro-/nanoscale friction and wear mechanisms of thin films using atomic force and friction force microscopy, Acta Mater. **46**, 3793–3804 (1998)

21.24 U. Krotil, T. Stifter, H. Waschipky, K. Weishaupt, S. Hild, O. Marti: Pulse force mode: A new method for the investigation of surface properties, Surf. Interface Anal. **27**, 336–340 (1999)

21.25 B. Bhushan, C. Dandavate: Thin-film friction and adhesion studies using atomic force microscopy, J. Appl. Phys. **87**, 1201–1210 (2000)

21.26 B. Bhushan: *Micro-/Nanotribology and Its Applications* (Kluwer, Dordrecht 1997)

21.27 B. Bhushan: *Principles and Applications of Tribology* (Wiley, New York 1999)

21.28 B. Bhushan: *Modern Tribology Handbook – Vol. 1: Principles of Tribology* (CRC, Boca Raton 2001)

21.29 B. Bhushan: *Introduction to Tribology* (Wiley, New York 2002)

21.30 M. Reinstädtler, U. Rabe, V. Scherer, U. Hartmann, A. Goldade, B. Bhushan, W. Arnold: On the nanoscale measurement of friction using atomic force microscope cantilever torsional resonances, Appl. Phys. Lett. **82**, 2604–2606 (2003)

21.31 N.A. Burnham, R.J. Colton: Measuring the nanomechanical properties and surface forces of materials using an atomic force microscope, J. Vac. Sci. Technol. A **7**, 2906–2913 (1989)

21.32 P. Maivald, H.J. Butt, S.A.C. Gould, C.B. Prater, B. Drake, J.A. Gurley, V.B. Elings, P.K. Hansma: Using force modulation to image surface elasticities with the atomic force microscope, Nanotechnology **2**, 103–106 (1991)

21.33 B. Bhushan, A.V. Kulkarni, W. Bonin, J.T. Wyrobek: Nano/picoindentation measurements using capacitive transducer in atomic force microscopy, Philos. Mag. A **74**, 1117–1128 (1996)

21.34 B. Bhushan, V.N. Koinkar: Nanoindentation hardness measurements using atomic force microscopy, Appl. Phys. Lett. **75**, 5741–5746 (1994)

21.35 D. DeVecchio, B. Bhushan: Localized surface elasticity measurements using an atomic force microscope, Rev. Sci. Instrum. **68**, 4498–4505 (1997)

21.36 S. Amelio, A.V. Goldade, U. Rabe, V. Scherer, B. Bhushan, W. Arnold: Measurements of mechanical properties of ultra-thin diamond-like carbon coatings using atomic force acoustic microscopy, Thin Solid Films **392**, 75–84 (2001)

21.37 D.M. Eigler, E.K. Schweizer: Positioning single atoms with a scanning tunnelling microscope, Nature **344**, 524–528 (1990)

21.38 A.L. Weisenhorn, J.E. MacDougall, J.A.C. Gould, S.D. Cox, W.S. Wise, J. Massie, P. Maivald, V.B. Elings, G.D. Stucky, P.K. Hansma: Imaging and manipulating of molecules on a zeolite surface with an atomic force microscope, Science **247**, 1330–1333 (1990)

21.39 I.W. Lyo, P. Avouris: Field-induced nanometer-to-atomic-scale manipulation of silicon surfaces with the STM, Science **253**, 173–176 (1991)

21.40 O.M. Leung, M.C. Goh: Orientation ordering of polymers by atomic force microscope tip-surface interactions, Science **225**, 64–66 (1992)

21.41 D.W. Abraham, H.J. Mamin, E. Ganz, J. Clark: Surface modification with the scanning tunneling microscope, IBM J. Res. Dev. **30**, 492–499 (1986)

21.42 R.M. Silver, E.E. Ehrichs, A.L. de Lozanne: Direct writing of submicron metallic features with a scanning tunnelling microscope, Appl. Phys. Lett. **51**, 247–249 (1987)

21.43 A. Kobayashi, F. Grey, R.S. Williams, M. Ano: Formation of nanometer-scale grooves in silicon with a scanning tunneling microscope, Science **259**, 1724–1726 (1993)

21.44 B. Parkinson: Layer-by-layer nanometer scale etching of two-dimensional substrates using the scanning tunneling microscope, J. Am. Chem. Soc. **112**, 7498–7502 (1990)

21.45 A. Majumdar, P.I. Oden, J.P. Carrejo, L.A. Nagahara, J.J. Graham, J. Alexander: Nanometer-scale lithography using the atomic force microscope, Appl. Phys. Lett. **61**, 2293–2295 (1992)

21.46 B. Bhushan: Micro-/nanotribology and its applications to magnetic storage devices and MEMS, Tribol. Int. **28**, 85–96 (1995)

21.47 L. Tsau, D. Wang, K.L. Wang: Nanometer scale patterning of silicon(100) surface by an atomic force microscope operating in air, Appl. Phys. Lett. **64**, 2133–2135 (1994)

21.48 E. Delawski, B.A. Parkinson: Layer-by-layer etching of two-dimensional metal chalcogenides with the atomic force microscope, J. Am. Chem. Soc. **114**, 1661–1667 (1992)

21.49 B. Bhushan, G.S. Blackman: Atomic force microscopy of magnetic rigid disks and sliders and its applications to tribology, ASME J. Tribol. **113**, 452–458 (1991)

21.50 O. Marti, B. Drake, P.K. Hansma: Atomic force microscopy of liquid-covered surfaces: atomic resolution images, Appl. Phys. Lett. **51**, 484–486 (1987)

21.51 B. Drake, C.B. Prater, A.L. Weisenhorn, S.A.C. Gould, T.R. Albrecht, C.F. Quate, D.S. Cannell, H.G. Hansma, P.K. Hansma: Imaging crystals, polymers and processes in water with the atomic force microscope, Science **243**, 1586–1589 (1989)

21.52 M. Binggeli, R. Christoph, H.E. Hintermann, J. Colchero, O. Marti: Friction force measurements on potential controlled graphite in an electrolytic environment, Nanotechnology **4**, 59–63 (1993)

21.53 G. Meyer, N.M. Amer: Novel optical approach to atomic force microscopy, Appl. Phys. Lett. **53**, 1045–1047 (1988)

21.54 J.H. Coombs, J.B. Pethica: Properties of vacuum tunneling currents: Anomalous barrier heights, IBM J. Res. Dev. **30**, 455–459 (1986)

21.55 M.D. Kirk, T. Albrecht, C.F. Quate: Low-temperature atomic force microscopy, Rev. Sci. Instrum. **59**, 833–835 (1988)

21.56 F.J. Giessibl, C. Gerber, G. Binnig: A low-temperature atomic force/scanning tunneling microscope for ultrahigh vacuum, J. Vac. Sci. Technol. B **9**, 984–988 (1991)

21.57 T.R. Albrecht, P. Grutter, D. Rugar, D.P.E. Smith: Low temperature force microscope with all-fiber interferometer, Ultramicroscopy **42–44**, 1638–1646 (1992)

21.58 H.J. Hug, A. Moser, T. Jung, O. Fritz, A. Wadas, I. Parashikor, H.J. Güntherodt: Low temperature magnetic force microscopy, Rev. Sci. Instrum. **64**, 2920–2925 (1993)

21.59 C. Basire, D.A. Ivanov: Evolution of the lamellar structure during crystallization of a semicrystalline-amorphous polymer blend: Time-resolved hot-stage SPM study, Phys. Rev. Lett. **85**, 5587–5590 (2000)

21.60 H. Liu, B. Bhushan: Investigation of nanotribological properties of self-assembled monolayers with alkyl and biphenyl spacer chains, Ultramicroscopy **91**, 185–202 (2002)

21.61 J. Foster, J. Frommer: Imaging of liquid crystal using a tunneling microscope, Nature **333**, 542–547 (1988)

21.62 D. Smith, H. Horber, C. Gerber, G. Binnig: Smectic liquid crystal monolayers on graphite observed by scanning tunneling microscopy, Science **245**, 43–45 (1989)

21.63 D. Smith, J. Horber, G. Binnig, H. Nejoh: Structure, registry and imaging mechanism of alkylcyanobiphenyl molecules by tunnelling microscopy, Nature **344**, 641–644 (1990)

21.64 Y. Andoh, S. Oguchi, R. Kaneko, T. Miyamoto: Evaluation of very thin lubricant films, J. Phys. D **25**, A71–A75 (1992)

21.65 Y. Martin, C.C. Williams, H.K. Wickramasinghe: Atomic force microscope-force mapping and profiling on a sub 100 scale, J. Appl. Phys. **61**, 4723–4729 (1987)

21.66 J.E. Stern, B.D. Terris, H.J. Mamin, D. Rugar: Deposition and imaging of localized charge on insulator surfaces using a force microscope, Appl. Phys. Lett. **53**, 2717–2719 (1988)

21.67 K. Yamanaka, H. Ogisco, O. Kolosov: Ultrasonic force microscopy for nanometer resolution subsurface imaging, Appl. Phys. Lett. **64**, 178–180 (1994)

21.68 K. Yamanaka, E. Tomita: Lateral force modulation atomic force microscope for selective imaging of friction forces, Jpn. J. Appl. Phys. **34**, 2879–2882 (1995)

21.69 U. Rabe, K. Janser, W. Arnold: Vibrations of free and surface-coupled atomic force microscope: Theory and experiment, Rev. Sci. Instrum. **67**, 3281–3293 (1996)

21.70 Y. Martin, H.K. Wickramasinghe: Magnetic imaging by force microscopy with 1000 Å resolution, Appl. Phys. Lett. **50**, 1455–1457 (1987)

21.71 D. Rugar, H.J. Mamin, P. Güthner, S.E. Lambert, J.E. Stern, I. McFadyen, T. Yogi: Magnetic force microscopy – General principles and application to longitudinal recording media, J. Appl. Phys. **63**, 1169–1183 (1990)

21.72 C. Schönenberger, S.F. Alvarado: Understanding magnetic force microscopy, Z. Phys. B **80**, 373–383 (1990)

21.73 U. Hartmann: Magnetic force microscopy, Annu. Rev. Mater. Sci. **29**, 53–87 (1999)

21.74 D.W. Pohl, W. Denk, M. Lanz: Optical stethoscopy-image recording with resolution lambda/20, Appl. Phys. Lett. **44**, 651–653 (1984)

21.75 E. Betzig, J.K. Troutman, T.D. Harris, J.S. Weiner, R.L. Kostelak: Breaking the diffraction barrier – optical microscopy on a nanometric scale, Science **251**, 1468–1470 (1991)

21.76 E. Betzig, P.L. Finn, J.S. Weiner: Combined shear force and near-field scanning optical microscopy, Appl. Phys. Lett. **60**, 2484 (1992)

21.77 P.F. Barbara, D.M. Adams, D.B. O'Connor: Characterization of organic thin film materials with near-field scanning optical microscopy (NSOM), Annu. Rev. Mater. Sci. **29**, 433–469 (1999)

21.78 C.C. Williams, H.K. Wickramasinghe: Scanning thermal profiler, Appl. Phys. Lett. **49**, 1587–1589 (1986)

21.79 C.C. Williams, H.K. Wickramasinghe: Microscopy of chemical-potential variations on an atomic scale, Nature **344**, 317–319 (1990)

21.80 A. Majumdar: Scanning thermal microscopy, Annu. Rev. Mater. Sci. **29**, 505–585 (1999)

21.81 O.E. Husser, D.H. Craston, A.J. Bard: Scanning electrochemical microscopy – High resolution deposition and etching of materials, J. Electrochem. Soc. **136**, 3222–3229 (1989)

21.82 Y. Martin, D.W. Abraham, H.K. Wickramasinghe: High-resolution capacitance measurement and potentiometry by force microscopy, Appl. Phys. Lett. **52**, 1103–1105 (1988)

21.83 M. Nonnenmacher, M.P. O'Boyle, H.K. Wickramasinghe: Kelvin probe force microscopy, Appl. Phys. Lett. **58**, 2921–2923 (1991)

21.84 J.M.R. Weaver, D.W. Abraham: High resolution atomic force microscopy potentiometry, J. Vac. Sci. Technol. B **9**, 1559–1561 (1991)

21.85 D. DeVecchio, B. Bhushan: Use of a nanoscale Kelvin probe for detecting wear precursors, Rev. Sci. Instrum. **69**, 3618–3624 (1998)

21.86 B. Bhushan, A.V. Goldade: Measurements and analysis of surface potential change during wear of single-crystal silicon (100) at ultralow loads using Kelvin probe microscopy, Appl. Surf. Sci. **157**, 373–381 (2000)

21.87 P.K. Hansma, B. Drake, O. Marti, S.A.C. Gould, C.B. Prater: The scanning ion-conductance microscope, Science **243**, 641–643 (1989)

21.88 C.B. Prater, P.K. Hansma, M. Tortonese, C.F. Quate: Improved scanning ion-conductance microscope using microfabricated probes, Rev. Sci. Instrum. **62**, 2634–2638 (1991)

21.89 J. Matey, J. Blanc: Scanning capacitance microscopy, J. Appl. Phys. **57**, 1437–1444 (1985)

21.90 C.C. Williams: Two-dimensional dopant profiling by scanning capacitance microscopy, Annu. Rev. Mater. Sci. **29**, 471–504 (1999)

21.91 D.T. Lee, J.P. Pelz, B. Bhushan: Instrumentation for direct, low frequency scanning capacitance microscopy, and analysis of position dependent stray capacitance, Rev. Sci. Instrum. **73**, 3523–3533 (2002)

21.92 P.K. Hansma, J. Tersoff: Scanning tunneling microscopy, J. Appl. Phys. **61**, R1–R23 (1987)

21.93 I. Giaever: Energy gap in superconductors measured by electron tunneling, Phys. Rev. Lett. **5**, 147–148 (1960)

21.94 D. Sarid, V. Elings: Review of scanning force microscopy, J. Vac. Sci. Technol. B **9**, 431–437 (1991)

21.95 U. Durig, O. Zuger, A. Stalder: Interaction force detection in scanning probe microscopy: Methods and applications, J. Appl. Phys. **72**, 1778–1797 (1992)

21.96 J. Frommer: Scanning tunneling microscopy and atomic force microscopy in organic chemistry, Angew. Chem. Int. Ed. **31**, 1298–1328 (1992)

21.97 H.J. Güntherodt, R. Wiesendanger (Eds.): *Scanning Tunneling Microscopy I: General Principles and Applications to Clean and Adsorbate-Covered Surfaces* (Springer, Berlin, Heidelberg 1992)

21.98 R. Wiesendanger, H.J. Güntherodt (Eds.): *Scanning Tunneling Microscopy II: Further Applications and Related Scanning Techniques* (Springer, Berlin, Heidelberg 1992)

21.99 D.A. Bonnell (Ed.): *Scanning Tunneling Microscopy and Spectroscopy – Theory, Techniques, and Applications* (VCH, New York 1993)

21.100 O. Marti, M. Amrein (Eds.): *STM and SFM in Biology* (Academic, San Diego 1993)

21.101 J.A. Stroscio, W.J. Kaiser (Eds.): *Scanning Tunneling Microscopy* (Academic, Boston 1993)

21.102 H.J. Güntherodt, D. Anselmetti, E. Meyer (Eds.): *Forces in Scanning Probe Methods* (Kluwer, Dordrecht 1995)

21.103 G. Binnig, H. Rohrer: Scanning tunnelling microscopy, Surf. Sci. **126**, 236–244 (1983)

21.104 B. Bhushan, J. Ruan, B.K. Gupta: A scanning tunnelling microscopy study of fullerene films, J. Phys. D **26**, 1319–1322 (1993)

21.105 R.L. Nicolaides, W.E. Yong, W.F. Packard, H.A. Zhou: Scanning tunneling microscope tip structures, J. Vac. Sci. Technol. A **6**, 445–447 (1988)

21.106 J.P. Ibe, P.P. Bey, S.L. Brandon, R.A. Brizzolara, N.A. Burnham, D.P. DiLella, K.P. Lee, C.R.K. Marrian, R.J. Colton: On the electrochemical etching of tips for scanning tunneling microscopy, J. Vac. Sci. Technol. A **8**, 3570–3575 (1990)

21.107 R. Kaneko, S. Oguchi: Ion-implanted diamond tip for a scanning tunneling microscopy, Jpn. J. Appl. Phys. **28**, 1854–1855 (1990)

21.108 F.J. Giessibl: Atomic resolution of the silicon(111)–(7×7) surface by atomic force microscopy, Science **267**, 68–71 (1995)

21.109 B. Anczykowski, D. Krüger, K.L. Babcock, H. Fuchs: Basic properties of dynamic force spectroscopy with the scanning force microscope in experiment and simulation, Ultramicroscopy **66**, 251–259 (1996)

21.110 T.R. Albrecht, C.F. Quate: Atomic resolution imaging of a nonconductor by atomic force microscopy, J. Appl. Phys. **62**, 2599–2602 (1987)

21.111 S. Alexander, L. Hellemans, O. Marti, J. Schneir, V. Elings, P.K. Hansma: An atomic-resolution atomic-force microscope implemented using an optical lever, J. Appl. Phys. **65**, 164–167 (1989)

21.112 G. Meyer, N.M. Amer: Optical-beam-deflection atomic force microscopy: The NaCl(001) surface, Appl. Phys. Lett. **56**, 2100–2101 (1990)

21.113 A.L. Weisenhorn, M. Egger, F. Ohnesorge, S.A.C. Gould, S.P. Heyn, H.G. Hansma, R.L. Sinsheimer, H.E. Gaub, P.K. Hansma: Molecular resolution images of Langmuir–Blodgett films and DNA by atomic force microscopy, Langmuir **7**, 8–12 (1991)

21.114 J. Ruan, B. Bhushan: Atomic-scale and microscale friction of graphite and diamond using friction force microscopy, J. Appl. Phys. **76**, 5022–5035 (1994)

21.115 D. Rugar, P.K. Hansma: Atomic force microscopy, Phys. Today **43**, 23–30 (1990)

21.116 D. Sarid: *Scanning Force Microscopy* (Oxford Univ. Press, Oxford 1991)

21.117 G. Binnig: Force microscopy, Ultramicroscopy **42–44**, 7–15 (1992)

21.118 E. Meyer: Atomic force microscopy, Surf. Sci. **41**, 3–49 (1992)

21.119 H.K. Wickramasinghe: Progress in scanning probe microscopy, Acta Mater. **48**, 347–358 (2000)

21.120 A.J. den Boef: The influence of lateral forces in scanning force microscopy, Rev. Sci. Instrum. **62**, 88–92 (1991)

21.121 M. Radmacher, R.W. Tillman, M. Fritz, H.E. Gaub: From molecules to cells: Imaging soft samples with the atomic force microscope, Science **257**, 1900–1905 (1992)

21.122 F. Ohnesorge, G. Binnig: True atomic resolution by atomic force microscopy through repulsive and attractive forces, Science **260**, 1451–1456 (1993)

21.123 G. Neubauer, S.R. Coben, G.M. McClelland, D. Horne, C.M. Mate: Force microscopy with a bidirectional capacitance sensor, Rev. Sci. Instrum. **61**, 2296–2308 (1990)

21.124 T. Goddenhenrich, H. Lemke, U. Hartmann, C. Heiden: Force microscope with capacitive displace-

21.125 U. Stahl, C.W. Yuan, A.L. Delozanne, M. Tortonese: Atomic force microscope using piezoresistive cantilevers and combined with a scanning electron microscope, Appl. Phys. Lett. **65**, 2878–2880 (1994)

21.126 R. Kassing, E. Oesterschulze: Sensors for scanning probe microscopy. In: *Micro-/Nanotribology and Its Applications*, ed. by B. Bhushan (Kluwer, Dordrecht 1997) pp. 35–54

21.127 C.M. Mate: Atomic-force-microscope study of polymer lubricants on silicon surfaces, Phys. Rev. Lett. **68**, 3323–3326 (1992)

21.128 S.P. Jarvis, A. Oral, T.P. Weihs, J.B. Pethica: A novel force microscope and point contact probe, Rev. Sci. Instrum. **64**, 3515–3520 (1993)

21.129 D. Rugar, H.J. Mamin, P. Güthner: Improved fiber-optical interferometer for atomic force microscopy, Appl. Phys. Lett. **55**, 2588–2590 (1989)

21.130 C. Schönenberger, S.F. Alvarado: A differential interferometer for force microscopy, Rev. Sci. Instrum. **60**, 3131–3135 (1989)

21.131 D. Sarid, D. Iams, V. Weissenberger, L.S. Bell: Compact scanning-force microscope using laser diode, Opt. Lett. **13**, 1057–1059 (1988)

21.132 N.W. Ashcroft, N.D. Mermin: *Solid State Physics* (Holt Reinhart and Winston, New York 1976)

21.133 G. Binnig, D.P.E. Smith: Single-tube three-dimensional scanner for scanning tunneling microscopy, Rev. Sci. Instrum. **57**, 1688 (1986)

21.134 S.I. Park, C.F. Quate: Digital filtering of STM images, J. Appl. Phys. **62**, 312 (1987)

21.135 J.W. Cooley, J.W. Tukey: An algorithm for machine calculation of complex Fourier series, Math. Comput. **19**, 297 (1965)

21.136 J. Ruan, B. Bhushan: Atomic-scale friction measurements using friction force microscopy: Part I – General principles and new measurement techniques, ASME J. Tribol. **116**, 378–388 (1994)

21.137 T.R. Albrecht, S. Akamine, T.E. Carver, C.F. Quate: Microfabrication of cantilever styli for the atomic force microscope, J. Vac. Sci. Technol. A **8**, 3386–3396 (1990)

21.138 O. Marti, S. Gould, P.K. Hansma: Control electronics for atomic force microscopy, Rev. Sci. Instrum. **59**, 836–839 (1988)

21.139 O. Wolter, T. Bayer, J. Greschner: Micromachined silicon sensors for scanning force microscopy, J. Vac. Sci. Technol. B **9**, 1353–1357 (1991)

21.140 E. Meyer, R. Overney, R. Luthi, D. Brodbeck: Friction force microscopy of mixed Langmuir–Blodgett films, Thin Solid Films **220**, 132–137 (1992)

21.141 H.J. Dai, J.H. Hafner, A.G. Rinzler, D.T. Colbert, R.E. Smalley: Nanotubes as nanoprobes in scanning probe microscopy, Nature **384**, 147–150 (1996)

21.142 J.H. Hafner, C.L. Cheung, A.T. Woolley, C.M. Lieber: Structural and functional imaging with carbon nanotube AFM probes, Prog. Biophys. Mol. Biol. **77**, 73–110 (2001)

21.143 G.S. Blackman, C.M. Mate, M.R. Philpott: Interaction forces of a sharp tungsten tip with molecular films on silicon surface, Phys. Rev. Lett. **65**, 2270–2273 (1990)

21.144 S.J. O'Shea, M.E. Welland, T. Rayment: Atomic force microscope study of boundary layer lubrication, Appl. Phys. Lett. **61**, 2240–2242 (1992)

21.145 J.P. Cleveland, S. Manne, D. Bocek, P.K. Hansma: A nondestructive method for determining the spring constant of cantilevers for scanning force microscopy, Rev. Sci. Instrum. **64**, 403–405 (1993)

21.146 D.W. Pohl: Some design criteria in STM, IBM J. Res. Dev. **30**, 417 (1986)

21.147 W.T. Thomson, M.D. Dahleh: *Theory of Vibration with Applications*, 5th edn. (Prentice Hall, Upper Saddle River 1998)

21.148 J. Colchero: Reibungskraftmikroskopie. Ph.D. Thesis (University of Konstanz, Konstanz 1993), in German

21.149 G.M. McClelland, R. Erlandsson, S. Chiang: Atomic force microscopy: General principles and a new implementation. In: *Review of Progress in Quantitative Nondestructive Evaluation*, Vol. 6B, ed. by D.O. Thompson, D.E. Chimenti (Plenum, New York 1987) pp. 1307–1314

21.150 Y.R. Shen: *The Principles of Nonlinear Optics* (Wiley, New York 1984)

21.151 T. Baumeister, S.L. Marks: *Standard Handbook for Mechanical Engineers*, 7th edn. (McGraw-Hill, New York 1967)

21.152 J. Colchero, O. Marti, H. Bielefeldt, J. Mlynek: Scanning force and friction microscopy, Phys. Status Solidi (a) **131**, 73–75 (1991)

21.153 R. Young, J. Ward, F. Scire: Observation of metal-vacuum-metal tunneling, field emission, and the transition region, Phys. Rev. Lett. **27**, 922 (1971)

21.154 R. Young, J. Ward, F. Scire: The topographiner: An instrument for measuring surface microtopography, Rev. Sci. Instrum. **43**, 999 (1972)

21.155 C. Gerber, O. Marti: Magnetostrictive positioner, IBM Tech. Discl. Bull. **27**, 6373 (1985)

21.156 R. Garcìa Cantù, M.A. Huerta Garnica: Long-scan imaging by STM, J. Vac. Sci. Technol. A **8**, 354 (1990)

21.157 C.J. Chen: In situ testing and calibration of tube piezoelectric scanners, Ultramicroscopy **42–44**, 1653–1658 (1992)

21.158 R.G. Carr: Finite element analysis of PZT tube scanner motion for scanning tunnelling microscopy, J. Microsc. **152**, 379–385 (1988)

21.159 C.J. Chen: Electromechanical deflections of piezoelectric tubes with quartered electrodes, Appl. Phys. Lett. **60**, 132 (1992)

21.160 N. Libioulle, A. Ronda, M. Taborelli, J.M. Gilles: Deformations and nonlinearity in scanning tunneling microscope images, J. Vac. Sci. Technol. B **9**, 655–658 (1991)

ment detection, J. Vac. Sci. Technol. A **8**, 383–387 (1990)

21.161 E.P. Stoll: Restoration of STM images distorted by time-dependent piezo driver aftereffects, Ultramicroscopy **42–44**, 1585–1589 (1991)

21.162 R. Durselen, U. Grunewald, W. Preuss: Calibration and applications of a high precision piezo scanner for nanometrology, Scanning **17**, 91–96 (1995)

21.163 J. Fu: In situ testing and calibrating of Z-piezo of an atomic force microscope, Rev. Sci. Instrum. **66**, 3785–3788 (1995)

21.164 R.C. Barrett, C.F. Quate: Optical scan-correction system applied to atomic force microscopy, Rev. Sci. Instrum. **62**, 1393 (1991)

21.165 R. Toledo-Crow, P.C. Yang, Y. Chen, M. Vaez-Iravani: Near-field differential scanning optical microscope with atomic force regulation, Appl. Phys. Lett. **60**, 2957–2959 (1992)

21.166 J.E. Griffith, G.L. Miller, C.A. Green: A scanning tunneling microscope with a capacitance-based position monitor, J. Vac. Sci. Technol. B **8**, 2023–2027 (1990)

21.167 A.E. Holman, C.D. Laman, P.M.L.O. Scholte, W.C. Heerens, F. Tuinstra: A calibrated scanning tunneling microscope equipped with capacitive sensors, Rev. Sci. Instrum. **67**, 2274–2280 (1996)

22. General and Special Probes in Scanning Microscopies

Jason Hafner, Edin (I-Chen) Chen, Ratnesh Lal, Sungho Jin

Scanning probe microscopy (SPM) provides nanometer-scale mapping of numerous sample properties in essentially any environment. This unique combination of high resolution and broad applicability has led to the application of SPM to many areas of science and technology, especially those interested in the structure and properties of materials at the nanometer scale. SPM images are generated through measurements of a tip–sample interaction. A well-characterized tip is the key element to data interpretation and is typically the limiting factor.

Commercially available atomic force microscopy (AFM) tips, integrated with force-sensing cantilevers, are microfabricated from silicon and silicon nitride by lithographic and anisotropic etching techniques. The performance of these tips can be characterized by imaging nanometer-scale standards of known dimension, and the resolution is found to roughly correspond to the tip radius of curvature, the tip aspect ratio, and the sample height. Although silicon and silicon nitride tips have a somewhat large radius of curvature, low aspect ratio, and limited lifetime due to wear, the widespread use of AFM today is due in large part to the broad availability of these tips. In some special cases, small asperities on the tip can provide resolution much higher than the tip radius of curvature for low-Z samples such as crystal surfaces and ordered protein arrays.

Several strategies have been developed to improve AFM tip performance. Oxide sharpening improves tip sharpness and enhances tip asperities. For high-aspect-ratio samples such as integrated circuits, silicon AFM tips can be modified by focused ion beam (FIB) milling. FIB tips reach 3° cone angles over lengths of several microns and can be fabricated at arbitrary angles.

Other high resolution and high-aspect-ratio tips are produced by electron-beam deposition

22.1	Atomic Force Microscopy	620
	22.1.1 Principles of Operation	620
	22.1.2 Standard Probe Tips	621
	22.1.3 Probe Tip Performance	622
	22.1.4 Oxide-Sharpened Tips	623
	22.1.5 Focused Ion Beam Tips	624
	22.1.6 Electron-Beam Deposition Tips	624
	22.1.7 Single- and Multiwalled Carbon Nanotube Tips	624
	22.1.8 Bent Carbon Nanotube Tips	628
	22.1.9 Low-Stiffness Cantilevers with Carbon Nanotube Tips	629
	22.1.10 Conductive Probe Tips	630
22.2	Scanning Tunneling Microscopy	630
	22.2.1 Mechanically Cut STM Tips	630
	22.2.2 Electrochemically Etched STM Tips	631
References		631

(EBD), in which a carbon spike is deposited onto the tip apex from the background gases in an electron microscope. Finally, carbon nanotubes have been employed as AFM tips. Their nanometer-scale diameter, long length, high stiffness, and elastic buckling properties make them possibly the ultimate tip material for AFM. Nanotubes can be manually attached to silicon or silicon nitride AFM tips or *grown* onto tips by chemical vapor deposition (CVD), which should soon make them widely available. In scanning tunneling microscopy (STM), the electron tunneling signal decays exponentially with tip–sample separation, so that in principle only the last few atoms contribute to the signal. STM tips are, therefore, not as sensitive to the nanoscale tip geometry and can be made by simple mechanical cutting or electrochemical etching of metal wires. In choosing tip materials, one prefers hard, stiff metals that will not oxidize or corrode in the imaging environment.

In scanning probe microscopy (SPM), an image is created by raster-scanning a sharp probe tip over a sample and measuring some highly localized tip–sample interaction as a function of position. SPMs are based on several interactions, the major types including scanning tunneling microscopy (STM), which measures an electronic tunneling current; atomic force microscopy (AFM), which measures force interactions; and near-field scanning optical microscopy (NSOM), which measures local optical properties by exploiting near-field effects (Fig. 22.1). These methods allow the characterization of many properties (structural, mechanical, electronic, optical) on essentially any material (metals, semiconductors, insulators, biomolecules) and in essentially any environment (vacuum, liquid, ambient air conditions). The unique combination of nanoscale resolution, previously the domain of electron microscopy, *and broad applicability* has led to the proliferation of SPM into virtually all areas of nanometer-scale science and technology.

Several enabling technologies have been developed for SPM, or borrowed from other techniques. Piezoelectric tube scanners allow accurate, subangstrom positioning of the tip or sample in three dimensions. Optical deflection systems and microfabricated cantilevers can detect forces in AFM down to the piconewton range. Sensitive electronics can measure STM currents < 1 pA. High-transmission fiber optics and sensitive photodetectors can manipulate and detect small optical signals of NSOM. Environmental control has been developed to allow SPM imaging in ultrahigh vacuum (UHV), cryogenic temperatures, at elevated temperatures, and in fluids. Vibration and drift have been controlled such that a probe tip can be held over a sin-

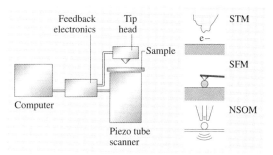

Fig. 22.1 A schematic of the components of a scanning probe microscope and the three types of signals observed: STM senses electron tunneling currents, AFM measures forces, and NSOM measures near-field optical properties via a subwavelength aperture

gle molecule for hours of observation. Microfabrication techniques have been developed for the mass production of probe tips, making SPMs commercially available and allowing the development of many new SPM modes and combinations with other characterization methods. However, of all this SPM development over the past 20 years, what has received the least attention is perhaps the most important aspect: the probe tip.

Interactions measured in SPMs occur at the tip–sample interface, which can range in size from a single atom to tens of nanometers. The size, shape, surface chemistry, and electronic and mechanical properties of the tip apex will directly influence the data signal and the interpretation of the image. Clearly, the better characterized the tip, the more useful the image information. In this chapter, the fabrication and performance of AFM and STM probes will be described.

22.1 Atomic Force Microscopy

AFM is the most widely used form of SPM, since it requires neither an electrically conductive sample, as in STM, nor an optically transparent sample or substrate, as in most NSOMs. Basic AFM modes measure the topography of a sample, with the only requirement being that the sample be deposited on a flat surface and rigid enough to withstand imaging. Since AFM can measure a variety of forces, including van der Waals forces, electrostatic forces, magnetic forces, adhesion forces, and friction forces, specialized modes of AFM can characterize the electrical, mechanical, and chemical properties of a sample in addition to its topography.

22.1.1 Principles of Operation

In AFM, a probe tip is integrated with a microfabricated force-sensing cantilever. A variety of silicon and silicon nitride cantilevers are commercially available with micrometer-scale dimensions, spring constants ranging from 0.01 to 100 N/m, and resonant frequencies ranging from 5 kHz to over 300 kHz. The cantilever deflection is detected by optical beam deflection, as illustrated in Fig. 22.2. A laser beam bounces off the back of the cantilever and is centered on a split photodiode. Cantilever deflections are

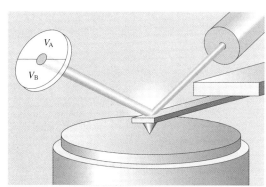

Fig. 22.2 An illustration of the optical beam deflection system that detects cantilever motion in the AFM. The voltage signal $V_A - V_B$ is proportional to the deflection

proportional to the difference signal $V_A - V_B$. Subangstrom deflections can be detected, and therefore forces down to tens of piconewtons can be measured. A more recently developed method of cantilever deflection measurement is through a piezoelectric layer on the cantilever that registers a voltage upon deflection [22.1].

A piezoelectric scanner rasters the sample under the tip while the forces are measured through deflections of the cantilever. To achieve more controlled imaging conditions, a feedback loop monitors the tip–sample force and adjusts the sample z-position to hold the force constant. The topographic image of the sample is then taken from the sample z-position data. The mode described is called the contact mode, in which the tip is deflected by the sample due to repulsive forces, or *contact*. It is generally only used for flat samples that can withstand lateral forces during scanning. To minimize lateral forces and sample damage, two alternating-current (AC) modes have been developed. In these, the cantilever is driven into AC oscillation near its resonant frequency (tens to hundreds of kHz) with desired amplitudes. When the tip approaches the sample, the oscillation is damped, and the reduced amplitude is the feedback signal, rather than the direct-current (DC) deflection. Again, topography is taken from the varying Z-position of the sample required to keep the tip oscillation amplitude constant. The two AC modes differ only in the nature of the interaction. In intermittent contact mode, also called tapping mode, the tip contacts the sample on each cycle, so the amplitude is reduced by ionic repulsion as in contact mode. In noncontact mode, long-range van der Waals forces reduce the amplitude by effectively shifting the spring constant experienced by the tip and changing its resonant frequency.

22.1.2 Standard Probe Tips

In early AFM work, cantilevers were made by hand from thin metal foils or small metal wires. Tips were created by gluing diamond fragments to the foil cantilevers or electrochemically etching the wires to a sharp point. Since these methods were labor intensive and not highly reproducible, they were not amenable to large-scale production. To address this problem, and the need for smaller cantilevers with higher resonant frequencies, batch fabrication techniques were developed (Fig. 22.3). Building on existing methods to batch-fabricate Si_3N_4 cantilevers, *Albrecht* et al. [22.2] etched an array of small square openings in an SiO_2 mask layer over a (100) silicon surface. The exposed square (100) regions were etched with KOH, an anisotropic etchant that terminates at the (111) planes, thus creating pyramidal etch pits in the silicon surface. The etch pit mask was then removed and another was applied to define the cantilever shapes with the pyramidal etch pits at the end. The Si wafer was then coated with a low-stress Si_3N_4 layer by low-pressure chemical vapor deposition (LPCVD). The Si_3N_4 fills the etch pit, using it as a mold to create a pyramidal tip. The silicon was later removed by etching to free the cantilevers and tips. Further steps resulting in the attachment of the cantilever to a macroscopic piece of glass are not described here. The resulting pyramidal tips were highly symmetric and had a tip radius of < 30 nm, as determined by scanning electron microscopy (SEM). This procedure has likely not changed significantly, since commercially available Si_3N_4 tips are still specified to have a radius of curvature of 30 nm.

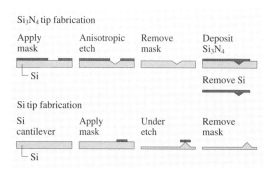

Fig. 22.3 A schematic overview of the fabrication of Si and Si_3N_4 tip fabrication as described in the text

Wolter et al. [22.3] developed methods to batch-fabricate single-crystal Si cantilevers with integrated tips. Microfabricated Si cantilevers were first prepared using previously described methods, and a small mask was formed at the end of the cantilever. The Si around the mask was etched by KOH, so that the mask was undercut. This resulted in a pyramidal silicon tip under the mask, which was then removed. Again, this partial description of the full procedure only describes tip fabrication. With some refinements the silicon tips were made in high yield with radii of curvature of less than 10 nm. Si tips are sharper than Si_3N_4 tips, because they are directly formed by the anisotropic etch in single-crystal Si, rather than using an etch pit as a mask for deposited material. Commercially available silicon probes are made by similar refined techniques and provide a typical radius of curvature of < 10 nm.

22.1.3 Probe Tip Performance

In atomic force microscopy the question of resolution can be a rather complicated issue. As an initial approximation, resolution is often considered strictly in geometrical terms that assume rigid tip–sample contact. The topographical image of a feature is broadened or narrowed by the size of the probe tip, so the resolution is approximately the width of the tip. Therefore, the resolution of AFM with standard commercially available tips is on the order of 5–10 nm. *Bustamante* and *Keller* [22.4] carried the geometrical model further by drawing an analogy to resolution in optical systems. Consider two sharp spikes separated by a distance d to be point objects imaged by AFM (Fig. 22.4). Assume the tip has a parabolic shape with an end radius R. The tip-broadened image of these spikes will appear as inverted parabolas. There will be a small depression between the images of depth Δz. The two spikes are considered *resolved* if Δz is larger than the instrumental noise in the z-direction. Defined in this manner, the resolution d, the minimum separation at which the spikes are resolved, is

$$d = 2\sqrt{2R(\Delta z)}, \tag{22.1}$$

where one must enter a minimal detectable depression for the instrument (Δz) to determine the resolution. So for a silicon tip with radius 5 nm and a minimum detectable Δz of 0.5 nm, the resolution is about 4.5 nm. However, the above model assumes the spikes are of equal height. *Bustamante* and *Keller* [22.4] went on to point out that, if the height of the spikes is not equal, the resolution will be affected. Assuming a height difference of Δh, the resolution becomes

$$d = \sqrt{2R}(\sqrt{\Delta z} + \sqrt{\Delta z + \Delta h}). \tag{22.2}$$

For a pair of spikes with a 2 nm height difference, the resolution drops to 7.2 nm for a 5 nm tip and 0.5 nm minimum detectable Δz. While geometrical considerations are a good starting point for defining resolution, they ignore factors such as the possible compression and deformation of the tip and sample. *Vesenka* et al. [22.5] confirmed a similar geometrical resolution model by imaging monodisperse gold nanoparticles with tips characterized by transmission electron microscopy (TEM).

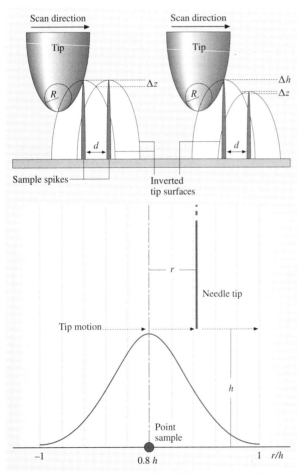

Fig. 22.4 The factors that determine AFM imaging resolution in contact mode (*top*) and noncontact mode (*bottom*) (after [22.4])

Noncontact AFM contrast is generated by long-range interactions such as van der Waals forces, so resolution will not simply be determined by geometry because the tip and sample are not in rigid contact. *Bustamante* and *Keller* [22.4] have derived an expression for the resolution in noncontact AFM for an idealized, infinitely thin *line* tip and a point particle as the sample (Fig. 22.4). Noncontact AFM is sensitive to the gradient of long-range forces, so the van der Waals force gradient was calculated as a function of position for the tip at height h above the surface. If the resolution d is defined as the full-width at half-maximum of this curve, the resolution is

$$d = 0.8h \, . \tag{22.3}$$

This shows that, even for an ideal geometry, the resolution is fundamentally limited in noncontact mode by the tip–sample separation. Under UHV conditions, the tip–sample separation can be made very small, so atomic resolution is possible on flat, crystalline surfaces. Under ambient conditions, however, the separation must be larger to keep the tip from being trapped in the ambient water layer on the surface. This larger separation can lead to a point where further improvements in tip sharpness do not improve resolution. It has been found that imaging 5 nm gold nanoparticles in noncontact mode with carbon nanotube tips of 2 nm diameter leads to particle widths of 12 nm, larger than the 7 nm width one would expect assuming rigid contact [22.8]. However, in tapping-mode operation, the geometrical definition of resolution is relevant, since the tip and sample come into rigid contact. When imaging 5 nm gold particles with 2 nm carbon nanotube tips in tapping mode, the expected 7 nm particle width is obtained [22.9].

The above descriptions of AFM resolution cannot explain the subnanometer resolution achieved on crystal surfaces [22.10] and ordered arrays of biomolecules [22.11] in contact mode with commercially available probe tips. Such tips have nominal radii of curvature ranging from 5 to 30 nm, an order of magnitude larger than the resolution achieved. A detailed model to explain the high resolution on ordered membrane proteins has been put forth by [22.6]. In this model, the larger part of the silicon nitride tip apex balances the tip–sample interaction through electrostatic forces, while a very small tip asperity interacts with the sample to provide contrast (Fig. 22.5). This model is supported by measurements at varying salt concentrations to vary the electrostatic interaction strength and the observation of defects in the ordered samples. However, the existence of such asperities has never been

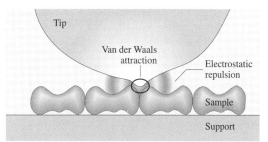

Fig. 22.5 A tip model to explain the high resolution obtained on ordered samples in contact mode (after [22.6])

confirmed by independent electron microscopy images of the tip. Another model, considered especially applicable to atomic resolution on crystal surfaces, assumes that the tip is in contact with a region of the sample much larger than the observed resolution, and that force components matching the periodicity of the sample are transmitted to the tip, resulting in an *averaged* image of the periodic lattice. Regardless of the mechanism, the structures determined are accurate and make this a highly valuable method for membrane proteins. However, this level of resolution should not be expected for most biological systems.

22.1.4 Oxide-Sharpened Tips

Both Si and Si_3N_4 tips with increased aspect ratio and reduced tip radius can be fabricated through oxide sharpening of the tip. If a pyramidal or cone-shaped silicon tip is thermally oxidized to SiO_2 at low temperature ($< 1050\,^\circ$C), Si–SiO_2 stress formation reduces the oxidation rate at regions of high curvature. The result is a sharper, higher-aspect-ratio cone of silicon at the high-curvature tip apex inside the outer pyramidal layer

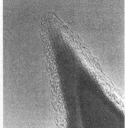

Fig. 22.6 Oxide sharpening of silicon tips. The *left image* shows a sharpened core of silicon in an outer layer of SiO_2. The *right image* is a higher magnification view of such a tip after the SiO_2 is removed (after [22.7])

of SiO$_2$ (Fig. 22.6). Etching the SiO$_2$ layer with HF then leaves tips with aspect ratios up to 10 : 1 and radii down to 1 nm [22.7], although 5–10 nm is the nominal specification for most commercially available tips. This oxide-sharpening technique can also be applied to Si$_3$N$_4$ tips by oxidizing the silicon etch pits that are used as molds. As with tip fabrication, oxide sharpening is not quite as effective for Si$_3$N$_4$. Si$_3$N$_4$ tips were reported to have an 11 nm radius of curvature [22.12], while commercially available oxide-sharpened Si$_3$N$_4$ tips have a nominal radius of < 20 nm.

22.1.5 Focused Ion Beam Tips

A common AFM application in integrated circuit manufacture and microelectromechanical systems (MEMS) is to image structures with very steep sidewalls such as trenches. To image these features accurately, one must consider the micrometer-scale tip structure, rather than the nanometer-scale structure of the tip apex. Since tip fabrication processes rely on anisotropic etchants, the cone half-angles of pyramidal tips are approximately 20°. Images of deep trenches taken with such tips display slanted sidewalls and may not reach the bottom of the trench due to the tip broadening effects. To image such samples more faithfully, high-aspect-ratio tips are fabricated by focused ion beam (FIB) machining a Si tip to produce a sharp spike at the tip apex. Commercially available FIB tips have half-cone angles of $< 3°$ over lengths of several micrometers, yielding aspect ratios of approximately 10 : 1. The radius of curvature at the tip end is similar to that of the tip before the FIB machining. Another consideration for high-aspect-ratio tips is the tip tilt. To ensure that the pyramidal tip is the lowest part of the tip–cantilever assembly, most AFM designs tilt the cantilever about 15° from parallel. Therefore, even an ideal *line tip* will not give an accurate image of high steep sidewalls, but will produce an image that depends on the scan angle. Due to the versatility of FIB machining, tips are available with the spikes at an angle to compensate for this effect.

22.1.6 Electron-Beam Deposition Tips

Another method of producing high-aspect-ratio tips for AFM is called electron-beam deposition (EBD). First developed for STM tips [22.13, 14], EBD tips were introduced for AFM by focusing an SEM onto the apex of a pyramidal tip arranged so that it pointed along the electron beam axis (Fig. 22.7). Carbon material was deposited by the dissociation of background gases in the

Fig. 22.7 A pyramidal tip before (*left*, 2 μm-scale bar) and after (*right*, 1 μm-scale bar) electron beam deposition (after [22.13])

SEM vacuum chamber. *Schiffmann* [22.15] systematically studied the following parameters and how they affected EBD tip geometry:

Deposition time: 0.5–8 min
Beam current: 3–300 pA
Beam energy: 1–30 keV
Working distance: 8–48 mm

EBD tips were cylindrical with end radii of 20–40 nm, lengths of 1–5 μm, and diameters of 100–200 nm. Like FIB tips, EBD tips were found to achieve improved imaging of steep features. By controlling the position of the focused beam, the tip geometry can be further controlled. Tips were fabricated with lengths over 5 μm and aspect ratios greater than 100 : 1, yet these were too fragile to use as AFM tips [22.13].

22.1.7 Single- and Multiwalled Carbon Nanotube Tips

Carbon nanotubes (CNTs), which were discovered in 1991, are composed of graphene sheets that are rolled up into tubes. Due to their high-aspect-ratio geometry, small tip diameter, and excellent mechanical properties, CNTs have become a promising candidate for new AFM probes to replace standard silicon or silicon nitride probes. CNT tips could offer high-resolution images, while the length of CNT tips allows the tracing of steep and deep features.

Structures of Carbon Nanotubes
CNTs are seamless cylinders formed by the honeycomb lattice of a single layer of crystalline graphite, called a graphene sheet. In general, CNTs are divided into two types, single-walled nanotubes (SWNTs) and

multiwalled nanotubes (MWNTs). Figure 22.8 shows the structures of CNTs explored by a high-resolution TEM [22.16, 17]. A SWNT is composed of only one rolled-up grapheme, whereas a MWNT consists of a number of concentric tubes. Multiwalled CNTs grown by the thermal CVD process generally exhibit concentric cylinder shape (Fig. 22.9a), while those grown by direct-current plasma-enhanced CVD (DC-PECVD) often exhibit a stacked cone structure (also known as herringbone- or bamboo-like structures, a cross-section of which is illustrated in Fig. 22.9b). Herringbone-like CNTs are also called carbon nanofibers (CNFs) since they are not made of perfect graphene tube cylinders.

Carbon Nanotube Probes by Attachment Approaches

CNTs have been attached onto AFM cantilever pyramid tips by various approaches. The first CNT AFM probes [22.18] were fabricated by techniques developed for assembling single-nanotube field-emission tips [22.19]. This process, illustrated in Fig. 22.10, used a purified MWNT material synthesized by the carbon arc procedure. The raw material must contain at least a few percent of long nanotubes ($> 10\,\mu$m), purified by oxidation to $\approx 1\%$ of its original mass. A torn edge of the purified material was attached to a micromanipulator by carbon tape and viewed under an optical microscope. Individual nanotubes and nanotube bundles were visible as filaments under dark-field illumination. A commercially available AFM tip was attached to another micromanipulator opposing the nanotube material. Glue was applied to the tip apex from high-vacuum carbon tape supporting the nanotube material. Nanotubes were then manually attached to the tip apex by micromanipulation. As assembled, MWNT tips were often too long for imaging due to thermal vibration during their use as AFM probes. Nanotubes tips were shortened by applying 10 V pulses to the tip while near a sputtered niobium surface. This process etched 100 nm lengths of nanotubes per pulse.

Since the nanotube orientation cannot be well controlled during manual attachment processes, the transfer procedure from the nanotube probe cartridge to the Si tips was operated under an electric field [22.20]. When applying a low voltage, the nanotube is attracted to the cantilever tip and aligned with the apex of the tip. This approach provides better control of the orientation of nanotube probes because of the electric-field alignment and electrostatic attraction of nanotube probes. When the nanotube is suitably aligned, the voltage is increased

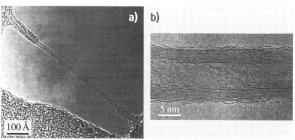

Fig. 22.8a,b The structure of carbon nanotubes. (a) TEM image of SWNTs (after [22.16]). (b) TEM image of MWNTs (after [22.17])

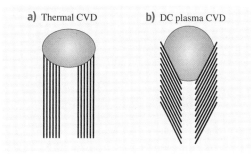

Fig. 22.9a,b Schematic structures of (a) tubelike carbon nanotubes and (b) stacked-cone nanotubes

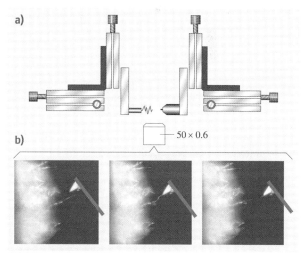

Fig. 22.10a,b Schematic drawing of the setup for manual assembly of carbon nanotube tips (a) and (b) optical microscopy images of the assembly process (the cantilever was drawn in for clarity)

to induce an arc discharge in which the nanotube is energetically disassociated and the formation of a carbide may occur at the contact site. Thus, the nanotube can be attached to the cantilever free from the cartridge. The mechanical attachment method has also been carried out in a SEM rather than an optical microscope [22.21]. This process allows selecting a single nanotube and attaching it to a specific site on the Si tip. This approach eliminates the need for pulse-etching, since short nanotubes can be attached to the tip, and the *glue* can be applied by EBD.

A method to attach CNTs onto AFM tips using magnetic-field alignment has been developed [22.23]. The experimental apparatus is designed to introduce a magnetic field onto a single AFM probe and a nanotube suspension. With this apparatus, the anisotropic properties of the CNT cause the nanotubes that come into contact with the probe tip to be preferentially oriented parallel to the tip direction and hence protrude down from the end. Another attachment method based on liquid deposition of CNTs onto AFM probes is the dielectrophoresis process [22.24, 25]. A Si AFM probe and a metal plate are used as electrodes to apply the AC electric field. A charge-coupled device (CCD) connected to a computer could be used to monitor the process. With in situ observation using the CCD image, the counterelectrode was slowly moved close to the AFM probe until the suspension surface touches its apex. The electrode was then gradually withdrawn until a CNT tip with the desired length was assembled. In this dielectrophoresis process, the length of the CNT probe is controlled by the distance that the counterelectrode is translated under the AC field.

Nanotube Probe Synthesis by Thermal CVD

The mechanical attachment approaches are tedious and time consuming since nanotube tips are made individually. So, these methods cannot be applied for mass production. The problems of manual assembly of nanotube probes discussed above can largely be solved by directly growing nanotubes onto AFM tips by metal-catalyzed chemical vapor deposition (CVD). Nanometer-scale metal catalyst particles are heated in a gas mixture containing hydrocarbon or CO. The gas molecules dissociate on the metal surface, and carbon is adsorbed into the catalyst particle. When this carbon precipitates, it nucleates a nanotube of similar diameter to the catalyst particle. Therefore, CVD allows control over nanotube size and structure, including the production of SWNTs [22.26] with radii as low as 3.5 Å [22.27].

Several key issues must be addressed to grow nanotube AFM tips by CVD:

1. The alignment of the nanotubes at the tip
2. The number of nanotubes that grow at the tip
3. The length of the nanotube tip.

Li et al. [22.28] found that nanotubes grow perpendicular to a porous surface containing embedded catalyst. This approach was exploited to fabricate nanotube tips by CVD [22.29] with the proper alignment, as illustrated in Fig. 22.11. A flattened area of $\approx 1-5\,\mu m^2$ was created on Si tips by scanning in contact mode at high load ($1\,\mu N$) on a hard, synthetic diamond surface. The tip was then anodized in HF to create $100\,nm$-diameter pores in this flat surface [22.30]. It is important to anodize only the last $20-40\,\mu m$ of the cantilever, which includes the tip, so that the rest of the cantilever is still reflective for use in the AFM. This was achieved by anodizing the tip in a small drop of HF under the view of an optical microscope. Next, iron was electrochemically deposited into the pores to form catalyst particles [22.31]. Tips prepared in this way were heated in low concentrations of ethylene at $800\,°C$, which is known to favor the growth of thin nanotubes [22.26]. When imaged by SEM, nanotubes were found to grow perpendicular to the surface from the pores as desired, and TEM revealed that the nanotubes were thin, individual, multiwalled nanotubes with typical radii of $3-5\,nm$.

These *pore-growth* CVD nanotube tips were typically several micrometers in length – too long for imaging – and were pulse-etched to usable length of $< 500\,nm$. The tips exhibited elastic buckling behavior and were very robust in imaging. The pore-growth method demonstrated the potential of CVD to simplify the fabrication of nanotube tips, although there were

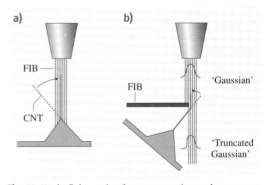

Fig. 22.11a,b Schematics for two experimental setup conditions using a focused ion beam for (**a**) aligning a nanotube tip and (**b**) bending the tip (after [22.22])

still limitations. In particular, the porous layer was difficult to prepare and rather fragile.

An alternative approach to CVD fabrication of nanotube tips involves direct growth of SWNTs on the surface of a pyramidal AFM tip [22.32, 33]. In this *surface-growth* approach, an alumina/iron/molybdenum powdered catalyst known to produce SWNT [22.26] was dispersed in ethanol at 1 mg/ml. Silicon tips were dipped in this solution and allowed to dry, leaving a sparse layer of ≈ 100 nm catalyst clusters on the tip. When CVD conditions were applied, single-walled nanotubes grew along the silicon tip surface. At a pyramid edge, nanotubes can either bend to align with the edge or protrude from the surface. If the energy required to bend the tube and follow the edge is less than the attractive nanotube surface energy, then the nanotube will follow the pyramid edge to the apex. Therefore, nanotubes were effectively steered toward the tip apex by the pyramid edges. At the apex, the nanotube protruded from the tip, since the energetic cost of bending around the sharp silicon tip was too high. The high aspect ratio at the oxide-sharpened silicon tip apex was critical for good nanotube alignment. These *surface-growth* nanotube tips exhibit a high aspect ratio and high-resolution imaging, as well as elastic buckling. This method has been expanded to include wafer-scale production of nanotube tips with high yields [22.34], yet one obstacle remains to the mass production of nanotube probe tips. Nanotubes protruding from the tip are several micrometers long, and since they are so thin, they must be etched to less than 100 nm.

Hybrid Nanotube Tip Fabrication: Pick-Up Tips

Another method of creating nanotube tips is something of a hybrid between assembly and CVD. The motivation was to create AFM probes that have an *individual* SWNT at the tip to achieve the ultimate imaging resolution. In order to synthesize isolated SWNT, isolated catalyst particles were formed by dipping a silicon wafer in an isopropyl alcohol solution of $Fe(NO_3)_3$. When heated in a CVD furnace, the iron became mobile and aggregated to form isolated iron particles. By controlling the reaction time, the SWNT lengths were kept shorter than their typical separation, so that the nanotubes never had a chance to form bundles. In the *pick-up tip* method, these isolated SWNT substrates were imaged by AFM with silicon tips in air [22.9]. When the tip encountered a vertical SWNT, the oscillation amplitude was damped, so the AFM pulled the sample away from the tip. This procedure pulled the SWNT into contact with the tip along its length, so that it became attached to the tip. This assembly process happened automatically when imaging in tapping mode; no special tip manipulation was required.

PECVD-Grown Nanotube Probe

The attachment methods are time consuming and often result in nonreproducible CNT configuration and placement. While thermal CVD approaches can potentially lead to wafer-scale production of AFM tips, the number, orientation, and length of CNTs are difficult to control. At the end of the fabrication, these processes usually require a one-at-a-time manipulation approach to remove extra CNTs and/or to shorten the remaining CNTs for SPM applications.

The key process for CVD-grown CNT probe fabrication is catalyst patterning, which determines the position, number, and diameter of the probe. Electrophoretically deposited or spin-coated colloidal catalyst particles on Si pyramid tips cannot provide reliable control of the position and number of catalyst particles. MWNT probes on tipless cantilevers have been fabricated based on conventional Si fabrication process in which the catalyst pattering was proceeded by typical e-beam lithography and lift-off of spin-coated poly(methyl methacrylate) (PMMA) layer, and plasma-enhanced chemical vapor deposition (PECVD) was used for CNT growth [22.35, 36]. The fabrication method described in [22.35] allows CNT tips to be grown directly on silicon cantilevers at the wafer scale. CNT tip locations and diameters are defined by e-beam lithography. CNT length and orientation are controlled by the growth conditions of the PECVD method. Therefore, there is no need to shorten the CNT after the growth. In PECVD, an electric field is present in the plasma discharge to direct the nanotubes to grow parallel to the electric field. A tilted probe is desirable as it compensates for the operating tilt angle of the AFM cantilever so that the probe itself is close to vertical for stable imaging.

A spin-coated PMMA layer cannot be uniformly conformal on the relatively small piece of tipless cantilevers or on the Si pyramid tip. For e-beam lithography-based processes, the patterned catalyst dots either have to be formed before the fabrication of the cantilevers (although then a protection layer is needed) [22.35] or lithography steps have to be applied twice to remove extra catalyst on commercial tipless-cantilever chips [22.36]. Therefore, the electron-beam-induced deposition (EBID) technique has been developed to make catalyst patterns for CNT probe fabrication [22.37, 38]. EBID is a simple and fast technique to make patterns and deposit materials simultaneously without using any e-beam resist. Its resist-free nature makes EBID a good choice for the fabrication of patterns on the edge of the substrate. A schematic diagram of probe fabrication based on EBID patterning and PECVD is shown in Fig. 22.12. No special carbonaceous precursor molecules were introduced, as the residual carbon-containing molecules naturally present in the SEM chamber were sufficient for EBID processing to form amorphous carbon dots on the cantilever surface. A single carbon dot with a diameter of ≈ 400 nm was deposited near the front-end edge of the cantilever by EBID. The carbon dot serves as a convenient etch mask for chemical etching of the catalyst film. The removal of the carbon dot mask after catalyst patterning was performed with oxygen reactive-ion etch, which exposed the catalyst island. The cantilever

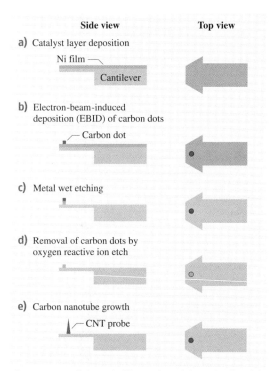

Fig. 22.12a–e Schematic illustration of the resist-free fabrication technique for a single CNT AFM tip

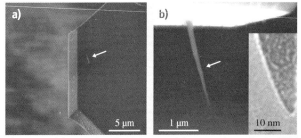

Fig. 22.13 (a) Top view SEM image of the very sharp single CNT probe. (b) Side view SEM image of the CNT probe (after [22.37]). The *arrow* indicates a very sharp, single CNT tip grown on the cantilever

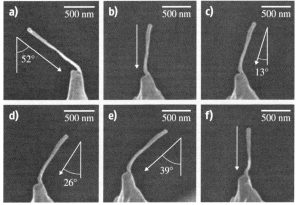

Fig. 22.14a–f SEM image of (a) metal-coated nanotube aligned at 52° with respect to the axis of the pyramidal tip. (b–f) SEM images of the same tip after being exposed to the ion beam incident along the direction of the *arrow* drawn in each image (after [22.39])

with the catalyst island was then transferred to the DC-PECVD system for subsequent growth of the CNT. Figure 22.13 shows SEM images of a CNT probe grown on a tipless cantilever.

22.1.8 Bent Carbon Nanotube Tips

The orientation of CNT tips can be manipulated by FIB treatments, utilizing the interaction between the ion beam and the CNT tip [22.22, 39]. Figure 22.11 shows a schematic of the process of aligning and bending the CNT by using FIB. The aligning and bending phenomena were observed in both as-grown CNT and metal-coated CNT tips. The aligning process is faster with larger values of beam current and acceleration voltage. Under the same voltage, a greater current or longer process time is needed for straightening compared with bending. By using this process, CNT tips can be aligned in any specified direction with precision of less than 1°. Precise control over the orientation of a metal-coated nanotube using a FIB is shown in Fig. 22.14. Figure 22.15 illustrates bending of the end of a CNT tip, which is expected to have potential applications for sidewall measurements in AFM imaging.

CNT tip bending can also be accomplished by changing the direction of the applied bias electric field during DC plasma-enhanced CVD growth [22.40, 41]. As depicted in Fig. 22.16, the nanotube tip can be bent either slightly, by $\approx 45°$ or by $\approx 90°$ using various electric-field angles during the growth process.

22.1.9 Low-Stiffness Cantilevers with Carbon Nanotube Tips

Direct growth of a CNT probe on a low-stiffness cantilever by PECVD is desirable for AFM imaging on soft or fragile materials. As introduced in Sect. 22.1.7, by combining an electron-beam lithography approach for catalyst patterning with PECVD for CNT growth, the location, length, and diameter of CNTs can be well

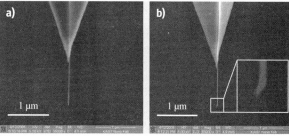

Fig. 22.15a,b Bending the end of the CNT with focused ion beam. (a) CNT as attached to a Si probe. (b) CNT end slightly bent after the FIB process toward the source (after [22.22])

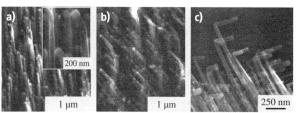

Fig. 22.16a–c Carbon nanotube bending using tilted bias electric field during plasma enhanced CVD growth. The nanotube tip can be bent (a) either slightly, (b) by $\approx 45°$, or (c) by $\approx 90°$ using various electric field angles during the growth process (after [22.40, 41])

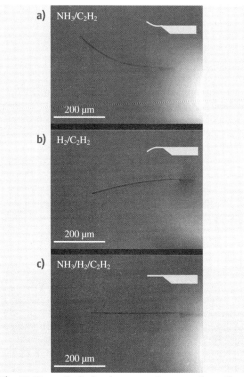

Fig. 22.17a–c Optical microscope images of cantilevers bending after plasma treatments with C_2H_2 gas and (a) NH_3 gas ($R = 1$). (b) H_2 gas ($R = 0$). (c) Mixed NH_3/H_2 gas ($R = 0.5$) (after [22.42])

controlled. The plasma-induced stresses and damages introduced during PECVD growth of nanotubes, however, result in severely bent cantilevers when a thin, low-stiffness cantilever is utilized as the substrate. If the bend is sufficiently large, the AFM laser spot focused at their end will be deflected off of the position-sensitive detector, rendering the cantilevers unusable for AFM measurements.

An in situ process to control the deflection of cantilever beams during CNT growth has been demonstrated by introducing hydrogen gas into the (acetylene + ammonia) feed gas and adjusting the ammonia-to-hydrogen flow ratio [22.42]. The total flow rate of NH_3 and H_2 was kept constant during growth, while the gas mix ratio (R), defined as $NH_3/(NH_3 + H_2)$, was varied in the range $0 \leq R \leq 1$. Figure 22.17 shows comparative, cross-sectional cantilever images for three different CNT growth conditions using different feed gas compositions. A large upward or downward bending of the cantilever is observed for $R = 1$ and $R = 0$, respectively. By employing a particular gas ratio of $R = 0.5$, a nearly flat cantilever beam can be obtained after PECVD growth of a CNT probe.

22.1.10 Conductive Probe Tips

Conductive AFM probes are useful for the study of electrical or ionic properties of nanostructures, especially for the investigation of biological nanofeatures such as ion channels and receptors, the key regulators of cellular homeostasis and sustenance. Disturbed ion-channel behavior in cell membranes such as in the transport of Ca^{2+}, K^+, Na^+ or Cl^- ions leads to a variety of channelopathies such as Alzheimer's, Parkinson's, cystic fibrosis, cardiac arrhythmias, and other systemic diseases. Real-time structure–activity relation of these channels and their (patho)physiological controls can be studied using conductive AFM. An integrated conductive AFM will allow simultaneous acquisition of structure and activity data and to correlate three-dimensional (3-D) nanostructure of individual ion channels and real-time transport of ions [22.43–45]. Either an intrinsically conductive and stable probe such as a carbon nanotube tip or a metal-coated silicon nitride tip can be utilized. The conductive AFM tip serves as one of the electrodes, measuring the ionic currents between the tip and a reference electrode, as illustrated in Fig. 22.18.

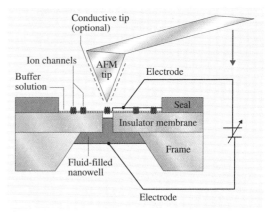

Fig. 22.18 Schematic illustration of the use of conductive AFM probe tip for ion channel conductivity study

22.2 Scanning Tunneling Microscopy

Scanning tunneling microscopy (STM) was the original scanning probe microscopy and generally produces the highest-resolution images, routinely achieving atomic resolution on flat, conductive surfaces. In STM, the probe tip consists of a sharpened metal wire that is held 0.3–1 nm from the sample. A potential difference of 0.1–1 V between the tip and sample leads to tunneling currents on the order of 0.1–1 nA. As in AFM, a piezo-scanner rasters the sample under the tip, and the z-position is adjusted to hold the tunneling current constant. The z-position data represents the *topography*, or in this case the surface of constant electron density. As with other SPMs, the tip properties and performance greatly depend on the experiment being carried out. Although it is nearly impossible to prepare a tip with known atomic structure, a number of factors are known to affect tip performance, and several preparation methods that produce good tips have been developed.

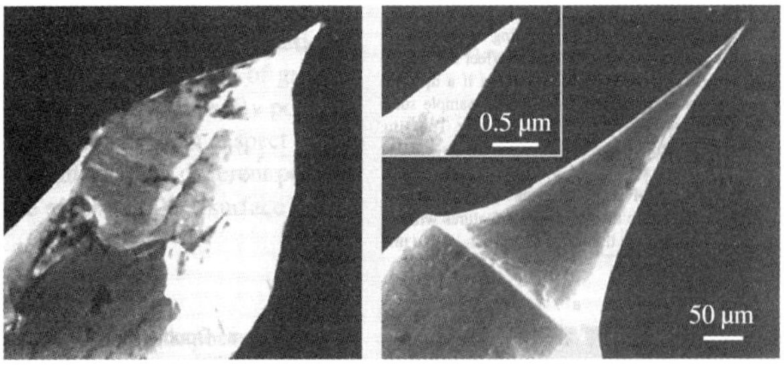

The nature of the sample being investigated and the scanning environment will affect the choice of the tip material and how the tip is fabricated. Factors to consider are mechanical properties – a hard material that will resist damage during tip–sample contact is desired. Chemical properties should also be considered – formation of oxides or other insulating contaminants will affect tip performance. Tungsten is a common tip material because it is very hard and will resist damage, but its use is limited to ultrahigh-vacuum (UHV) conditions, since it readily oxidizes. For imaging under ambient conditions an inert tip material such as platinum or gold is preferred. Platinum is typically alloyed with iridium to increase its stiffness.

22.2.1 Mechanically Cut STM Tips

STM tips can be fabricated by simple mechanical procedures such as grinding or cutting metal wires. Such tips are not formed with highly reproducible shapes and have a large opening angle and a large radius of cur-

Fig. 22.19 A mechanically cut STM tip (*left*) and an electrochemically etched STMtip (*right*) (after [22.46])

vature in the range of 0.1–1 μm (Fig. 22.19a). They are not useful for imaging samples with surface roughness above a few nanometers. However, on atomically flat samples, mechanically cut tips can achieve atomic resolution due to the nature of the tunneling signal, which drops exponentially with tip–sample separation. Since mechanically cut tips contain many small asperities on the larger tip structure, atomic resolution is easily achieved as long as one atom of the tip is just a few angstroms lower than all of the others.

22.2.2 Electrochemically Etched STM Tips

For samples with more than a few nanometers of surface roughness, the tip structure in the nanometer size range becomes an issue. Electrochemical etching can provide tips with reproducible and desirable shapes and sizes (Fig. 22.19), although the exact atomic structure of the tip apex is still not well controlled. The parameters of electrochemical etching depend greatly on the tip material and the desired tip shape. The following is an entirely general description. A fine metal wire (0.1–1 mm diameter) of the tip material is immersed in an appropriate electrochemical etchant solution. A bias voltage of 1–10 V is applied between the tip and a counterelectrode such that the tip is etched. Due to the enhanced etch rate at the electrolyte–air interface, a neck is formed in the wire. This neck is eventually etched thin enough that it cannot support the weight of the part of the wire suspended in the solution, and it breaks to form a sharp tip. The widely varying parameters and methods will be not be covered in detail here, but many recipes can be found in the literature for common tip materials [22.47–51].

References

22.1   R. Linnemann, T. Gotszalk, I.W. Rangelow, P. Dumania, E. Oesterschulze: Atomic force microscopy and lateral force microscopy using piezoresistive cantilevers, J. Vac. Sci. Technol. B **14**(2), 856–860 (1996)

22.2   T.R. Albrecht, S. Akamine, T.E. Carver, C.F. Quate: Microfabrication of cantilever styli for the atomic force microscope, J. Vac. Sci. Technol. A **8**(4), 3386–3396 (1990)

22.3   O. Wolter, T. Bayer, J. Greschner: Micromachined silicon sensors for scanning force microscopy, J. Vac. Sci. Technol. B **9**(2), 1353–1357 (1991)

22.4   C. Bustamante, D. Keller: Scanning force microscopy in biology, Phys. Today **48**(12), 32–38 (1995)

22.5   J. Vesenka, S. Manne, R. Giberson, T. Marsh, E. Henderson: Colloidal gold particles as an incompressible atomic force microscope imaging standard for assessing the compressibility of biomolecules, Biophys. J. **65**, 992–997 (1993)

22.6   D.J. Müller, D. Fotiadis, S. Scheuring, S.A. Müller, A. Engel: Electrostatically balanced subnanometer imaging of biological specimens by atomic force microscope, Biophys. J. **76**(2), 1101–1111 (1999)

22.7   R.B. Marcus, T.S. Ravi, T. Gmitter, K. Chin, D.J. Liu, W. Orvis, D.R. Ciarlo, C.E. Hunt, J. Trujillo: Formation of silicon tips with < 1 nm radius, Appl. Phys. Lett. **56**(3), 236–238 (1990)

22.8   J.H. Hafner, C.L. Cheung, C.M. Lieber: unpublished results (2001)

22.9   J.H. Hafner, C.L. Cheung, T.H. Oosterkamp, C.M. Lieber: High-yield assembly of individual single-walled carbon nanotube tips for scanning probe microscopies, J. Phys. Chem. B **105**(4), 743–746 (2001)

22.10   F. Ohnesorge, G. Binnig: True atomic resolution by atomic force microscopy through repulsive and attractive forces, Science **260**, 1451–1456 (1993)

22.11   D.J. Müller, D. Fotiadis, A. Engel: Mapping flexible protein domains at subnanometer resolution with the atomic force microscope, FEBS Letters **430**(1/2), 105–111 (1998), Special Issue SI

22.12   S. Akamine, R.C. Barrett, C.F. Quate: Improved atomic force microscope images using microcantilevers with sharp tips, Appl. Phys. Lett. **57**(3), 316–318 (1990)

22.13   D.J. Keller, C. Chih-Chung: Imaging steep, high structures by scanning force microscopy with electron beam deposited tips, Surf. Sci. **268**, 333–339 (1992)

22.14   T. Ichihashi, S. Matsui: In situ observation on electron beam induced chemical vapor deposition by transmission electron microscopy, J. Vac. Sci. Technol. B **6**(6), 1869–1872 (1988)

22.15   K.I. Schiffmann: Investigation of fabrication parameters for the electron-beam-induced deposition of contamination tips used in atomic force microscopy, Nanotechnology **4**, 163–169 (1993)

22.16   D.S. Bethune, C.H. Kiang, M.S. de Vries, G. Gorman, R. Savoy, J. Vazquez, R. Beyers: Cobalt-catalysed growth of carbon nanotubes with single-atomic-layer walls, Nature **363**(6430), 605–607 (1993)

22.17   E.T. Thostenson, Z. Ren, T.W. Chou: Advances in the science and technology of carbon nanotubes and their composites: A review, Compos. Sci. Technol. **61**(13), 1899–1912 (2001)

22.18   H.J. Dai, J.H. Hafner, A.G. Rinzler, D.T. Colbert, R.E. Smalley: Nanotubes as nanoprobes in scanning probe microscopy, Nature **384**(6605), 147–150 (1996)

22.19 A.G. Rinzler, Y.H. Hafner, P. Nikolaev, L. Lou, S.G. Kim, D. Tomanek, D.T. Colbert, R.E. Smalley: Unraveling nanotubes: Field emission from atomic wire, Science **269**, 1550 (1995)

22.20 R. Stevens, C. Nguyen, A. Cassell, L. Delzeit, M. Meyyappan, J. Han: Improved fabrication approach for carbon nanotube probe devices, Appl. Phys. Lett. **77**, 3453–3455 (2000)

22.21 H. Nishijima, S. Kamo, S. Akita, Y. Nakayama, K.I. Hohmura, S.H. Yoshimura, K. Takeyasu: Carbon-nanotube tips for scanning probe microscopy: Preparation by a controlled process and observation of deoxyribonucleic acid, Appl. Phys. Lett. **74**, 4061–4063 (1999)

22.22 B.C. Park, K.Y. Jung, W.Y. Song, O. Beom-Hoan, S.J. Ahn: Bending of a carbon nanotube in vacuum using a focused ion beam, Adv. Mater. **18**, 95–98 (2006)

22.23 A. Hall, W.G. Matthews, R. Superfine, M.R. Falvo, S. Washburna: Simple and efficient method for carbon nanotube attachment to scanning probes and other substrates, Appl. Phys. Lett. **82**, 2506–2508 (2003)

22.24 J. Tang, G. Yang, Q. Zhang, A. Parhat, B. Maynor, J. Liu, L.C. Qin, O. Zhou: Rapid and reproducible fabrication of carbon nanotube AFM probes by dielectrophoresis, Nano Lett. **5**, 11–14 (2005)

22.25 J.-E. Kim, J.-K. Park, C.-S. Han: Use of dielectrophoresis in the fabrication of an atomic force microscope tip with a carbon nanotube: Experimental investigation, Nanotechnology **17**, 2937–2941 (2006)

22.26 J.H. Hafner, M.J. Bronikowski, B.R. Azamian, P. Nikolaev, A.G. Rinzler, D.T. Colbert, K.A. Smith, R.E. Smalley: Catalytic growth of single-wall carbon nanotubes from metal particles, Chem. Phys. Lett. **296**(1/2), 195–202 (1998)

22.27 P. Nikolaev, M.J. Bronikowski, R.K. Bradley, F. Rohmund, D.T. Colbert, K.A. Smith, R.E. Smalley: Gas-phase catalytic growth of single-walled carbon nanotubes from carbon monoxide, Chem. Phys. Lett. **313**(1/2), 91–97 (1999)

22.28 W.Z. Li, S.S. Xie, L.X. Qian, B.H. Chang, B.S. Zou, W.Y. Zhou, R.A. Zhao, G. Wang: Large-scale synthesis of aligned carbon nanotubes, Science **274**(5293), 1701–1703 (1996)

22.29 J.H. Hafner, C.L. Cheung, C.M. Lieber: Growth of nanotubes for probe microscopy tips, Nature **398**(6730), 761–762 (1999)

22.30 V. Lehmann: The physics of macroporous silicon formation, Thin Solid Films **255**, 1–4 (1995)

22.31 F. Ronkel, J.W. Schultze, R. Arensfischer: Electrical contact to porous silicon by electrodeposition of iron, Thin Solid Films **276**(1–2), 40–43 (1996)

22.32 J.H. Hafner, C.L. Cheung, C.M. Lieber: Direct growth of single-walled carbon nanotube scanning probe microscopy tips, J. Am. Chem. Soc. **121**(41), 9750–9751 (1999)

22.33 E.B. Cooper, S.R. Manalis, H. Fang, H. Dai, K. Matsumoto, S.C. Minne, T. Hunt, C.F. Quate: Terabit-per-square-inch data storage with the atomic force microscope, Appl. Phys. Lett. **75**(22), 3566–3568 (1999)

22.34 E. Yenilmez, Q. Wang, R.J. Chen, D. Wang, H. Dai: Wafer scale production of carbon nanotube scanning probe tips for atomic force microscopy, Appl. Phys. Lett. **80**(12), 2225–2227 (2002)

22.35 Q. Ye, A.M. Cassell, H.B. Liu, K.J. Chao, J. Han, M. Meyyappan: Large-scale fabrication of carbon nanotube probe tips for atomic force microscopy critical dimension imaging applications, Nano Lett. **4**, 1301–1308 (2004)

22.36 H. Cui, S.V. Kalinin, X. Yang, D.H. Lowndes: Growth of carbon nanofibers on tipless cantilevers for high resolution topography and magnetic force imaging, Nano Lett. **4**, 2157–2161 (2004)

22.37 I.-C. Chen, L.-H. Chen, X.-R. Ye, C. Daraio, S. Jin, C.A. Orme, A. Quist, R. Lal: Extremely sharp carbon nanocone probes for atomic force microscopy imaging, Appl. Phys. Lett. **88**, 153102 (2006)

22.38 I.-C. Chen, L.-H. Chen, C.A. Orme, A. Quist, R. Lal, S. Jin: Fabrication of high-aspect-ratio carbon nanocone probes by electron beam induced deposition patterning, Nanotechnology **17**, 4322 (2006)

22.39 Z.F. Deng, E. Yenilmez, A. Reilein, J. Leu, H. Dai, K.A. Moler: Nanotube manipulation with focused ion beam, Appl. Phys. Lett. **88**, 023119 (2006)

22.40 J.F. AuBuchon, L.-H. Chen, S. Jin: Control of carbon capping for regrowth of aligned carbon nanotubes, J. Phys. Chem. B **109**, 6044–6048 (2005)

22.41 J.F. AuBuchon, L.-H. Chen, A.I. Gapin, S. Jin: electric-field-guided growth of carbon nanotubes during DC plasma-enhanced CVD, Chem. Vap. Depos. **12**(6), 370–374 (2006)

22.42 I.-C. Chen, L.-H. Chen, C.A. Orme, S. Jin: Control of curvature in highly compliant probe cantilevers during carbon nanotube growth, Nano Lett. **7**(10), 3035–3040 (2007)

22.43 A. Quist, I. Doudevski, H. Lin, R. Azimova, D. Ng, B. Frangione, B. Kagan, J. Ghiso, R. Lal: Amyloid ion channels: A common structural link for protein-misfolding disease, Proc. Natl. Acad. Sci. USA **102**, 10427 (2005)

22.44 A.P. Quist, A. Chand, S. Ramachandran, C. Daraio, S. Jin, R. Lal: AFM imaging and electrical recording of lipid bilayers supported over microfabricated silicon chip nanopores: A lab on-chip system for lipid membrane and ion channels, Langmuir **23**(3), 1375 (2007)

22.45 J. Thimm, A. Mechler, H. Lin, S.K. Rhee, R. Lal: Calcium dependent open-closed conformations and interfacial energy maps of reconstituted individual hemichannels, J. Biol. Chem. **280**, 10646 (2005)

22.46 A. Stemmer, A. Hefti, U. Aebi, A. Engel: Scanning tunneling and transmission electron microscopy on identical areas of biological specimens, Ultramicroscopy **30**(3), 263 (1989)

22.47 J.H. Hafner, C.L. Cheung, A.T. Woolley, C.M. Lieber: Structural and functional imaging with carbon nanotube AFM probes, Prog. Biophys. Mol. Biol. **77**(1), 73–110 (2001)

22.48 R. Nicolaides, L. Yong, W.E. Packard, W.F. Zhou, H.A. Blackstead, K.K. Chin, J.D. Dow, J.K. Furdyna, M.H. Wei, R.C.J. Jaklevic, W. Kaiser, A.R. Pelton, M.V. Zeller, J. Bellina Jr.: Scanning tunneling microscope tip structures, J. Vac. Sci. Technol. A **6**(2), 445–447 (1988)

22.49 J.P. Ibe, P.P. Bey, S.L. Brandow, R.A. Brizzolara, N.A. Burnham, D.P. DiLella, K.P. Lee, C.R.K. Marrian, R.J. Colton: On the electrochemical etching of tips for scanning tunneling microscopy, J. Vac. Sci. Technol. A **8**, 3570–3575 (1990)

22.50 L. Libioulle, Y. Houbion, J.-M. Gilles: Very sharp platinum tips for scanning tunneling microscopy, Rev. Sci. Instrum. **66**(1), 97–100 (1995)

22.51 A.J. Nam, A. Teren, T.A. Lusby, A.J. Melmed: Benign making of sharp tips for STM and FIM: Pt, Ir, Au, Pd, and Rh, J. Vac. Sci. Technol. B **13**(4), 1556–1559 (1995)

23. Noncontact Atomic Force Microscopy and Related Topics

Franz J. Giessibl, Yasuhiro Sugawara, Seizo Morita, Hirotaka Hosoi, Kazuhisa Sueoka, Koichi Mukasa, Akira Sasahara, Hiroshi Onishi

Scanning probe microscopy (SPM) methods such as scanning tunneling microscopy (STM) and noncontact atomic force microscopy (NC-AFM) are the basic technologies for nanotechnology and also for future bottom-up processes. In Sect. 23.1, the principles of AFM such as its operating modes and the NC-AFM frequency-modulation method are fully explained. Then, in Sect. 23.2, applications of NC-AFM to semiconductors, which make clear its potential in terms of spatial resolution and function, are introduced. Next, in Sect. 23.3, applications of NC-AFM to insulators such as alkali halides, fluorides and transition-metal oxides are introduced. Lastly, in Sect. 23.4, applications of NC-AFM to molecules such as carboxylate ($RCOO^-$) with R = H, CH_3, $C(CH_3)_3$ and CF_3 are introduced. Thus, NC-AFM can observe atoms and molecules on various kinds of surfaces such as semiconductors, insulators and metal oxides with atomic or molecular resolution. These sections are essential to understand the state of the art and future possibilities for NC-AFM, which is the second generation of atom/molecule technology.

23.1 **Atomic Force Microscopy (AFM)** 636
    23.1.1 Imaging Signal in AFM 636
    23.1.2 Experimental Measurement and Noise 637
    23.1.3 Static AFM Operating Mode 637
    23.1.4 Dynamic AFM Operating Mode 638
    23.1.5 The Four Additional Challenges Faced by AFM 638
    23.1.6 Frequency-Modulation AFM (FM-AFM) 639
    23.1.7 Relation Between Frequency Shift and Forces 640
    23.1.8 Noise in Frequency Modulation AFM: Generic Calculation 641
    23.1.9 Conclusion 641

23.2 **Applications to Semiconductors** 641
    23.2.1 Si(111)-(7×7) Surface 642
    23.2.2 Si(100)-(2×1) and Si(100)-(2×1):H Monohydride Surfaces 643
    23.2.3 Metal Deposited Si Surface 645

23.3 **Applications to Insulators** 647
    23.3.1 Alkali Halides, Fluorides and Metal Oxides 647
    23.3.2 Atomically Resolved Imaging of a NiO(001) Surface 652

23.4 **Applications to Molecules** 654
    23.4.1 Why Molecules and Which Molecules? 654
    23.4.2 Mechanism of Molecular Imaging ... 654
    23.4.3 Perspectives 657

References 658

The scanning tunneling microscope (STM) is an atomic tool based on an electric method that measures the tunneling current between a conductive tip and a conductive surface. It can electrically observe individual atoms/molecules. It can characterize or analyze the electronic nature around surface atoms/molecules. In addition, it can manipulate individual atoms/molecules. Hence, the STM is the first generation of atom/molecule technology. On the other hand, the atomic force microscopy (AFM) is a unique atomic tool based on a mechanical method that can even deal with insulator surfaces. Since the invention of noncontact AFM (NC-AFM) in 1995, the NC-AFM and NC-AFM-based methods have rapidly developed into powerful surface tools on the atomic/molecular scales, because NC-AFM has the following characteristics: (1) it has true atomic resolution, (2) it can measure atomic force (so-called atomic force spectroscopy), (3) it can observe even insulators, and (4) it can measure mechanical responses such as elastic deformation. Thus, NC-AFM is the second generation of atom/molecule technology. Scanning probe microscopy (SPM) such as STM and NC-AFM is the basic technology for nanotechnology and also for future bottom-up processes.

In Sect. 23.1, the principles of NC-AFM will be fully introduced. Then, in Sect. 23.2, applications to semiconductors will be presented. Next, in Sect. 23.3, applications to insulators will be described. And, in Sect. 23.4, applications to molecules will be introduced. These sections are essential to understanding the state of the art and future possibilities for NC-AFM.

23.1 Atomic Force Microscopy (AFM)

The atomic force microscope (AFM), invented by *Binnig* [23.1] and introduced in 1986 by *Binnig* et al. [23.2] is an offspring of the scanning tunneling microscope (STM) [23.3]. The STM is covered in several books and review articles, e.g. [23.4–9]. Early in the development of STM it became evident that relatively strong forces act between a tip in close proximity to a sample. It was found that these forces could be put to good use in the atomic force microscope (AFM). Detailed information about the noncontact AFM can be found in [23.10–12].

23.1.1 Imaging Signal in AFM

Figure 23.1 shows a sharp tip close to a sample. The potential energy between the tip and the sample V_{ts} causes a z-component of the tip–sample force $F_{ts} = -\partial V_{ts}/\partial z$. Depending on the mode of operation, the AFM uses F_{ts}, or some entity derived from F_{ts}, as the imaging signal.

Unlike the tunneling current, which has a very strong distance dependence, F_{ts} has long- and short-range contributions. We can classify the contributions by their range and strength. In vacuum, there are van-der-Waals, electrostatic and magnetic forces with a long range (up to 100 nm) and short-range chemical forces (fractions of nm).

The van-der-Waals interaction is caused by fluctuations in the electric dipole moment of atoms and their mutual polarization. For a spherical tip with radius R next to a flat surface (z is the distance between the plane connecting the centers of the surface atoms and the center of the closest tip atom) the van-der-Waals potential is given by [23.13]

$$V_{vdW} = -\frac{A_H}{6z} \,. \tag{23.1}$$

The Hamaker constant A_H depends on the type of materials (atomic polarizability and density) of the tip and sample and is of the order of 1 eV for most solids [23.13].

When the tip and sample are both conductive and have an electrostatic potential difference $U \neq 0$, electrostatic forces are important. For a spherical tip with radius R, the force is given by [23.14]

$$F_{electrostatic} = -\frac{\pi \varepsilon_0 R U^2}{z} \,. \tag{23.2}$$

Chemical forces are more complicated. Empirical model potentials for chemical bonds are the Morse potential (see e.g. [23.13])

$$V_{Morse} = -E_{bond}\left(2\mathrm{e}^{-\kappa(z-\sigma)} - \mathrm{e}^{-2\kappa(z-\sigma)}\right) \tag{23.3}$$

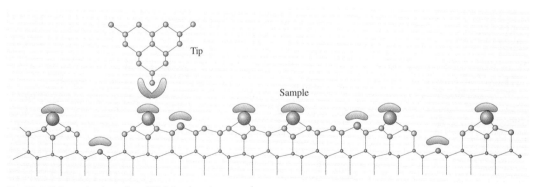

Fig. 23.1 Schematic view of an AFM tip close to a sample

and the Lennard-Jones potential [23.13]

$$V_{\text{Lennard-Jones}} = -E_{\text{bond}}\left(2\frac{\sigma^6}{z^6} - \frac{\sigma^{12}}{z^{12}}\right). \quad (23.4)$$

These potentials describe a chemical bond with bonding energy E_{bond} and equilibrium distance σ. The Morse potential has an additional parameter: a decay length κ.

23.1.2 Experimental Measurement and Noise

Forces between the tip and sample are typically measured by recording the deflection of a cantilever beam that has a tip mounted on its end (Fig. 23.2). Today's microfabricated silicon cantilevers were first created in the group of *Quate* [23.15–17] and at IBM [23.18].

The cantilever is characterized by its spring constant k, eigenfrequency f_0 and quality factor Q.

For a rectangular cantilever with dimensions w, t and L (Fig. 23.2), the spring constant k is given by [23.6]

$$k = \frac{E_Y w t^3}{4L^3}, \quad (23.5)$$

where E_Y is the Young's modulus. The eigenfrequency f_0 is given by [23.6]

$$f_0 = 0.162 \frac{t}{L^2}\sqrt{\frac{E}{\rho}}, \quad (23.6)$$

where ρ is the mass density of the cantilever material. The Q-factor depends on the damping mechanisms present in the cantilever. For micromachined cantilevers operated in air, Q is typically a few hundred, while Q can reach hundreds of thousands in vacuum.

In the first AFM, the deflection of the cantilever was measured with an STM; the back side of the cantilever

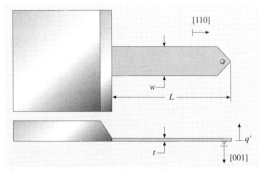

Fig. 23.2 Top view and side view of a microfabricated silicon cantilever (schematic)

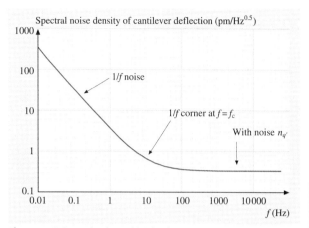

Fig. 23.3 Schematic view of $1/f$ noise apparent in force detectors. Static AFMs operate in a frequency range from 0.01 Hz to a few hundred Hz, while dynamic AFMs operate at frequencies around 10 kHz to a few hundred kHz. The noise of the cantilever deflection sensor is characterized by the $1/f$ corner frequency f_c and the constant deflection noise density $n_{q'}$ for the frequency range where white noise dominates

was metalized, and a tunneling tip was brought close to it to measure the deflection [23.2]. Today's designs use optical (interferometer, beam-bounce) or electrical methods (piezoresistive, piezoelectric) to measure the cantilever deflection. A discussion of the various techniques can be found in [23.19], descriptions of piezoresistive detection schemes are found in [23.17, 20] and piezoelectric methods are explained in [23.21–24].

The quality of the cantilever deflection measurement can be expressed in a schematic plot of the deflection noise density versus frequency as in Fig. 23.3.

The noise density has a $1/f$ dependence for low frequency and merges into a constant noise density (white noise) above the $1/f$ corner frequency.

23.1.3 Static AFM Operating Mode

In the static mode of operation, the force translates into a deflection $q' = F_{\text{ts}}/k$ of the cantilever, yielding images as maps of $z(x, y, F_{\text{ts}} = \text{const.})$. The noise level of the force measurement is then given by the cantilever's spring constant k times the noise level of the deflection measurement. In this respect, a small value for k increases force sensitivity. On the other hand, instabilities are more likely to occur with soft cantilevers (Sect. 23.1.1). Because the deflection of the cantilever should be significantly larger than the deformation of the tip and sample, the cantilever should be much softer than the bonds between the bulk atoms in the tip and sample. Interatomic force constants in solids are in the range 10–100 N/m; in biological samples, they can be as small as 0.1 N/m. Thus, typical values for k in the static mode are 0.01–5 N/m.

Even though it has been demonstrated that atomic resolution is possible with static AFM, the method can only be applied in certain cases. The detrimental effects of $1/f$-noise can be limited by working at low temperatures [23.25], where the coefficients of thermal expansion are very small or by building the AFM using a material with a low thermal-expansion coefficient [23.26]. The long-range attractive forces have to be canceled by immersing the tip and sample in a liquid [23.26] or by partly compensating the attractive force by pulling at the cantilever after jump-to-contact has occurred [23.27]. *Jarvis* et al. have canceled the long-range attractive force with an electromagnetic force applied to the cantilever [23.28]. Even with these restrictions, static AFM does not produce atomic resolution on reactive surfaces like silicon, as the chemical bonding of the AFM tip and sample poses an unsurmountable problem [23.29, 30].

23.1.4 Dynamic AFM Operating Mode

In the dynamic operation modes, the cantilever is deliberately vibrated. There are two basic methods of dynamic operation: amplitude-modulation (AM) and frequency-modulation (FM) operation. In AM-

AFM [23.31], the actuator is driven by a fixed amplitude A_{drive} at a fixed frequency f_{drive} where f_{drive} is close to f_0. When the tip approaches the sample, elastic and inelastic interactions cause a change in both the amplitude and the phase (relative to the driving signal) of the cantilever. These changes are used as the feedback signal. While the AM mode was initially used in a noncontant mode, it was later implemented very successfully at a closer distance range in ambient conditions involving repulsive tip–sample interactions.

The change in amplitude in AM mode does not occur instantaneously with a change in the tip–sample interaction, but on a timescale of $\tau_{\text{AM}} \approx 2Q/f_0$ and the AM mode is slow with high-Q cantilevers. However, the use of high Q-factors reduces noise. *Albrecht* et al. found a way to combine the benefits of high Q and high speed by introducing the frequency-modulation (FM) mode [23.32], where the change in the eigenfrequency settles on a timescale of $\tau_{\text{FM}} \approx 1/f_0$.

Using the FM mode, the resolution was improved dramatically and finally atomic resolution [23.33, 34] was obtained by reducing the tip–sample distance and working in vacuum. For atomic studies in vacuum, the FM mode (Sect. 23.1.6) is now the preferred AFM technique. However, atomic resolution in vacuum can also be obtained with the AM mode, as demonstrated by *Erlandsson* et al. [23.35].

23.1.5 The Four Additional Challenges Faced by AFM

Some of the inherent AFM challenges are apparent by comparing the tunneling current and tip–sample force as a function of distance (Fig. 23.4).

The tunneling current is a monotonic function of the tip–sample distance and has a very sharp distance dependence. In contrast, the tip–sample force has long- and short-range components and is not monotonic.

Jump-to-Contact and Other Instabilities

If the tip is mounted on a soft cantilever, the initially attractive tip–sample forces can cause a sudden jump-to-contact when approaching the tip to the sample. This instability occurs in the quasistatic mode if [23.36, 37]

$$k < \max\left(-\frac{\partial^2 V_{\text{ts}}}{\partial z^2}\right) = k_{\text{ts}}^{\max} . \tag{23.7}$$

Jump-to-contact can be avoided even for soft cantilevers by oscillating at a large enough amplitude A [23.38]

$$kA > \max(-F_{\text{ts}}) . \tag{23.8}$$

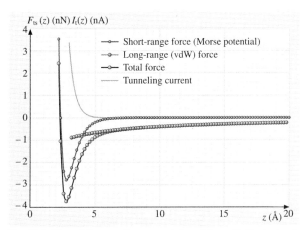

Fig. 23.4 Plot of the tunneling current I_t and force F_{ts} (typical values) as a function of the distance z between the front atom and surface atom layer

If hysteresis occurs in the $F_{ts}(z)$-relation, energy ΔE_{ts} needs to be supplied to the cantilever for each oscillation cycle. If this energy loss is large compared to the intrinsic energy loss of the cantilever, amplitude control can become difficult. An additional approximate criterion for k and A is then

$$\frac{kA^2}{2} \geq \frac{\Delta E_{ts} Q}{2\pi}. \qquad (23.9)$$

Contribution of Long-Range Forces

The force between the tip and sample is composed of many contributions: electrostatic, magnetic, van-der-Waals and chemical forces in vacuum. All of these force types except for the chemical forces have strong long-range components which conceal the atomic force components. For imaging by AFM with atomic resolution, it is desirable to filter out the long-range force contributions and only measure the force components which vary on the atomic scale. While there is no way to discriminate between long- and short-range forces in static AFM, it is possible to enhance the short-range contributions in dynamic AFM by proper choice of the oscillation amplitude A of the cantilever.

Noise in the Imaging Signal

Measuring the cantilever deflection is subject to noise, especially at low frequencies ($1/f$ noise). In static AFM, this noise is particularly problematic because of the approximate $1/f$ dependence. In dynamic AFM, the low-frequency noise is easily discriminated when using a bandpass filter with a center frequency around f_0.

Nonmonotonic Imaging Signal

The tip–sample force is not monotonic. In general, the force is attractive for large distances and, upon decreasing the distance between tip and sample, the force turns repulsive (Fig. 23.4). Stable feedback is only possible on a monotonic subbranch of the force curve.

Frequency-modulation AFM helps to overcome challenges. The nonmonotonic imaging signal in AFM is a remaining complication for FM-AFM.

23.1.6 Frequency-Modulation AFM (FM-AFM)

In FM-AFM, a cantilever with eigenfrequency f_0 and spring constant k is subject to controlled positive feedback such that it oscillates with a constant amplitude A [23.32], as shown in Fig. 23.5.

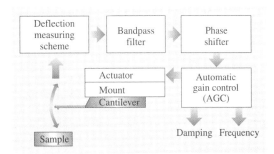

Fig. 23.5 Block diagram of a frequency-modulation force sensor

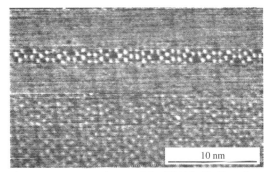

Fig. 23.6 First AFM image of the Si(111)-(7×7) surface. Parameters: $k = 17\,\mathrm{Nm}$, $f_0 = 114\,\mathrm{kHz}$, $Q = 28\,000$, $A = 34\,\mathrm{nm}$, $\Delta f = -70\,\mathrm{Hz}$, $V_t = 0\,\mathrm{V}$

Experimental Set-Up

The deflection signal is phase-shifted, routed through an automatic gain control circuit and fed back to the actuator. The frequency f is a function of f_0, its quality factor Q, and the phase shift ϕ between the mechanical excitation generated at the actuator and the deflection of the cantilever. If $\phi = \pi/2$, the loop oscillates at $f = f_0$. Three physical observables can be recorded: (1) a change in the resonance frequency Δf, (2) the control signal of the automatic gain control unit as a measure of the tip–sample energy dissipation, and (3) an average tunneling current (for conducting cantilevers and tips).

Applications

FM-AFM was introduced by *Albrecht* and coworkers in magnetic force microscopy [23.32]. The noise level and imaging speed was enhanced significantly compared to amplitude-modulation techniques. Achieving atomic resolution on the Si(111)-(7×7) surface has been an important step in the development of the STM [23.39] and, in 1994, this surface was imaged by AFM with true atomic resolution for the first time [23.33] (Fig. 23.6).

The initial parameters which provided true atomic resolution (see caption of Fig. 23.6) were found empirically. Surprisingly, the amplitude necessary to obtain good results was very large compared to atomic dimensions. It turned out later that the amplitudes had to be so large to fulfill the stability criteria listed in Sect. 23.1.5. Cantilevers with $k \approx 2000\,\mathrm{N/m}$ can be operated with amplitudes in the Å-range [23.24].

23.1.7 Relation Between Frequency Shift and Forces

The cantilever (spring constant k, effective mass m^*) is a macroscopic object and its motion can be described by classical mechanics. Figure 23.7 shows the deflection $q'(t)$ of the tip of the cantilever: it oscillates with an amplitude A at a distance $q(t)$ from a sample.

Generic Calculation

The Hamiltonian of the cantilever is

$$H = \frac{p^2}{2m^*} + \frac{kq'^2}{2} + V_{ts}(q) \tag{23.10}$$

where $p = m^*\,dq'/dt$. The unperturbed motion is given by

$$q'(t) = A\cos(2\pi f_0 t) \tag{23.11}$$

and the frequency is

$$f_0 = \frac{1}{2\pi}\sqrt{\frac{k}{m^*}}. \tag{23.12}$$

If the force gradient $k_{ts} = -\partial F_{ts}/\partial z = \partial^2 V_{ts}/\partial z^2$ is constant during the oscillation cycle, the calculation of the frequency shift is trivial

$$\Delta f = \frac{f_0}{2k}k_{ts}. \tag{23.13}$$

However, in classic FM-AFM k_{ts} varies over orders of magnitude during one oscillation cycle and a perturbation approach, as shown below, has to be employed for the calculation of the frequency shift.

Hamilton–Jacobi Method

The first derivation of the frequency shift in FM-AFM was achieved in 1997 [23.38] using canonical perturbation theory [23.40]. The result of this calculation is

$$\Delta f = -\frac{f_0}{kA^2}\langle F_{ts}q'\rangle$$

$$= -\frac{f_0}{kA^2}\int_0^{1/f_0} F_{ts}(d + A + q'(t))q'(t)\,dt. \tag{23.14}$$

The applicability of first-order perturbation theory is justified because, in FM-AFM, E is typically in the range of several keV, while V_{ts} is of the order of a few eV. DÃ¼rig [23.41] found a generalized algorithm that even allows one to reconstruct the tip–sample potential if not only the frequency shift, but the higher harmonics of the cantilever oscillation are known.

A Descriptive Expression for Frequency Shifts as a Function of the Tip–Sample Forces

With integration by parts, the complicated expression (23.14) is transformed into a very simple expression that resembles (23.13) [23.42]

$$\Delta f = \frac{f_0}{2k}\int_{-A}^{A} k_{ts}(z-q')\frac{\sqrt{A^2 - q'^2}}{\frac{\pi}{2}kA^2}\,dq'. \tag{23.15}$$

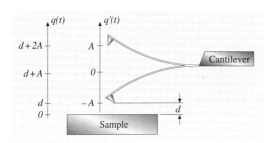

Fig. 23.7 Schematic view of an oscillating cantilever and definition of geometric terms

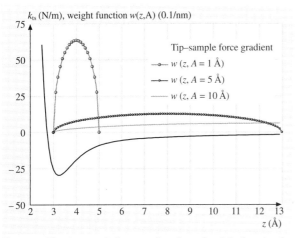

Fig. 23.8 The tip–sample force gradient k_{ts} and weight function for the calculation of the frequency shift

This expression is closely related to (23.13): the constant k_{ts} is replaced by a weighted average, where the weight function $w(q', A)$ is a semicircle with radius A divided by the area of the semicircle $\pi A^2/2$ (Fig. 23.8). For $A \to 0$, $w(q', A)$ is a representation of Dirac's delta function and the trivial zero-amplitude result of (23.13) is immediately recovered. The frequency shift results from a convolution between the tip–sample force gradient and weight function. This convolution can easily be reversed with a linear transformation and the tip–sample force can be recovered from the curve of frequency shift versus distance [23.42].

The dependence of the frequency shift on amplitude confirms an empirical conjecture: small amplitudes increase the sensitivity to short-range forces. Adjusting the amplitude in FM-AFM is comparable to tuning an optical spectrometer to a passing wavelength. When short-range interactions are to be probed, the amplitude should be in the range of the short-range forces. While using amplitudes in the Å-range has been elusive with conventional cantilevers because of the instability problems described in Sect. 23.1.5, cantilevers with a stiffness of the order of 1000 N/m like those introduced in [23.23] are well suited for small-amplitude operation.

23.1.8 Noise in Frequency Modulation AFM: Generic Calculation

The vertical noise in FM-AFM is given by the ratio between the noise in the imaging signal and the slope of the imaging signal with respect to z

$$\delta z = \frac{\delta \Delta f}{\left| \frac{\partial \Delta f}{\partial z} \right|}. \tag{23.16}$$

Figure 23.9 shows a typical curve of frequency shift versus distance. Because the distance between the tip and sample is measured indirectly through the frequency shift, it is clearly evident from Fig. 23.9 that the noise in the frequency measurement $\delta \Delta f$ translates into vertical noise δz and is given by the ratio between $\delta \Delta f$ and the slope of the frequency shift curve $\Delta f(z)$ (23.16). Low vertical noise is obtained for a low-

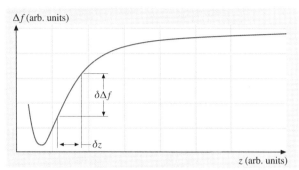

Fig. 23.9 Plot of the frequency shift Δf as a function of the tip–sample distance z. The noise in the tip–sample distance measurement is given by the noise of the frequency measurement $\delta \Delta f$ divided by the slope of the frequency shift curve

noise frequency measurement and a steep slope of the frequency-shift curve.

The frequency noise $\delta \Delta f$ is typically inversely proportional to the cantilever amplitude A [23.32, 43]. The derivative of the frequency shift with distance is constant for $A \ll \lambda$ where λ is the range of the tip–sample interaction and proportional to $A^{-1.5}$ for $A \gg \lambda$ [23.38]. Thus, minimal noise occurs if [23.44]

$$A_{\text{optimal}} \approx \lambda \tag{23.17}$$

for chemical forces, $\lambda \approx 1$ Å. However, for stability reasons, (Sect. 23.1.5) extremely stiff cantilevers are needed for small-amplitude operation. The excellent noise performance of the stiff cantilever and the small-amplitude technique has been verified experimentally [23.24].

23.1.9 Conclusion

Dynamic force microscopy, and in particular frequency-modulation atomic force microscopy has matured into a viable technique that allows true atomic resolution of conducting and insulating surfaces and spectroscopic measurements on individual atoms [23.10, 45]. Even true atomic resolution in lateral force microscopy is now possible [23.46]. Challenges remain in the chemical composition and structural arrangement of the AFM tip.

23.2 Applications to Semiconductors

For the first time, corner holes and adatoms on the Si(111)-(7×7) surface have been observed in very local areas by *Giessible* using pure noncontact AFM in ultra-high vacuum (UHV) [23.33]. This was the breakthrough of true atomic-resolution imaging on a well-defined clean surface using the noncontact AFM. Since then,

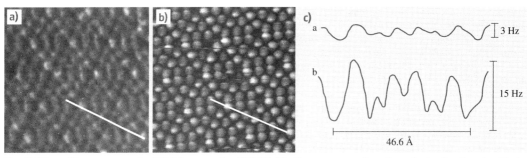

Fig. 23.10a–c Noncontact-mode AFM images of a Si(111)-(7×7) reconstructed surface obtained using the Si tips (**a**) without and (**b**) with a dangling bond. The scan area is 99 Å × 99 Å. (**c**) The cross-sectional profiles along the long diagonal of the 7 × 7 unit cell indicated by the *white lines* in (**a**) and (**b**)

Si(111)-(7×7) [23.34, 35, 45, 47], InP(110) [23.48] and Si(100)-(2×1) [23.34] surfaces have been successively resolved with true atomic resolution. Furthermore, thermally induced motion of atoms or atomic-scale point defects on a InP(110) surface have been observed at room temperature [23.48]. In this section we will describe typical results of atomically resolved noncontact AFM imaging of semiconductor surfaces.

23.2.1 Si(111)-(7×7) Surface

Figure 23.10 shows the atomic-resolution images of the Si(111)-(7×7) surface [23.49]. Here, Fig. 23.10a

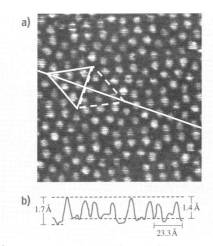

Fig. 23.11 (**a**) Noncontact mode AFM image with contrast of inequivalent adatoms and (**b**) a cross-sectional profile indicated by the *white line*. The halves of the 7 × 7 unit cell surrounded by the *solid line* and *broken line* correspond to the faulted and unfaulted halves, respectively. The scan area is 89 Å × 89 Å

(type I) was obtained using the Si tip without dangling, which is covered with an inert oxide layer. Figure 23.10b (type II) was obtained using the Si tip with a dangling bond, on which the Si atoms were deposited due the mechanical soft contact between the tip and the Si surface. The variable frequency shift mode was used. We can see not only adatoms and corner holes but also missing adatoms described by the dimer–adatom–stacking (DAS) fault model. We can see that the image contrast in Fig. 23.10b is clearly stronger than that in Fig. 23.10a.

Interestingly, by using the Si tip with a dangling bond, we observed contrast between inequivalent halves and between inequivalent adatoms of the 7 × 7 unit cell. Namely, as shown in Fig. 23.11a, the faulted halves (surrounded with a solid line) are brighter than the unfaulted halves (surrounded with a broken line). Here, the positions of the faulted and unfaulted halves were determined from the step direction. From the cross-sectional profile along the long diagonal of the 7 × 7 unit cell in Fig. 23.11b, the heights of the corner adatoms are slightly higher than those of the adjacent center adatoms in the faulted and unfaulted halves of the unit cell. The measured corrugation are in the following decreasing order: Co-F > Ce-F > Co-U > Ce-U, where Co-F and Ce-F indicate the corner and center adatoms in faulted halves, and Co-U and Ce-U indicate the corner and center adatoms in unfaulted halves, respectively. Averaging over several units, the corrugation height differences are estimated to be 0.25 Å, 0.15 Å and 0.05 Å for Co-F, Ce-F and Co-U, respectively, with respect to to Ce-U. This tendency, that the heights of the corner adatoms are higher than those of the center adatoms, is consistent with the experimental results using a silicon tip [23.47], although they could not determine the faulted and unfaulted halves of the unit cell in the measured

AFM images. However, this tendency is completely contrary to the experimental results using a tungsten tip [23.35]. This difference may originate from the difference between the tip materials, which seems to affect the interaction between the tip and the reactive sample surface. Another possibility is that the tip is in contact with the surface during the small fraction of the oscillating cycle in their experiments [23.35].

We consider that the contrast between inequivalent adatoms is not caused by tip artifacts for the following reasons: (1) each adatom, corner hole and defect was clearly observed, (2) the apparent heights of the adatoms are the same whether they are located adjacent to defects or not, and (3) the same contrast in several images for the different tips has been observed.

It should be noted that the corrugation amplitude of adatoms ≈ 1.4 Å in Fig. 23.11b is higher than that of $0.8-1.0$ Å obtained with the STM, although the depth of the corner holes obtained with noncontact AFM is almost the same as that observed with STM. Moreover, in noncontact-mode AFM images, the corrugation amplitude of adatoms was frequently larger than the depth of the corner holes. The origin of such large corrugation of adatoms may be due to the effect of the chemical interaction, but is not yet clear.

The atom positions, surface energies, dynamic properties and chemical reactivities on the Si(111)-(7×7) reconstructed surface have been extensively investigated theoretically and experimentally. From these investigations, the possible origins of the contrast between inequivalent adatoms in AFM images are the followings: the true atomic heights that correspond to the adatom core positions, the stiffness (spring constant) of interatomic bonding with the adatoms corresponding to the frequencies of the surface mode, the charge on the adatom, and the chemical reactivity of the adatoms. Table 23.1 summarizes the decreasing orders of the inequivalent adatoms for individual property. From Table 23.1, we can see that the calculated adatom heights and the stiffness of interatomic bonding cannot explain the AFM data, while the amount of charge of adatom and the chemical reactivity of adatoms can explain the our data. The contrast due to the amount of charge of adatom means that the AFM image is originated from the difference of the vdW or electrostatic physical interactions between the tip and the valence electrons at the adatoms. The contrast due to the chemical reactivity of adatoms means that the AFM image is originated from the difference of covalent bonding chemical interaction between the atoms at the tip apex and dangling bond of adatoms. Thus, we can see there are two possible interactions which explain the strong contrast between inequivalent adatoms of 7×7 unit cell observed using the Si tip with dangling bond.

The weak-contrast image in Fig. 23.10a is due to vdW and/or electrostatic force interactions. On the other hand, the strong-contrast images in Figs. 23.10b and 23.11a are due to a covalent bonding formation between the AFM tip with Si atoms and Si adatoms. These results indicate the capability of the noncontact-mode AFM to image the variation in chemical reactivity of Si adatoms. In the future, by controlling an atomic species at the tip apex, the study of chemical reactivity on an atomic scale will be possible using noncontact AFM.

23.2.2 Si(100)-(2×1) and Si(100)-(2×1):H Monohydride Surfaces

In order to investigate the imaging mechanism of the noncontact AFM, a comparative study between a reactive surface and an insensitive surface using the same tip is very useful. Si(100)-(2×1):H monohydride surface is a Si(100)-(2×1) reconstructed surface that is terminated by a hydrogen atom. It does not reconstruct as metal is deposited on the semiconductor surface. The surface structure hardly changes. Thus, the Si(100)-(2×1):H monohydride surface is one of most useful surface for a model system to investigate the imaging mechanism, experimentally and theoretically. Furthermore, whether the interaction between a very small atom such as hy-

Table 23.1 Comparison between the adatom heights observed in an AFM image and the variety of properties for inequivalent adatoms

	Decreasing order	Agreement
AFM image	Co-F > Ce-F > Co-U > Ce-U	–
Calculated height	Co-F > Co-U > Ce-F > Ce-U	×
Stiffness of interatomic bonding	Ce-U > Co-U > Ce-F > Co-F	×
Amount of charge of adatom	Co-F > Ce-F > Co-U > Ce-U	○
Calculated chemical reactivity	Faulted > unfaulted	○
Experimental chemical reactivity	Co-F > Ce-F > Co-U > Ce-U	○

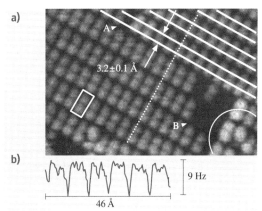 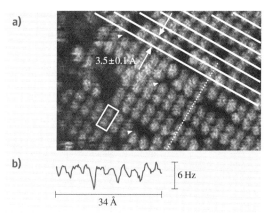

Fig. 23.12 (a) Noncontact AFM image of a Si(001)(2×1) reconstructed surface. The scan area was 69×46 Å. One 2×1 unit cell is outlined with a *box*. *White rows* are superimposed to show the bright spots arrangement. The distance between the bright spots on the dimer row is 3.2 ± 0.1 Å. On the *white arc*, the alternative bright spots are shown. (b) Cross-sectional profile indicated by the *white dotted line*

Fig. 23.13 (a) Noncontact AFM image of Si(001)-(2×1):H surface. The scan area was 69×46 Å. One 2×1 unit cell is outlined with a *box*. *White rows* are superimposed to show the bright spots arrangement. The distance between the bright spots on the dimer row is 3.5 ± 0.1 Å. (b) Cross-sectional profile indicated by the *white dotted line*

drogen and a tip apex is observable with noncontact AFM is interested. Here, we show noncontact AFM images measured on a Si(100)-(2×1) reconstructed surface with a dangling bond and on a Si(100)-(2×1):H monohydride surface on which the dangling bond is terminated by a hydrogen atom [23.50].

Figure 23.12a shows the atomic-resolution image of the Si(100)-(2×1) reconstructed surface. Pairs of bright spots arranged in rows with a 2×1 symmetry were observed with clear contrast. Missing pairs of bright spots were also observed, as indicated by arrows. Furthermore, the pairs of bright spots are shown by the white dashed arc and appear to be the stabilize-buckled asymmetric dimer structure. Furthermore, the distance between the pairs of bright spots is 3.2 ± 0.1 Å.

Figure 23.13a shows the atomic-resolution image of the Si(100)-(2×1):H monohydride surface. Pairs of bright spots arranged in rows were observed. Missing paired bright spots as well as those paired in rows and single bright spots were observed, as indicated by arrows. Furthermore, the distance between paired bright spots is 3.5 ± 0.1 Å. This distance of 3.5 ± 0.1 Å is 0.2 Å larger than that of the Si(100)-(2×1) reconstructed surface. Namely, it is found that the distance between bright spots increases in size due to the hydrogen termination.

The bright spots in Fig. 23.12 do not merely image the silicon-atom site, because the distance between the bright spots forming the dimer structure of Fig. 23.12a, 3.2 ± 0.1 Å, is lager than the distance between silicon atoms of every dimer structure model. (The maximum is the distance between the upper silicones in an asymmetric dimer structure 2.9 Å.) This seems to be due to the contribution to the imaging of the chemical bonding interaction between the dangling bond from the apex of the silicon tip and the dangling bond on the Si(100)-(2×1) reconstructed surface. Namely, the chemical bonding interaction operates strongly, with strong direction dependence, between the dangling bond pointing out of the silicon dimer structure on the Si(100)-(2×1) reconstructed surface and the dangling bond pointing out of the apex of the silicon tip; a dimer structure is obtained with a larger separation than between silicones on the surface.

The bright spots in Fig. 23.13 seem to be located at hydrogen atom sites on the Si(100)-(2×1):H monohydride surface, because the distance between the bright spots forming the dimer structure (3.5 ± 0.1 Å) approximately agrees with the distance between the hydrogens, i. e., 3.52 Å. Thus, the noncontact AFM atomically resolved the individual hydrogen atoms on the topmost layer. On this surface, the dangling bond is terminated by a hydrogen atom, and the hydrogen atom on the topmost layer does not have chemical reactivity. Therefore, the interaction between the hydrogen atom on the topmost layer and the apex of the silicon tip does not contribute to the chemical bonding interaction with strong direction dependence as on the silicon surface, and the bright spots in the noncontact AFM image correspond to the hydrogen atom sites on the topmost layer.

23.2.3 Metal Deposited Si Surface

In this section, we will introduce the comparative study of force interactions between a Si tip and a metal-deposited Si surface, and between a metal adsorbed Si tip and a metal-deposited Si surface [23.51, 52]. As for the metal-deposited Si surface, Si(111)-($\sqrt{3} \times \sqrt{3}$)-Ag (hereafter referred to as $\sqrt{3}$-Ag) surface was used.

For the $\sqrt{3}$-Ag surface, the honeycomb-chained trimer (HCT) model has been accepted as the appropriate model. As shown in Fig. 23.14, this structure contains a Si trimer in the second layer, 0.75 Å below the Ag trimer in the topmost layer. The topmost Ag atoms and lower Si atoms form covalent bonds. The interatomic distances between the nearest-neighbor Ag atoms forming the Ag trimer and between the lower Si atoms forming the Si trimer are 3.43 and 2.31 Å, respectively. The apexes of the Si trimers and Ag trimers face the [11$\bar{2}$] direction and the direction tilted a little to the [$\bar{1}\bar{1}$2] direction, respectively.

In Fig. 23.15, we show the noncontact AFM images measured using a normal Si tip at a frequency shift of (a) −37 Hz, (b) −43 Hz and (c) −51 Hz, respectively. These frequency shifts correspond to tip–sample distances of about 0–3 Å. We defined the zero position of the tip–sample distance, i.e., the contact point, as the point at which the vibration amplitude began to decrease. The rhombus indicates the $\sqrt{3} \times \sqrt{3}$ unit cell. When the tip approached the surface, the contrast of the noncontact AFM images become strong and the pattern changed remarkably. That is, by approaching the tip toward the sample surface, the hexagonal pattern, the trefoil-like pattern composed of three dark lines, and the triangle pattern can be observed sequentially. In Fig. 23.15a, the distance between the bright

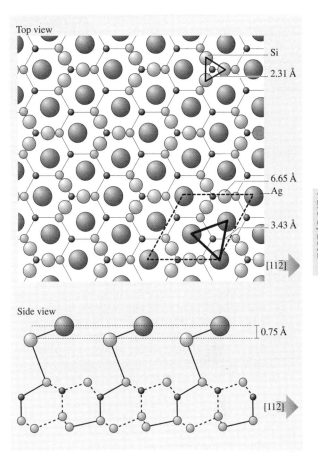

Fig. 23.14 HCT model for the structure of the Si(111)-($\sqrt{3} \times \sqrt{3}$)-Ag surface. *Black closed circle, gray closed circle, open circle*, and *closed circle with horizontal line* indicate Ag atom at the topmost layer, Si atom at the second layer, Si atom at the third layer, and Si atom at the fourth layer, respectively. The *rhombus* indicates the $\sqrt{3} \times \sqrt{3}$ unit cell. The *thick, large, solid triangle* indicates an Ag trimer. The *thin, small, solid triangle* indicates a Si trimer

spots is 3.9 ± 0.2 Å. In Fig. 23.15c, the distance between the bright spots is 3.0 ± 0.2 Å, and the direction of the

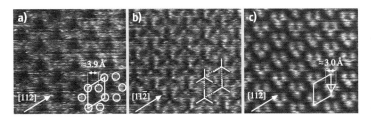

Fig. 23.15a–c Noncontact AFM images obtained at frequency shifts of (**a**) −37 Hz, (**b**) −43 Hz, and (**c**) −51 Hz on a Si(111)-($\sqrt{3} \times \sqrt{3}$)-Ag surface. This distance dependence was obtained with a Si tip. The scan area is 38 Å × 34 Å. A *rhombus* indicates the $\sqrt{3} \times \sqrt{3}$ unit cell

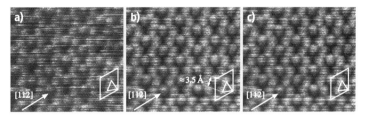

Fig. 23.16a–c Noncontact AFM images obtained at frequency shifts of (a) -4.4 Hz, (b) -6.9 Hz, and (c) -9.4 Hz on a Si(111)-$(\sqrt{3} \times \sqrt{3})$-Ag surface. This distance dependence was obtained with the Ag-adsorbed tip. The scan area is 38 Å × 34 Å

apex of all the triangles composed of three bright spots is $[11\bar{2}]$.

In Fig. 23.16, we show the noncontact AFM images measured by using Ag-absorbed tip at a frequency shift of (a) -4.4 Hz, (b) -6.9 Hz and (c) -9.4 Hz, respectively. The tip–sample distances Z are roughly estimated to be $Z = 1.9$, 0.6 and ≈ 0 Å (in the noncontact region), respectively. When the tip approached the surface, the pattern of the noncontact AFM images did not change, although the contrast become clearer. A triangle pattern can be observed. The distance between the bright spots is 3.5 ± 0.2 Å. The direction of the apex of all the triangles composed of three bright spots is tilted a little from the $[\bar{1}\bar{1}2]$ direction.

Thus, noncontact AFM images measured on Si(111)-$(\sqrt{3} \times \sqrt{3})$-Ag surface showed two types of distance dependence in the image patterns depending on the atom species on the apex of the tip.

By using the normal Si tip with a dangling bond, in Fig. 23.15a, the measured distance between the bright spot of 3.9 ± 0.2 Å agrees with the distance of 3.84 Å between the centers of the Ag trimers in the HCT model within the experimental error. Furthermore, the hexagonal pattern composed of six bright spots also agrees with the honeycomb structure of the Ag trimer in HCT model. So the most appropriate site corresponding to the bright spots in Fig. 23.15a is the site of the center of Ag trimers. In Fig. 23.15c, the measured distance of 3.0 ± 0.2 Å between the bright spots forming the triangle pattern agrees with neither the distance between the Si trimer of 2.31 Å nor the distance between the Ag trimer of 3.43 Å in the HCT model, while the direction of the apex of the triangles composed of three bright spots agrees with the $[11\bar{2}]$ direction of the apex of the Si trimer in the HCT model. So the most appropriate site corresponding to the bright spots in Fig. 23.15c is the intermediate site between the Si atoms and Ag atoms. On the other hand, by using the Ag-adsorbed tip, the measured distance between the bright spots of 3.5 ± 0.2 Å in Fig. 23.16 agrees with the distance of 3.43 Å between the nearest-neighbor Ag atoms forming the Ag trimer in the topmost layer in the HCT model within the experimental error. Furthermore, the direction of the apex of the triangles composed of three bright spots also agrees with the direction of the apex of the Ag trimer, i.e., tilted $[\bar{1}\bar{1}2]$, in the HCT model. So, the most appropriate site corresponding to the bright spots in Fig. 23.16 is the site of individual Ag atoms forming the Ag trimer in the topmost layer.

It should be noted that, by using the noncontact AFM with a Ag-adsorbed tip, for the first time, the individual Ag atom on the $\sqrt{3}$-Ag surface could be resolved in real space, although by using the noncontact AFM with an Si tip, it could not be resolved. So far, the $\sqrt{3}$-Ag surface has been observed by a scanning tunneling microscope (STM) with atomic resolution. However, the STM can also measure the local charge density of states near the Fermi level on the surface. From first-principle calculations, it was proven that unoccupied surface states are densely distributed around

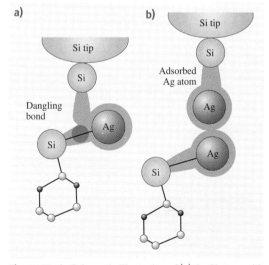

Fig. 23.17a,b Schematic illustration of (a) the Si atom with dangling bond and (b) the Ag-adsorbed tip above the Si–Ag covalent bond on a Si(111)-$(\sqrt{3} \times \sqrt{3})$-Ag surface

the center of the Ag trimer. As a result, bright contrast is obtained at the center of the Ag trimer with the STM.

Finally, we consider the origin of the atomic-resolution imaging of the individual Ag atoms on the $\sqrt{3}$-Ag surface. Here, we discuss the difference between the force interactions when using the Si tip and the Ag-adsorbed tip. As shown in Fig. 23.17a, when using the Si tip, there is a dangling bond pointing out of the topmost Si atom on the apex of the Si tip. As a result, the force interaction is dominated by physical bonding interactions, such as the Coulomb force, far from the surface and by chemical bonding interaction very close to the surface. Namely, if a reactive Si tip with a dangling bond approaches a surface, at distances far from the surface the Coulomb force acts between the electron localized on the dangling bond pointing out of the topmost Si atom on the apex of the tip, and the positive charge distributed around the center of the Ag trimer. At distances very close to the surface, the chemical bonding interaction will occur due to the onset of orbital hybridization between the dangling bond pointing out of the topmost Si atom on the apex of the Si tip and a Si−Ag covalent bond on the surface. Hence, the individual Ag atoms will not be resolved and the image pattern will change depending on the tip–sample distance. On the other hand, as shown in Fig. 23.17b, by using the Ag-adsorbed tip, the dangling bond localized out of topmost Si atom on the apex of the Si tip is terminated by the adsorbed Ag atom. As a result, even at very close tip–sample distances, the force interaction is dominated by physical bonding interactions such as the vdW force. Namely, if the Ag-adsorbed tip approaches the surface, the vdW force acts between the Ag atom on the apex of the tip and the Ag or Si atom on the surface. Ag atoms in the topmost layer of the $\sqrt{3}$-Ag surface are located higher than the Si atoms in the lower layer. Hence, the individual Ag atoms (or their nearly true topography) will be resolved, and the image pattern will not change even at very small tip–sample distances. It should be emphasized that there is a possibility to identify or recognize atomic species on a sample surface using noncontact AFM if we can control the atomic species at the tip apex.

23.3 Applications to Insulators

Insulators such as alkali halides, fluorides, and metal oxides are key materials in many applications, including optics, microelectronics, catalysis, and so on. Surface properties are important in these technologies, but they are usually poorly understood. This is due to their low conductivity, which makes it difficult to investigate them using electron- and ion-based measurement techniques such as low-energy electron diffraction, ion-scattering spectroscopy, and scanning tunneling microscopy (STM). Surface imaging by noncontact atomic force microscopy (NC-AFM) does not require a sample surface with a high conductivity because NC-AFM detects a force between the tip on the cantilever and the surface of the sample. Since the first report of atomically resolved NC-AFM on a Si(111)-(7×7) surface [23.33], several groups have succeeded in obtaining *true* atomic resolution images of insulators, including defects, and it has been shown that NC-AFM is a powerful new tool for atomic-scale surface investigation of insulators.

In this section we will describe typical results of atomically resolved NC-AFM imaging of insulators such as alkali halides, fluorides and metal oxides. For the alkali halides and fluorides, we will focus on contrast formation, which is the most important issue for interpreting atomically resolved images of binary compounds on the basis of experimental and theoretical results. For the metal oxides, typical examples of atomically resolved imaging will be exhibited and the difference between the STM and NC-AFM images will be demonstrated. Also, theoretical studies on the interaction between realistic Si tips and representative oxide surfaces will be shown. Finally, we will describe an antiferromagnetic NiO(001) surface imaged with a ferromagnetic tip to explore the possibility of detecting short-range magnetic interactions using the NC-AFM.

23.3.1 Alkali Halides, Fluorides and Metal Oxides

The surfaces of alkali halides were the first insulating materials to be imaged by NC-AFM with *true* atomic resolution [23.53]. To date, there have been reports on atomically resolved images of (001) cleaved surfaces for single-crystal NaF, RbBr, LiF, KI, NaCl, [23.54], KBr [23.55] and thin films of NaCl(001) on Cu(111) [23.56]. In this section we describe the contrast formation of alkali halides surfaces on the basis of experimental and theoretical results.

Alkali Halides

In experiments on alkali halides, the symmetry of the observed topographic images indicates that the protrusions exhibit only one type of ions, either the positive or negatively charged ions. This leads to the conclusion that the atomic contrast is dominantly caused by electrostatic interactions between a charged atom at the apex of the tip and the surface ions, i.e. long-range forces between the macroscopic tip and the sample, such as the van der Waals force, are modulated by an alternating short-range electrostatic interaction with the surface ions. Theoretical work employing the atomistic simulation technique has revealed the mechanism for contrast formation on an ionic surface [23.57]. A significant part of the contrast is due to the displacement of ions in the force field, not only enhancing the atomic corrugations, but also contributing to the electrostatic potential by forming dipoles at the surface. The experimentally observed atomic corrugation height is determined by the interplay of the long- and short-range forces. In the case of NaCl, it has been experimentally demonstrated that a blunter tip produces a lager corrugation when the tip–sample distance is shorter [23.54]. This result shows that the increased long-range forces induced by a blunter tip allow for more stable imaging closer to the surface. The stronger electrostatic short-range interaction and lager ion displacement produce a more pronounced atomic corrugation. At steps and kinks on an NaCl thin film on Cu(111), the corrugation amplitude of atoms with low coordination number has been observed to increase by a factor of up to two more than that of atomically flat terraces [23.56]. The low coordination number of the ions results in an enhancement of the electrostatic potential over the site and an increase in the displacement induced by the interaction with the tip.

Theoretical study predicts that the image contrast depends on the chemical species at the apex of the tip. *Bennewitz* et al. [23.56] have performed the calculations using an MgO tip terminated by oxygen and an Mg ion. The magnitude of the atomic contrast for the Mg-terminated tip shows a slight increase in comparison with an oxygen-terminated tip. The atomic contrast with the oxygen-terminated tip is dominated by the attractive electrostatic interaction between the oxygen on the tip apex and the Na ion, but the Mg-terminated tip attractively interacts with the Cl ion. In other words, these results demonstrated that the species of the ion imaged as the bright protrusions depends on the polarity of the tip apex.

These theoretical results emphasized the importance of the atomic species at the tip apex for the alkali halide (001) surface, while it is not straightforward to define the nature of the tip apex experimentally because of the high symmetry of the surface structure. However, there are a few experiments exploring the possibilities to determine the polarity of the tip apex. *Bennewitz* et al. [23.58] studied imaging of surfaces of a mixed alkali halide crystal, which was designed to observe the chemically inhomogeneous surface. The mixed crystal is composed of 60% KCl and 40% KBr, with the Cl and Br ions interfused randomly in the crystal. The image of the cleaved $KCl_{0.6}Br_{0.4}(001)$ surface indicates that only one type of ion is imaged as protrusions, as if it were a pure alkali halide crystal. However, the amplitude of the atomic corrugation varies strongly between the positions of the ions imaged as depressions. This variation in the corrugations corresponds to the constituents of the crystal, i.e. the Cl and Br ions, and it is concluded that the tip apex is negatively charged. Moreover, the deep depressions can be assigned to Br ions by comparing the number with the relative density of anions. The difference between Cl and Br anions with different masses is enhanced in the damping signal measured simultaneously with the topographic image [23.59]. The damping is recorded as an increase in the excitation amplitude necessary to maintain the oscillation amplitude of the cantilever in the constant-amplitude mode [23.56]. Although the dissipation phenomena on an atomic scale are a subject under discussion, any dissipative interaction must generally induce energy losses in the cantilever oscillation [23.60, 61]. The measurement of energy dissipation has the potential to enable chemical discrimination on an atomic scale. Recently, a new procedure for species recognition on a alkali halide surface was proposed [23.62]. This method is based on a comparison between theoretical results and the site-specific measurement of frequency versus distance. The differences in the force curves measured at the typical sites, such as protrusion, depression, and their bridge position, are compared to the corresponding differences obtained from atomistic simulation. The polarity of the tip apex can be determined, leading to the identification of the surface species. This method is applicable to highly symmetric surfaces and is useful for determining the sign of the tip polarity.

Fluorides

Fluorides are important materials for the progress of an atomic-scale-resolution NC-AFM imaging of insulators. There are reports in the literature of surface images for single-crystal BaF_2, SrF_2 [23.63], CaF_2 [23.64–66] and a CaF bilayer on Si(111) [23.67]. Surfaces of fluorite-type crystals are prepared by cleaving along the (111) planes. Their structure is more complex than the structure of alkali halides, which have a rock-salt structure. The complexity is of great interest for atomic-resolution imaging using NC-AFM and also for theoretical predictions of the interpretation of the atomic-scale contrast information.

The first atomically resolved images of a $CaF_2(111)$ surface were obtained in topographic mode [23.65], and the surface ions mostly appear as spherical caps. *Barth et al.* [23.68] have found that the $CaF_2(111)$ surface images obtained by using the constant-height mode, in which the frequency shift is recorded with a very low loop gain, can be categorized into two contrast patterns. In the first of these the ions appear as triangles and in the second they have the appearance of circles, similar to the contrast obtained in a topographic image. Theoretical studies demonstrated that these two different contrast patterns could be explained as a result of imaging with tips of different polarity [23.68–70]. When imaging with a positively charged (cation-terminated) tip, the triangular pattern appears. In this case, the contrast is dominated by the strong short-range electrostatic attraction between the positive tip and the negative F ions. The cross section along the [121] direction of the triangular image shows two maxima: one is a larger peak over the F(I) ions located in the topmost layer and the other is a smaller peak at the position of the F(III) ions in the third layer. The minima appear at the position of the Ca ions in the second layer. When imaging with a negatively charged (anion-terminated) tip, the spherical image contrast appears and the main periodicity is created by the Ca ions between the topmost and the third F ion layers. In the cross section along the [121] direction, the large maxima correspond to the Ca sites because of the strong attraction of the negative tip and the minima appear at the sites of maximum repulsion over the F(I) ions. At a position between two F ions, there are smaller maxima. This reflects the weaker repulsion over the F(III) ion sites compared to the protruding F(I) ion sites and a slower decay in the contrast on one side of the Ca ions.

The triangular pattern obtained with a positively charged tip appears at relatively large tip–sample distance, as shown in Fig. 23.18a. The cross section along the [121] direction, experiment results and theoretical studies both demonstrate the large-peak and small-shoulder characteristic for the triangular pattern image (Fig. 23.18d). When the tip approaches the surface more closely, the triangular pattern of the experimental images is more vivid (Fig. 23.18b), as predicted in the

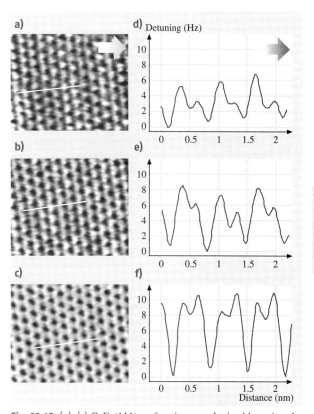

Fig. 23.18 (a)–(c) $CaF_2(111)$ surface images obtained by using the constant-height mode. From (a) to (c) the frequency shift was lowered. The *white lines* represent the positions of the cross section. (d)–(f) The cross section extracted from the Fourier-filtered images of (a)–(c). The *white* and *black arrows* represent the scanning direction. The images and the cross sections are from [23.68]

theoretical works. As the tip approaches, the amplitude of the shoulder increases until it is equal to that of the main peak, and this feature gives rise to the honeycomb pattern image, as shown in Fig. 23.18c. Moreover, theoretical results predict that the image contrast changes again when the tip apex is in close proximity to surface. Recently, *Giessibl* and *Reichling* [23.71] achieved atomic imaging in the repulsive region and proved experimentally the predicted change of the image contrast. As described here, there is good correspondence in the distance dependency of the image obtained by experimental and theoretical investigations.

From detailed theoretical analysis of the electrostatic potential [23.72], it was suggested that the change in displacement of the ions due to the proximity of the tip plays an important role in the formation of the image contrast. Such a drastic change in image contrast, depending on both the polarity of the terminated tip atom and on the tip–sample distance, is inherent to the fluoride (111) surface, and this image-contrast feature cannot be seen on the (001) surface of alkali halides with a simple crystal structure.

The results of careful experiments show another feature: that the cross sections taken along the three equivalent [121] directions do not yield identical results [23.68]. It is thought that this can be attributed to the asymmetry of the nanocluster at the tip apex, which leads to different interactions in the equivalent directions. A better understanding of the asymmetric image contrast may require more complicated modeling of the tip structure. In fact, it should be mentioned that perfect tips on an atomic scale can occasionally be obtained. These tips do yield identical results in forward and backward scanning, and cross sections in the three equivalent directions taken with this tip are almost identical [23.74].

The fluoride (111) surface is an excellent standard surface for calibrating tips on an atomic scale. The polarity of the tip-terminated atom can be determined from the image contrast pattern (spherical or triangular pattern). The irregularities in the tip structure can be detected, since the surface structure is highly symmetric. Therefore, once such a tip has been prepared, it can be used as a calibrated tip for imaging unknown surfaces.

The polarity and shape of the tip apex play an important role in interpreting NC-AFM images of alkali halide and fluorides surfaces. It is expected that the achievement of good correlation between experimental and theoretical studies will help to advance surface imaging of insulators by NC-AFM.

Metal Oxides

Most of the metal oxides that have attracted strong interest for their technological importance are insulating. Therefore, in the case of atomically resolved imaging of metal oxide surfaces by STM, efforts to increase the conductivity of the sample are needed, such as, the introduction of anions or cations defects, doping with other atoms and surface observations during heating of the sample. However, in principle, NC-AFM provides the possibility of observing nonconductive metal oxides without these efforts. In cases where the conductivity of the metal oxides is high enough for a tunneling current to flow, it should be noted that most surface images obtained by NC-AFM and STM are not identical.

Since the first report of atomically resolved images on a $TiO_2(110)$ surface with oxygen point defects [23.75], they have also been reported on rutile $TiO_2(100)$ [23.76–78], anatase $TiO_2(001)$ thin film on $SrTiO_3(100)$ [23.79] and on $LaAO_3(001)$ [23.80], $SnO_2(110)$ [23.81], $NiO(001)$ [23.82, 83], $SrTiO_3(100)$ [23.84], $CeO_2(111)$ [23.85] and $MoO_3(010)$ [23.86] surfaces. Also, *Barth* and *Reichling* have succeeded in obtaining atomically resolved NC-AFM images of a clean α-$Al_2O_3(0001)$ surface [23.73] and of a UHV cleaved $MgO(001)$ [23.87] surface, which are impossible to investigate using STM. In this section we describe typical results of the imaging of metal oxides by NC-AFM.

The α-$Al_2O_3(0001)$ surface exists in several ordered phases that can reversibly be transformed into each other by thermal treatments and oxygen exposure. It

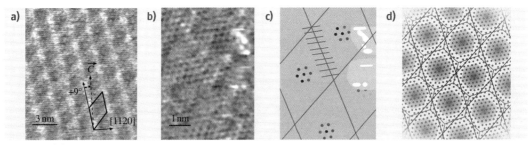

Fig. 23.19 (a) Image of the high-temperature, reconstructed clean α-Al_2O_3 surface obtained by using the constant-height mode. The *rhombus* represents the unit cell of the $(\sqrt{31} \times \sqrt{31})R + 9°$ reconstructed surface. (b) Higher-magnification image of (a). Imaging was performed at a reduced tip–sample distance. (c) Schematic representation of the indicating regions of hexagonal order in the center of reconstructed rhombi. (d) Superposition of the hexagonal domain with reconstruction rhombi found by NC-AFM imaging. Atoms in the *gray shaded regions* are well ordered. The images and the schematic representations are from [23.73]

is known that the high-temperature phase has a large $(\sqrt{31} \times \sqrt{31})R \pm 9°$ unit cell. However, the details of the atomic structure of this surface have not been revealed, and two models have been proposed. *Barth* and *Reichling* [23.73] have directly observed this reconstructed α-$Al_2O_3(0001)$ surface by NC-AFM. They confirmed that the dominant contrast of the low-magnification image corresponds to a rhombic grid representing a unit cell of $(\sqrt{31} \times \sqrt{31})R + 9°$, as shown in Fig. 23.19a. Also, more details of the atomic structures were determined from the higher-magnification image (Fig. 23.19b), which was taken at a reduced tip–sample distance. In this atomically resolved image, it was revealed that each side of the rhombus is intersected by ten atomic rows, and that a hexagonal arrangement of atoms exists in the center of the rhombi (Fig. 23.19c). This feature agrees with the proposed surface structure that predicts order in the center of the hexagonal surface domains and disorder at the domain boundaries. Their result is an excellent demonstration of the capabilities of the NC-AFM for the atomic-scale surface investigation of insulators.

The atomic structure of the $SrTiO_3(100)$-$(\sqrt{5} \times \sqrt{5})R26.6°$ surface, as well as that of $Al_2O_3(0001)$ can be determined on the basis of the results of NC-AFM imaging [23.84]. $SrTiO_3$ is one of the perovskite oxides, and its (100) surface exhibits the many different kinds of reconstructed structures. In the case of the $(\sqrt{5} \times \sqrt{5})R26.6°$ reconstruction, the oxygen vacancy–Ti^{3+}–oxygen model (where the terminated surface is TiO_2 and the observed spots are related to oxygen vacancies) was proposed from the results of STM imaging. As shown in Fig. 23.20, *Kubo* and *Nozoye* [23.84] have performed measurements using both STM and NC-AFM, and have found that the size of the bright spots as observed by NC-AFM is always smaller than that for STM measurement, and that the dark spots,

which are not observed by STM, are arranged along the [001] and [010] directions in the NC-AFM image. A theoretical simulation of the NC-AFM image using first-principles calculations shows that the bright and dark spots correspond to Sr and oxygen atoms, respectively. It has been proposed that the structural model of the reconstructed surface consists of an ordered Sr adatom located at the oxygen fourfold site on the TiO_2-terminated layer (Fig. 23.20c).

Because STM images are related to the spatial distribution of the wave functions near the Fermi level, atoms without a local density of states near the Fermi level are generally invisible even on conductive materials. On the other hand, the NC-AFM image reflects the strength of the tip–sample interaction force originating from chemical, electrostatic and other interactions. Therefore, even STM and NC-AFM images obtained using an identical tip and sample may not be identical generally. The simultaneous imaging of a metal oxide surface enables the investigation of a more detailed surface structure. The images of a $TiO_2(110)$ surface simultaneously obtained with STM and NC-AFM [23.78] are a typical example. The STM image shows that the dangling-bond states at the tip apex overlap with the dangling bonds of the 3d states protruding from the Ti atom, while the NC-AFM primarily imaged the uppermost oxygen atom.

Recently, calculations of the interaction of a Si tip with metal oxides surfaces, such as $Al_2O_3(0001)$, $TiO_2(110)$, and $MgO(001)$, were reported [23.88, 89]. Previous simulations of AFM imaging of alkali halides and fluorides assume that the tip would be oxides or contaminated and hence have been performed with a model of ionic oxide tips. In the case of imaging a metal oxide surface, pure Si tips are appropriate for a more realistic tip model because the tip is sputtered for

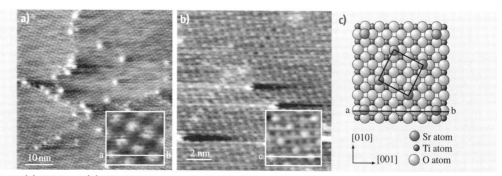

Fig. 23.20 (a) STM and **(b)** NC-AFM images of a $SrTiO_3(100)$ surface. **(c)** A proposed model of the $SrTiO_3(100)$-$(\sqrt{5} \times \sqrt{5})R26.6°$ surface reconstruction. The images and the schematic representations are from [23.84]

cleaning in many experiments. The results of ab initio calculations for a Si tip with a dangling bond demonstrate that the balance between polarization of the tip and covalent bonding between the tip and the surface should determine the tip–surface force. The interaction force can be related to the nature of the surface electronic structure. For wide-gap insulators with a large valence-band offset that prevents significant electron-density transfer between the tip and the sample, the force is dominated by polarization of the tip. When the gap is narrow, the charge transfer increase and covalent bonding dominates the tip–sample interaction. The forces over anions (oxygen ions) in the surface are larger than over cations (metal ions), as they play a more significant role in charge transfer. This implies that a pure Si tip would always show the brightest contrast over the highest anions in the surface. In addition, *Foster* et al. [23.88] suggested the method of using applied voltage, which controls the charge transfer, during an AFM measurement to define the nature of tip apex.

The collaboration between experimental and theoretical studies has made great progress in interpreting the imaging mechanism for binary insulators surface and reveals that a well-defined tip with atomic resolution is preferable for imaging a surface. As described previously, a method for the evaluation of the nature of the tip has been developed. However, the most desirable solution would be the development of suitable techniques for well-defined tip preparation and a few attempts at controlled production of Si tips have been reported [23.24, 90, 91].

23.3.2 Atomically Resolved Imaging of a NiO(001) Surface

The transition metal oxides, such as NiO, CoO, and FeO, feature the simultaneous existence of an energy gap and unpaired electrons, which gives rise to a variety of magnetic property. Such magnetic insulators are widely used for the exchange biasing for magnetic and spintronic devices. NC-AFM enables direct surface imaging of magnetic insulators on an atomic scale. The forces detected by NC-AFM originate from several kinds of interaction between the surface and the tip, including magnetic interactions in some cases. Theoretical studies predict that short-range magnetic interactions such as the exchange interaction should enable the NC-AFM to image magnetic moments on an atomic scale. In this section, we will describe imaging of the antiferromagnetic NiO(001) surface using a ferromagnetic tip. Also, theoretical studies of the exchange force interaction between a magnetic tip and a sample will be described.

Theoretical Studies of the Exchange Force

In the system of a magnetic tip and sample, the interaction detected by NC-AFM includes the short-range magnetic interaction in addition to the long-range magnetic dipole interaction. The energy of the short-range interaction depends on the electron spin states of the atoms on the apex of the tip and the sample surface, and the energy difference between spin alignments (parallel or antiparallel) is referred to as the exchange interaction energy. Therefore, the short-range magnetic interaction leads to the atomic-scale magnetic contrast, depending on the local energy difference between spin alignments.

In the past, extensive theoretical studies on the short-range magnetic interaction between a ferromagnetic tip and a ferromagnetic sample have been performed by a simple calculation [23.92], a tight-binding approximation [23.93] and first-principles calculations [23.94]. In the calculations performed by *Nakamura* et al. [23.94], three-atomic-layer Fe(001) films are used as a model for the tip and sample. The exchange force is defined as the difference between the forces in each spin configuration of the tip and sample (parallel and antiparallel). The result of this calculation demonstrates that the amplitude of the exchange force is measurable for AFM (about 0.1 nN). Also, they forecasted that the discrimination of the exchange force would enable direct imaging of the magnetic moments on an atomic scale. *Foster* and *Shluger* [23.95] have theoretically investigated the interaction between a spin-polarized H atom and a Ni atom on a NiO(001) surface. They demonstrated that the difference in magnitude in the exchange interaction between opposite-spin Ni ions in a NiO surface could be sufficient to be measured in a low-temperature NC-AFM experiment. Recently, first-principles calculation of the interaction of a ferromagnetic Fe tip with an NiO surface has demonstrated that it should be feasible to measure the difference in exchange force between opposite-spin Ni ions [23.96].

Atomically Resolved Imaging Using Noncoated and Fe-Coated Si Tips

The detection of the exchange interaction is a challenging task for NC-AFM applications. An antiferromagnetic insulator NiO single crystal that has regularly aligned atom sites with alternating electron spin states is one of the best candidates to prove the feasibility of

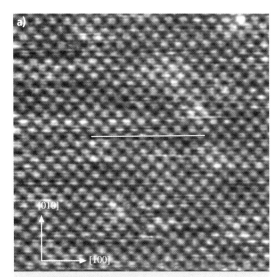

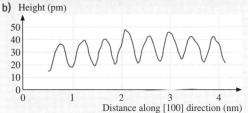

Fig. 23.21 (a) Atomically resolved image obtained with an Fe-coated tip. (b) Shows the cross sections of the middle part in (a). Their corrugations are about 30 pm

sembles that of the alkali halide (001) surface. The symmetry of the image reveals that only one type of atom appears to be at the maximum. From this image, it seems difficult to distinguish which of the atoms are observed as protrusions. The theoretical works indicate that a metal tip interacts strongly with the oxygen atoms on the MgO(001) surface [23.95]. From this result, it is presumed that the bright protrusions correspond to the oxygen atoms. However, it is still questionable which of the atoms are visible with a Fe-coated tip.

If the short-range magnetic interaction is included in the atomic image, the corrugation amplitude of the atoms should depend on the direction of the spin over the atom site. From the results of first-principles calculations [23.94], the contribution of the short-range magnetic interaction to the measured corrugation amplitude is expected to be about a few percent of the total interaction. Discrimination of such small perturbations is therefore needed. In order to reduce the noise, the corrugation amplitude was added on the basis of the periodicity of the NC-AFM image. In addition, the topographical asymmetry, which is the index characterizing the difference in atomic corrugation amplitude, has been defined [23.101]. The result shows that the value of the topographical asymmetry calculated from the image obtained with an Fe-coated Si tip depends on the direction of summing of the corrugation amplitude, and that the dependency corresponds to the antiferromagnetic spin ordering of the NiO(001) surface [23.101, 102]. Therefore, this result implies that the dependency of the topographical asymmetry originates in the short-range magnetic interaction. However, in some cases the topographic asymmetry with uncoated Si tips has a finite value [23.103]. The possibility that the asymmetry includes the influence of the structure of tip apex and of the relative orientation between the surface and tip cannot be excluded. In addition, it is suggested that the absence of unambiguous exchange contrast is due to the fact that surface ion instabilities occur at tip–sample distances that are small enough for a magnetic interact [23.100]. Another possibility is that the magnetic properties of the tips are not yet fully controlled because the topographic asymmetries obtained by Fe- and Ni-coated tips show no significant difference [23.103]. In any cases, a careful comparison is needed to evaluate the exchange interaction included in an atomic image.

From the aforementioned theoretical works, it is presumed that a metallic tip has the capability to image an oxygen atom as a bright protrusion. Recently, the magnetic properties of the NiO(001) surface were

detecting the exchange force for the following reason. NiO has an antiferromagnetic AF$_2$ structure as the most stable below the Néel temperature of 525 K. This well-defined magnetic structure, in which Ni atoms on the (001) surface are arranged in a checkerboard pattern, leads to the simple interpretation of an image containing the atomic-scale contrast originating in the exchange force. In addition, a clean surface can easily be prepared by cleaving.

Figure 23.21a shows an atomically resolved image of a NiO(001) surface with a ferromagnetic Fe-coated tip [23.97]. The bright protrusions correspond to atoms spaced about 0.42 nm apart, consistent with the expected periodic arrangement of the NiO(001) surface. The corrugation amplitude is typically 30 pm, which is comparable to the value previously reported [23.82, 83, 98–100], as shown in Fig. 23.21b. The atomic-resolution image (Fig. 23.21b), in which there is one maximum and one minimum within the unit cell, reinvestigated by first-principles electronic-structure calculations [23.104]. It was shown that the surface oxygen has finite spin magnetic moment, which originates from symmetry breaking. We must take into account the possibility that a metal atom at the ferromagnetic tip apex may interact with a Ni atom on the second layer through a magnetic interaction mediated by the electrons in an oxygen atom on the surface.

The measurements presented here demonstrate the feasibility of imaging magnetic structures on an atomic scale by NC-AFM. In order to realize explicit detection of exchange force, further experiments and a theoretical study are required. In particular, the development of a tip with well-defined atomic structure and magnetic properties is essential for *exchange force microscopy*.

23.4 Applications to Molecules

In the future, it is expected that electronic, chemical, and medical devices will be downsized to the nanometer scale. To achieve this, visualizing and assembling individual molecular components is of fundamental importance. Topographic imaging of nonconductive materials, which is beyond the range of scanning tunneling microscopes, is a challenge for atomic force microscopy (AFM). Nanometer-sized domains of surfactants terminated with different functional groups have been identified by lateral force microscopy (LFM) [23.106] and by chemical force microscopy (CFM) [23.107] as extensions of AFM. At a higher resolution, a periodic array of molecules, Langmuir–Blodgett films [23.108] for example, was recognized by AFM. However, it remains difficult to visualize an isolated molecule, molecule vacancy, or the boundary of different periodic domains, with a microscope with the tip in contact.

23.4.1 Why Molecules and Which Molecules?

Access to individual molecules has not been a trivial task even for noncontact atomic force microscopy (NC-AFM). The force pulling the tip into the surface is less sensitive to the gap width (r), especially when chemically stable molecules cover the surface. The attractive potential between two stable molecules is shallow and exhibits r^{-6} decay [23.13].

High-resolution topography of formate (HCOO$^-$) [23.109] was first reported in 1997 as a molecular adsorbate. The number of imaged molecules is now increasing because of the technological importance of molecular interfaces. To date, the following studies on molecular topography have been published: C$_{60}$ [23.105, 110], DNAs [23.111, 112], adenine and thymine [23.113], alkanethiols [23.113, 114], a perylene derivative (PTCDA) [23.115], a metal porphyrin (Cu-TBPP) [23.116], glycine sulfate [23.117], polypropylene [23.118], vinylidene fluoride [23.119], and a series of carboxylates (RCOO$^-$) [23.120–126]. Two of these

Fig. 23.22 The constant frequency-shift topography of domain boundaries on a C$_{60}$ multilayered film deposited on a Si(111) surface based on [23.105]. Image size: 35×35 nm^2

are presented in Figs. 23.22 and 23.23 to demonstrate the current stage of achievement. The proceedings of the annual NC-AFM conference represent a convenient opportunity for us to update the list of molecules imaged.

23.4.2 Mechanism of Molecular Imaging

A systematic study of carboxylates (RCOO$^-$) with R = H, CH$_3$, C(CH$_3$)$_3$, C≡CH, and CF$_3$ revealed that the van der Waals force is responsible for the molecule-dependent microscope topography despite its long-range (r^{-6}) nature. Carboxylates adsorbed on the (110) surface of rutile TiO$_2$ have been extensively studied as a prototype for organic materials interfaced with an inorganic metal oxide [23.127]. A carboxylic acid molecule (RCOOH) dissociates on this surface to a carboxylate (RCOO$^-$) and a proton (H$^+$) at room temperature, as illustrated in Fig. 23.24. The pair

Fig. 23.23 The constant frequency-shift topography of a DNA helix on a mica surface based on [23.111]. Image size: 43×43 nm^2. The image revealed features with a spacing of 3.3 nm, consistent with the helix turn of B-DNA ▶

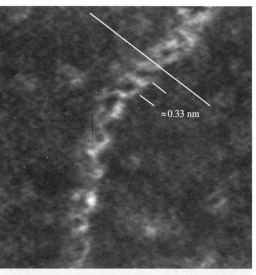

of negatively charged oxygen atoms in the RCOO$^-$ coordinate two positively charged Ti atoms on the surface. The adsorbed carboxylates create a long-range ordered monolayer. The lateral distances of the adsorbates in the ordered monolayer are regulated at 0.65 and 0.59 nm along the [110] and [001] directions. By scanning a mixed monolayer containing different carboxylates, the microscope topography of the terminal groups can be quantitatively compared while minimizing tip-dependent artifacts.

Figure 23.25 presents the observed constant frequency-shift topography of four carboxylates terminated by different alkyl groups. On the formate-covered surface of panel (a), individual formates (R = H) were resolved as protrusions of uniform brightness. The dark holes represent unoccupied surface sites. The cross section in the lower panel shows that the accuracy of the height measurement was 0.01 nm or better. Brighter particles appeared in the image when the formate monolayer was exposed to acetic acid (CH$_3$COOH) as shown in panel (b). Some formates

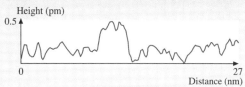

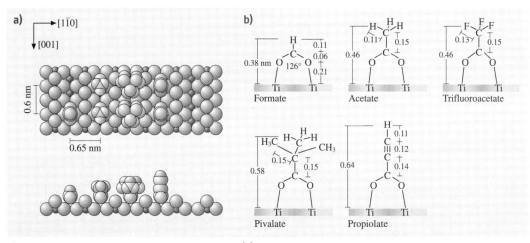

Fig. 23.24a,b The carboxylates and TiO$_2$ substrate. (**a**) Top and side view of the ball model. *Small shaded* and *large shaded balls* represent Ti and O atoms in the substrate. Protons yielded in the dissociation reaction are not shown. (**b**) Atomic geometry of formate, acetate, pivalate, propiolate, and trifluoroacetate adsorbed on the TiO$_2$(110) surface. The O–Ti distance and O–C–O angle of the formate were determined in the quantitative analysis using photoelectron diffraction [23.128]

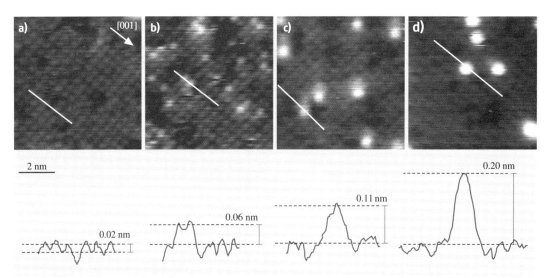

Fig. 23.25a–d The constant frequency-shift topography of carboxylate monolayers prepared on the TiO$_2$(110) surface based on [23.121, 123, 125]. Image size: 10×10 nm^2. (**a**) Pure formate monolayer; (**b**) formate–acetate mixed layer; (**c**) formate–pivalate mixed layer; (**d**) formate–propiolate mixed layer. Cross sections determined on the *lines* are shown in the *lower panel*

were exchanged with acetates (R = CH$_3$) impinging from the gas phase [23.129]. Because the number of brighter spots increased with exposure time to acetic acid, the brighter particle was assigned to the acetate [23.121]. Twenty-nine acetates and 188 formates were identified in the topography. An isolated acetate and its surrounding formates exhibited an image height difference of 0.06 nm. Pivalate is terminated by bulky R = (CH$_3$)$_3$. Nine bright pivalates were surrounded by formates of ordinary brightness in the image of panel (c) [23.123]. The image height difference of an isolated pivalate over the formates was 0.11 nm. Propiolate with C≡CH is a needle-like adsorbate of single-atom diameter. That molecule exhibited in panel (d) a microscope topography 0.20 nm higher than that of the formate [23.125].

The image topography of formate, acetate, pivalate, and propiolate followed the order of the size of the alkyl groups. Their physical topography can be assumed based on the C−C and C−H bond lengths in the corresponding RCOOH molecules in the gas phase [23.130], and is illustrated in Fig. 23.24. The top hydrogen atom of the formate is located 0.38 nm above the surface plane containing the Ti atom pair, while three equivalent hydrogen atoms of the acetate are more elevated at 0.46 nm. The uppermost H atoms in the pivalate are raised by 0.58 nm relative to the Ti plane. The H atom terminating the triple-bonded carbon chain in the propiolate is at 0.64 nm. Figure 23.26 summarizes the observed image heights relative to the formate, as a function of the physical height of the topmost H atoms given in the model. The straight line fitted the four observations [23.122]. When the horizontal axis was scaled with other properties (molecular weight, the number of atoms in a molecule, or the number of electrons in valence states), the correlation became poor.

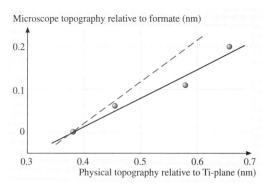

Fig. 23.26 The constant frequency-shift topography of the alkyl-substituted carboxylates as a function of their physical topography given in the model of Fig. 23.3 based on [23.123]

On the other hand, if the tip apex traced the contour of a molecule composed of hard-sphere atoms, the image topography would reproduce the physical topography in a one-to-one ratio, as shown by the broken line in Fig. 23.26. However, the slope of the fitted line was 0.7. A slope of less than unity is interpreted as the long-range nature of the tip–molecule force. The observable frequency shift reflects the sum of the forces between the tip apex and individual molecules. When the tip passes above a tall molecule embedded in short molecules, it is pulled up to compensate for the increased force originating from the tall molecule. Forces between the lifted tip and the short molecules are reduced due to the increased tip–surface distance. Feedback regulation pushes down the probe to restore the lost forces.

This picture predicts that microscope topography is sensitive to the lateral distribution of the molecules, and that was in fact the case. Two-dimensionally clustered acetates exhibited enhanced image height over an isolated acetate [23.121]. The tip–molecule force therefore remained nonzero at distances over the lateral separation of the carboxylates on this surface (0.59–0.65 nm). Chemical bond interactions cannot be important across such a wide tip–molecule gap, whereas atom-scale images of Si(111)(7×7) are interpreted with the fractional formation of tip–surface chemical bonds [23.24, 45, 49]. Instead, the attractive component of the van der Waals force is probable responsible for the observed molecule-dependent topography. The absence of the tip–surface chemical bond is reasonable on the carboxylate-covered surface terminated with stable C−H bonds.

The attractive component of the van der Waals force contains electrostatic terms caused by permanent-dipole/permanent-dipole coupling, permanent-dipole/induced-dipole coupling, and induced-dipole/induced-dipole coupling (dispersion force). The four carboxylates examined are equivalent in terms of their permanent electric dipole, because the alkyl groups are nonpolar. The image contrast of one carboxylate relative to another is thus ascribed to the dispersion force and/or the force created by the coupling between the permanent dipole on the tip and the induced dipole on the molecule. If we further assume that the Si tip used exhibits the smallest permanent dipole, the dispersion force remains dominant to create the NC-AFM topography dependent on the nonpolar groups of atoms. A numerical simulation based on this assumption [23.125] successfully reproduced the propiolate topography of Fig. 23.25d. A calculation that does not include quantum chemical treatment is expected to work, unless the tip approaches the surface too closely, or the molecule possesses a dangling bond.

In addition to the contribution of the dispersion force, the permanent dipole moment of molecules may perturb the microscope topography through electrostatic coupling with the tip. Its possible role was demonstrated by imaging a fluorine-substituted acetate. The strongly polarized C−F bonds were expected to perturb the electrostatic field over the molecule. The constant frequency-shift topography of acetate ($R = CH_3$) and trifluoroacetate ($R = CF_3$) was indeed sensitive to the fluorine substitution. The acetate was observed to be 0.05 nm higher than the trifluoroacetate [23.122], although the F atoms in the trifluoroacetate as well as the H atoms in the acetate were lifted by 0.46 nm from the surface plane, as illustrated in Fig. 23.24.

23.4.3 Perspectives

The experimental results summarized in this section prove the feasibility of using NC-AFM to identify individual molecules. A systematic study on the constant frequency-shift topography of carboxylates with $R = CH_3$, $C(CH_3)_3$, $C\equiv CH$, and CF_3 has revealed the mechanism behind the high-resolution imaging of the chemically stable molecules. The dispersion force is primarily responsible for the molecule-dependent topography. The permanent dipole moment of the imaged molecule, if it exists, perturbs the topography through the electrostatic coupling with the tip. A tiny calculation containing empirical force fields works when simulating the microscope topography.

These results make us optimistic about analyzing physical and chemical properties of nanoscale supramolecular assemblies constructed on a solid surface. If the accuracy of topographic measurement is developed by one more order of magnitude, which is not an unrealistic target, it may be possible to identify structural isomers, chiral isomers, and conformational isomers of a molecule. Kelvin probe force microscopy (KPFM), an extension of NC-AFM, provides a nanoscale analysis of molecular electronic properties [23.118, 119]. Force spectroscopy with chemically modified tips seems promising for the detection of a selected chemical force. Operation in a liquid atmosphere [23.131] is required for the observation of biochemical materials in their natural environment.

References

23.1 G. Binnig: Atomic force microscope, method for imaging surfaces with atomic resolution, US Patent 4724318 (1986)

23.2 G. Binnig, C.F. Quate, C. Gerber: Atomic force microscope, Phys. Rev. Lett. **56**, 930–933 (1986)

23.3 G. Binnig, H. Rohrer, C. Gerber, E. Weibel: Surface studies by scanning tunneling microscopy, Phys. Rev. Lett. **49**, 57–61 (1982)

23.4 G. Binnig, H. Rohrer: The scanning tunneling microscope, Sci. Am. **253**, 50–56 (1985)

23.5 G. Binnig, H. Rohrer: In touch with atoms, Rev. Mod. Phys. **71**, S320–S330 (1999)

23.6 C.J. Chen: *Introduction to Scanning Tunneling Microscopy* (Oxford Univ. Press, Oxford 1993)

23.7 H.-J. Güntherodt, R. Wiesendanger (Eds.): *Scanning Tunneling Microscopy I–III* (Springer, Berlin, Heidelberg 1991)

23.8 J.A. Stroscio, W.J. Kaiser (Eds.): *Scanning Tunneling Microscopy* (Academic, Boston 1993)

23.9 R. Wiesendanger: *Scanning Probe Microscopy and Spectroscopy: Methods and Applications* (Cambridge Univ. Press, Cambridge 1994)

23.10 S. Morita, R. Wiesendanger, E. Meyer (Eds.): *Noncontact Atomic Force Microscopy* (Springer, Berlin, Heidelberg 2002)

23.11 R. Garcia, R. Perez: Dynamic atomic force microscopy methods, Surf. Sci. Rep. **47**, 197–301 (2002)

23.12 F.J. Giessibl: Advances in atomic force microscopy, Rev. Mod. Phys. **75**, 949–983 (2003)

23.13 J. Israelachvili: *Intermolecular and Surface Forces*, 2nd edn. (Academic, London 1991)

23.14 L. Olsson, N. Lin, V. Yakimov, R. Erlandsson: A method for in situ characterization of tip shape in AC-mode atomic force microscopy using electrostatic interaction, J. Appl. Phys. **84**, 4060–4064 (1998)

23.15 S. Akamine, R.C. Barrett, C.F. Quate: Improved atomic force microscopy images using cantilevers with sharp tips, Appl. Phys. Lett. **57**, 316–318 (1990)

23.16 T.R. Albrecht, S. Akamine, T.E. Carver, C.F. Quate: Microfabrication of cantilever styli for the atomic force microscope, J. Vac. Sci. Technol. A **8**, 3386–3396 (1990)

23.17 M. Tortonese, R.C. Barrett, C. Quate: Atomic resolution with an atomic force microscope using piezoresistive detection, Appl. Phys. Lett. **62**, 834–836 (1993)

23.18 O. Wolter, T. Bayer, J. Greschner: Micromachined silicon sensors for scanning force microscopy, J. Vac. Sci. Technol. **9**, 1353–1357 (1991)

23.19 D. Sarid: *Scanning Force Microscopy*, 2nd edn. (Oxford Univ. Press, New York 1994)

23.20 F.J. Giessibl, B.M. Trafas: Piezoresistive cantilevers utilized for scanning tunneling and scanning force microscope in ultrahigh vacuum, Rev. Sci. Instrum. **65**, 1923–1929 (1994)

23.21 P. Güthner, U.C. Fischer, K. Dransfeld: Scanning near-field acoustic microscopy, Appl. Phys. B **48**, 89–92 (1989)

23.22 K. Karrai, R.D. Grober: Piezoelectric tip-sample distance control for near field optical microscopes, Appl. Phys. Lett. **66**, 1842–1844 (1995)

23.23 F.J. Giessibl: High-speed force sensor for force microscopy and profilometry utilizing a quartz tuning fork, Appl. Phys. Lett. **73**, 3956–3958 (1998)

23.24 F.J. Giessibl, S. Hembacher, H. Bielefeldt, J. Mannhart: Subatomic features on the silicon (111)-(7×7) surface observed by atomic force microscopy, Science **289**, 422–425 (2000)

23.25 F. Giessibl, C. Gerber, G. Binnig: A low-temperature atomic force/scanning tunneling microscope for ultrahigh vacuum, J. Vac. Sci. Technol. B **9**, 984–988 (1991)

23.26 F. Ohnesorge, G. Binnig: True atomic resolution by atomic force microscopy through repulsive and attractive forces, Science **260**, 1451–1456 (1993)

23.27 F.J. Giessibl, G. Binnig: True atomic resolution on KBr with a low-temperature atomic force microscope in ultrahigh vacuum, Ultramicroscopy **42–44**, 281–286 (1992)

23.28 S.P. Jarvis, H. Yamada, H. Tokumoto, J.B. Pethica: Direct mechanical measurement of interatomic potentials, Nature **384**, 247–249 (1996)

23.29 L. Howald, R. Lüthi, E. Meyer, P. Güthner, H.-J. Güntherodt: Scanning force microscopy on the Si(111)7×7 surface reconstruction, Z. Phys. B **93**, 267–268 (1994)

23.30 L. Howald, R. Lüthi, E. Meyer, H.-J. Güntherodt: Atomic-force microscopy on the Si(111)7×7 surface, Phys. Rev. B **51**, 5484–5487 (1995)

23.31 Y. Martin, C.C. Williams, H.K. Wickramasinghe: Atomic force microscope – force mapping and profiling on a sub 100-Å scale, J. Appl. Phys. **61**, 4723–4729 (1987)

23.32 T.R. Albrecht, P. Grütter, H.K. Horne, D. Rugar: Frequency modulation detection using high-Q cantilevers for enhanced force microscope sensitivity, J. Appl. Phys. **69**, 668–673 (1991)

23.33 F.J. Giessibl: Atomic resolution of the silicon (111)-(7×7) surface by atomic force microscopy, Science **267**, 68–71 (1995)

23.34 S. Kitamura, M. Iwatsuki: Observation of silicon surfaces using ultrahigh-vacuum noncontact atomic force microscopy, Jpn. J. Appl. Phys. **35**, 668–L671 (1995)

23.35 R. Erlandsson, L. Olsson, P. Mårtensson: Inequivalent atoms and imaging mechanisms in AC-mode atomic-force microscopy of Si(111)7×7, Phys. Rev. B **54**, R8309–R8312 (1996)

23.36 N. Burnham, R.J. Colton: Measuring the nanomechanical and surface forces of materials using an atomic force microscope, J. Vac. Sci. Technol. A **7**, 2906–2913 (1989)

23.37 D. Tabor, R.H.S. Winterton: Direct measurement of normal and related van der Waals forces, Proc. R. Soc. Lond. A **312**, 435 (1969)

23.38 F.J. Giessibl: Forces and frequency shifts in atomic resolution dynamic force microscopy, Phys. Rev. B **56**, 16011–16015 (1997)

23.39 G. Binnig, H. Rohrer, C. Gerber, E. Weibel: 7×7 reconstruction on Si(111) resolved in real space, Phys. Rev. Lett. **50**, 120–123 (1983)

23.40 H. Goldstein: *Classical Mechanics* (Addison Wesley, Reading 1980)

23.41 U. Dürig: Interaction sensing in dynamic force microscopy, New J. Phys. **2**, 5.1–5.12 (2000)

23.42 F.J. Giessibl: A direct method to calculate tip-sample forces from frequency shifts in frequency-modulation atomic force microscopy, Appl. Phys. Lett. **78**, 123–125 (2001)

23.43 U. Dürig, H.P. Steinauer, N. Blanc: Dynamic force microscopy by means of the phase-controlled oscillator method, J. Appl. Phys. **82**, 3641–3651 (1997)

23.44 F.J. Giessibl, H. Bielefeldt, S. Hembacher, J. Mannhart: Calculation of the optimal imaging parameters for frequency modulation atomic force microscopy, Appl. Surf. Sci. **140**, 352–357 (1999)

23.45 M.A. Lantz, H.J. Hug, R. Hoffmann, P.J.A. van Schendel, P. Kappenberger, S. Martin, A. Baratoff, H.-J. Güntherodt: Quantitative measurement of short-range chemical bonding forces, Science **291**, 2580–2583 (2001)

23.46 F.J. Giessibl, M. Herz, J. Mannhart: Friction traced to the single atom, Proc. Natl. Acad. Sci. USA **99**, 12006–12010 (2002)

23.47 N. Nakagiri, M. Suzuki, K. Oguchi, H. Sugimura: Site discrimination of adatoms in Si(111)-7×7 by noncontact atomic force microscopy, Surf. Sci. Lett. **373**, L329–L332 (1997)

23.48 Y. Sugawara, M. Ohta, H. Ueyama, S. Morita: Defect motion on an InP(110) surface observed with noncontact atomic force microscopy, Science **270**, 1646–1648 (1995)

23.49 T. Uchihashi, Y. Sugawara, T. Tsukamoto, M. Ohta, S. Morita: Role of a covalent bonding interaction in noncontact-mode atomic-force microscopy on Si(111)7×7, Phys. Rev. B **56**, 9834–9840 (1997)

23.50 K. Yokoyama, T. Ochi, A. Yoshimoto, Y. Sugawara, S. Morita: Atomic resolution imaging on Si(100)2×1 and Si(100)2×1-H surfaces using a non-contact atomic force microscope, Jpn. J. Appl. Phys. **39**, L113–L115 (2000)

23.51 Y. Sugawara, T. Minobe, S. Orisaka, T. Uchihashi, T. Tsukamoto, S. Morita: Non-contact AFM images measured on Si(111)$\sqrt{3}\times\sqrt{3}$-Ag and Ag(111) surfaces, Surf. Interface Anal. **27**, 456–461 (1999)

23.52 K. Yokoyama, T. Ochi, Y. Sugawara, S. Morita: Atomically resolved Ag imaging on Si(111)$\sqrt{3}\times\sqrt{3}$-Ag surface with noncontact atomic force microscope, Phys. Rev. Lett. **83**, 5023–5026 (1999)

23.53 M. Bammerlin, R. Lüthi, E. Meyer, A. Baratoff, J. Lü, M. Guggisberg, C. Gerber, L. Howald, H.-J. Güntherodt: True atomic resolution on the surface of an insulator via ultrahigh vacuum dynamic force microscopy, Probe Microsc. J. **1**, 3–7 (1997)

23.54 M. Bammerlin, R. Lüthi, E. Meyer, A. Baratoff, J. Lü, M. Guggisberg, C. Loppacher, C. Gerber, H.-J. Güntherodt: Dynamic SFM with true atomic resolution on alkali halide surfaces, Appl. Phys. A **66**, S293–S294 (1998)

23.55 R. Hoffmann, M.A. Lantz, H.J. Hug, P.J.A. van Schendel, P. Kappenberger, S. Martin, A. Baratoff, H.-J. Güntherodt: Atomic resolution imaging and force versus distance measurements on KBr(001) using low temperature scanning force microscopy, Appl. Surf. Sci. **188**, 238–244 (2002)

23.56 R. Bennewitz, A.S. Foster, L.N. Kantotovich, M. Bammerlin, C. Loppacher, S. Schär, M. Guggisberg, E. Meyer, A.L. Shluger: Atomically resolved edges and kinks of NaCl islands on Cu(111): Experiment and theory, Phys. Rev. B **62**, 2074–2084 (2000)

23.57 A.I. Livshits, A.L. Shluger, A.L. Rohl, A.S. Foster: Model of noncontact scanning force microscopy on ionic surfaces, Phys. Rev. **59**, 2436–2448 (1999)

23.58 R. Bennewitz, O. Pfeiffer, S. Schär, V. Barwich, E. Meyer, L.N. Kantorovich: Atomic corrugation in nc-AFM of alkali halides, Appl. Surf. Sci. **188**, 232–237 (2002)

23.59 R. Bennewitz, S. Schär, E. Gnecco, O. Pfeiffer, M. Bammerlin, E. Meyer: Atomic structure of alkali halide surfaces, Appl. Phys. A **78**, 837–841 (2004)

23.60 M. Gauthier, L. Kantrovich, M. Tsukada: Theory of energy dissipation into surface viblationsal. In: *Noncontact Atomic Force Microscopy*, ed. by S. Morita, R. Wiesendanger, E. Meyer (Springer, Berlin, Heidelberg 2002) pp. 371–394

23.61 H.J. Hug, A. Baratoff: Measurement of dissipation induced by tip–sample interactions. In: *Noncontact Atomic Force Microscopy*, ed. by S. Morita, R. Wiesendanger, E. Meyer (Springer, Berlin, Heidelberg 2002) pp. 395–431

23.62 R. Hoffmann, L.N. Kantorovich, A. Baratoff, H.J. Hug, H.-J. Güntherodt: Sublattice identification in scanning force microscopy on alkali halide surfaces, Phys. Rev. B **92**, 146103-1–146103-4 (2004)

23.63 C. Barth, M. Reichling: Resolving ions and vacancies at step edges on insulating surfaces, Surf. Sci. **470**, L99–L103 (2000)

23.64 R. Bennewitz, M. Reichling, E. Matthias: Force microscopy of cleaved and electron-irradiated CaF_2(111) surfaces in ultra-high vacuum, Surf. Sci. **387**, 69–77 (1997)

23.65 M. Reichling, C. Barth: Scanning force imaging of atomic size defects on the $CaF_2(111)$ surface, Phys. Rev. Lett. **83**, 768–771 (1999)

23.66 M. Reichling, M. Huisinga, S. Gogoll, C. Barth: Degradation of the $CaF_2(111)$ surface by air exposure, Surf. Sci. **439**, 181–190 (1999)

23.67 A. Klust, T. Ohta, A.A. Bostwick, Q. Yu, F.S. Ohuchi, M.A. Olmstead: Atomically resolved imaging of a CaF bilayer on Si(111): Subsurface atoms and the image contrast in scanning force microscopy, Phys. Rev. B **69**, 035405-1–035405-5 (2004)

23.68 C. Barth, A.S. Foster, M. Reichling, A.L. Shluger: Contrast formation in atomic resolution scanning force microscopy of $CaF_2(111)$: Experiment and theory, J. Phys. Condens. Matter **13**, 2061–2079 (2001)

23.69 A.S. Foster, C. Barth, A.L. Shulger, M. Reichling: Unambiguous interpretation of atomically resolved force microscopy images of an insulator, Phys. Rev. Lett. **86**, 2373–2376 (2001)

23.70 A.S. Foster, A.L. Rohl, A.L. Shluger: Imaging problems on insulators: What can be learnt from NC-AFM modeling on CaF_2?, Appl. Phys. A **72**, S31–S34 (2001)

23.71 F.J. Giessibl, M. Reichling: Investigating atomic details of the $CaF_2(111)$ surface with a qPlus sensor, Nanotechnology **16**, S118–S124 (2005)

23.72 A.S. Foster, C. Barth, A.L. Shluger, R.M. Nieminen, M. Reichling: Role of tip structure and surface relaxation in atomic resolution dynamic force microscopy: $CaF_2(111)$ as a reference surface, Phys. Rev. B **66**, 235417-1–235417-10 (2002)

23.73 C. Barth, M. Reichling: Imaging the atomic arrangements on the high-temperature reconstructed α-Al_2O_3 surface, Nature **414**, 54–57 (2001)

23.74 M. Reichling, C. Barth: Atomically resolution imaging on fluorides. In: *Noncontact Atomic Force Microscopy*, ed. by S. Morita, R. Wiesendanger, E. Meyer (Springer, Berlin, Heidelberg 2002) pp.109–123

23.75 K. Fukui, H. Ohnishi, Y. Iwasawa: Atom-resolved image of the $TiO_2(110)$ surface by noncontact atomic force microscopy, Phys. Rev. Lett. **79**, 4202–4205 (1997)

23.76 H. Raza, C.L. Pang, S.A. Haycock, G. Thornton: Non-contact atomic force microscopy imaging of $TiO_2(100)$ surfaces, Appl. Surf. Sci. **140**, 271–275 (1999)

23.77 C.L. Pang, H. Raza, S.A. Haycock, G. Thornton: Imaging reconstructed $TiO_2(100)$ surfaces with non-contact atomic force microscopy, Appl. Surf. Sci. **157**, 223–238 (2000)

23.78 M. Ashino, T. Uchihashi, K. Yokoyama, Y. Sugawara, S. Morita, M. Ishikawa: STM and atomic-resolution noncontact AFM of an oxygen-deficient $TiO_2(110)$ surface, Phys. Rev. B **61**, 13955–13959 (2000)

23.79 R.E. Tanner, A. Sasahara, Y. Liang, E.I. Altmann, H. Onishi: Formic acid adsorption on anatase $TiO_2(001)$-(1×4) thin films studied by NC-AFM and STM, J. Phys. Chem. B **106**, 8211–8222 (2002)

23.80 A. Sasahara, T.C. Droubay, S.A. Chambers, H. Uetsuka, H. Onishi: Topography of anatase TiO_2 film synthesized on $LaAlO_3(001)$, Nanotechnology **16**, S18–S21 (2005)

23.81 C.L. Pang, S.A. Haycock, H. Raza, P.J. Møller, G. Thornton: Structures of the 4×1 and 1×2 reconstructions of $SnO_2(110)$, Phys. Rev. B **62**, R7775–R7778 (2000)

23.82 H. Hosoi, K. Sueoka, K. Hayakawa, K. Mukasa: Atomic resolved imaging of cleaved NiO(100) surfaces by NC-AFM, Appl. Surf. Sci. **157**, 218–221 (2000)

23.83 W. Allers, S. Langkat, R. Wiesendanger: Dynamic low-temperature scanning force microscopy on nickel oxide (001), Appl. Phys. A **72**, S27–S30 (2001)

23.84 T. Kubo, H. Nozoye: Surface Structure of $SrTiO3(100)$-$(\sqrt{5}\times\sqrt{5})$-R $26.6°$, Phys. Rev. Lett. **86**, 1801–1804 (2001)

23.85 K. Fukui, Y. Namai, Y. Iwasawa: Imaging of surface oxygen atoms and their defect structures on $CeO_2(111)$ by noncontact atomic force microscopy, Appl. Surf. Sci. **188**, 252–256 (2002)

23.86 S. Suzuki, Y. Ohminami, T. Tsutsumi, M.M. Shoaib, M. Ichikawa, K. Asakura: The first observation of an atomic scale noncontact AFM image of $MoO_3(010)$, Chem. Lett. **32**, 1098–1099 (2003)

23.87 C. Barth, C.R. Henry: Atomic resolution imaging of the (001) surface of UHV cleaved MgO by dynamic scanning force microscopy, Phys. Rev. Lett. **91**, 196102-1–196102-4 (2003)

23.88 A.S. Foster, A.Y. Gal, J.M. Airaksinen, O.H. Pakarinen, Y.J. Lee, J.D. Gale, A.L. Shluger, R.M. Nieminen: Towards chemical identification in atomic-resolution noncontact AFM imaging with silicon tips, Phys. Rev. B **68**, 195420-1–195420-8 (2003)

23.89 A.S. Foster, A.Y. Gal, J.D. Gale, Y.J. Lee, R.M. Nieminen, A.L. Shluger: Interaction of silicon dangling bonds with insulating surfaces, Phys. Rev. Lett. **92**, 036101-1–036101-4 (2004)

23.90 T. Eguchi, Y. Hasegawa: High resolution atomic force microscopic imaging of the Si(111)-(7×7) surface: Contribution of short-range force to the images, Phys. Rev. Lett. **89**, 266105-1–266105-4 (2002)

23.91 T. Arai, M. Tomitori: A Si nanopillar grown on a Si tip by atomic force microscopy in ultrahigh vacuum for a high-quality scanning probe, Appl. Phys. Lett. **86**, 073110-1–073110-3 (2005)

23.92 K. Mukasa, H. Hasegawa, Y. Tazuke, K. Sueoka, M. Sasaki, K. Hayakawa: Exchange interaction between magnetic moments of ferromagnetic sample and tip: Possibility of atomic-resolution images of exchange interactions using exchange force microscopy, Jpn. J. Appl. Phys. **33**, 2692–2695 (1994)

23.93 H. Ness, F. Gautier: Theoretical study of the interaction between a magnetic nanotip and a magnetic surface, Phys. Rev. B **52**, 7352–7362 (1995)

23.94 K. Nakamura, H. Hasegawa, T. Oguchi, K. Sueoka, K. Hayakawa, K. Mukasa: First-principles calculation of the exchange interaction and the exchange force between magnetic Fe films, Phys. Rev. B **56**, 3218–3221 (1997)

23.95 A.S. Foster, A.L. Shluger: Spin-contrast in non-contact SFM on oxide surfaces: Theoretical modeling of NiO(001) surface, Surf. Sci. **490**, 211–219 (2001)

23.96 T. Oguchi, H. Momida: Electronic structure and magnetism of antiferromagnetic oxide surface – First-principles calculations, J. Surf. Sci. Soc. Jpn. **26**, 138–143 (2005)

23.97 H. Hosoi, M. Kimura, K. Sueoka, K. Hayakawa, K. Mukasa: Non-contact atomic force microscopy of an antiferromagnetic NiO(100) surface using a ferromagnetic tip, Appl. Phys. A **72**, S23–S26 (2001)

23.98 H. Hölscher, S.M. Langkat, A. Schwarz, R. Wiesendanger: Measurement of three-dimensional force fields with atomic resolution using dynamic force spectroscopy, Appl. Phys. Lett. **81**, 4428–4430 (2002)

23.99 S.M. Langkat, H. Hölscher, A. Schwarz, R. Wiesendanger: Determination of site specific interaction forces between an iron coated tip and the NiO(001) surface by force field spectroscopy, Surf. Sci. **527**, 12–20 (2003)

23.100 R. Hoffmann, M.A. Lantz, H.J. Hug, P.J.A. van Schendel, P. Kappenberger, S. Martin, A. Baratoff, H.-J. Güntherodt: Atomic resolution imaging and frequency versus distance measurement on NiO(001) using low-temperature scanning force microscopy, Phys. Rev. B **67**, 085402-1–085402-6 (2003)

23.101 H. Hosoi, K. Sueoka, K. Hayakawa, K. Mukasa: Atomically resolved imaging of a NiO(001) surface. In: *Noncontact Atomic Force Microscopy*, ed. by S. Morita, R. Wiesendanger, E. Meyer (Springer, Berlin, Heidelberg 2002) pp. 125–134

23.102 K. Sueoka, A. Subagyo, H. Hosoi, K. Mukasa: Magnetic imaging with scanning force microscopy, Nanotechnology **15**, S691–S698 (2004)

23.103 H. Hosoi, K. Sueoka, K. Mukasa: Investigations on the topographic asymmetry of non-contact atomic force microscopy images of NiO(001) surface observed with a ferromagnetic tip, Nanotechnology **15**, 505–509 (2004)

23.104 H. Momida, T. Oguchi: First-principles studies of antiferromagnetic MnO and NiO surfaces, J. Phys. Soc. Jpn. **72**, 588–593 (2003)

23.105 K. Kobayashi, H. Yamada, T. Horiuchi, K. Matsushige: Structures and electrical properties of fullerene thin films on Si(111)-7×7 surface investigated by noncontact atomic force microscopy, Jpn. J. Appl. Phys. **39**, 3821–3829 (2000)

23.106 R.M. Overney, E. Meyer, J. Frommer, D. Brodbeck, R. Lüthi, L. Howald, H.-J. Güntherodt, M. Fujihira, H. Takano, Y. Gotoh: Friction measurements on phase-separated thin films with amodified atomic force microscope, Nature **359**, 133–135 (1992)

23.107 D. Frisbie, L.F. Rozsnyai, A. Noy, M.S. Wrighton, C.M. Lieber: Functional group imaging by chemical force microscopy, Science **265**, 2071–2074 (1994)

23.108 E. Meyer, L. Howald, R.M. Overney, H. Heinzelmann, J. Frommer, H.-J. Güntherodt, T. Wagner, H. Schier, S. Roth: Molecular-resolution images of Langmuir–Blodgett films using atomic force microscopy, Nature **349**, 398–400 (1992)

23.109 K. Fukui, H. Onishi, Y. Iwasawa: Imaging of individual formate ions adsorbed on TiO_2(110) surface by non-contact atomic force microscopy, Chem. Phys. Lett. **280**, 296–301 (1997)

23.110 K. Kobayashi, H. Yamada, T. Horiuchi, K. Matsushige: Investigations of C_{60} molecules deposited on Si(111) by noncontact atomic force microscopy, Appl. Surf. Sci. **140**, 281–286 (1999)

23.111 T. Uchihashi, M. Tanigawa, M. Ashino, Y. Sugawara, K. Yokoyama, S. Morita, M. Ishikawa: Identification of B-form DNA in an ultrahigh vacuum by noncontact-mode atomic force microscopy, Langmuir **16**, 1349–1353 (2000)

23.112 Y. Maeda, T. Matsumoto, T. Kawai: Observation of single- and double-strand DNA using non-contact atomic force microscopy, Appl. Surf. Sci. **140**, 400–405 (1999)

23.113 T. Uchihashi, T. Ishida, M. Komiyama, M. Ashino, Y. Sugawara, W. Mizutani, K. Yokoyama, S. Morita, H. Tokumoto, M. Ishikawa: High-resolution imaging of organic monolayers using noncontact AFM, Appl. Surf. Sci. **157**, 244–250 (2000)

23.114 T. Fukuma, K. Kobayashi, T. Horiuchi, H. Yamada, K. Matsushige: Alkanethiol self-assembled monolayers on Au(111) surfaces investigated by non-contact AFM, Appl. Phys. A **72**, S109–S112 (2001)

23.115 B. Gotsmann, C. Schmidt, C. Seidel, H. Fuchs: Molecular resolution of an organic monolayer by dynamic AFM, Eur. Phys. J. B **4**, 267–268 (1998)

23.116 C. Loppacher, M. Bammerlin, M. Guggisberg, E. Meyer, H.-J. Güntherodt, R. Lüthi, R. Schlittler, J.K. Gimzewski: Forces with submolecular resolution between the probing tip and Cu-TBPP molecules on Cu(100) observed with a combined AFM/STM, Appl. Phys. A **72**, S105–S108 (2001)

23.117 L.M. Eng, M. Bammerlin, C. Loppacher, M. Guggisberg, R. Bennewitz, R. Lüthi, E. Meyer, H.-J. Güntherodt: Surface morphology, chemical contrast, and ferroelectric domains in TGS bulk single crystals differentiated with UHV non-contact force microscopy, Appl. Surf. Sci. **140**, 253–258 (1999)

23.118 S. Kitamura, K. Suzuki, M. Iwatsuki: High resolution imaging of contact potential difference using a novel ultrahigh vacuum non-contact atomic

23.119 H. Yamada, T. Fukuma, K. Umeda, K. Kobayashi, K. Matsushige: Local structures and electrical properties of organic molecular films investigated by non-contact atomic force microscopy, Appl. Surf. Sci. **188**, 391–398 (2000)

23.120 K. Fukui, Y. Iwasawa: Fluctuation of acetate ions in the (2×1)-acetate overlayer on TiO_2(110)-(1×1) observed by noncontact atomic force microscopy, Surf. Sci. **464**, L719–L726 (2000)

23.121 A. Sasahara, H. Uetsuka, H. Onishi: Singlemolecule analysis by non-contact atomic force microscopy, J. Phys. Chem. B **105**, 1–4 (2001)

23.122 A. Sasahara, H. Uetsuka, H. Onishi: NC-AFM topography of HCOO and CH_3COO molecules co-adsorbed on TiO_2(110), Appl. Phys. A **72**, S101–S103 (2001)

23.123 A. Sasahara, H. Uetsuka, H. Onishi: Image topography of alkyl-substituted carboxylates observed by noncontact atomic force microscopy, Surf. Sci. **481**, L437–L442 (2001)

23.124 A. Sasahara, H. Uetsuka, H. Onishi: Noncontact atomic force microscope topography dependent on permanent dipole of individual molecules, Phys. Rev. B **64**, 121406 (2001)

23.125 A. Sasahara, H. Uetsuka, T. Ishibashi, H. Onishi: A needle-like organic molecule imaged by non-contact atomic force microscopy, Appl. Surf. Sci. **188**, 265–271 (2002)

23.126 H. Onishi, A. Sasahara, H. Uetsuka, T. Ishibashi: Molecule-dependent topography determined by noncontact atomic force microscopy: Carboxylates on TiO_2(110), Appl. Surf. Sci. **188**, 257–264 (2002)

23.127 H. Onishi: Carboxylates adsorbed on TiO_2(110). In: *Chemistry of Nano-molecular Systems*, ed. by T. Nakamura (Springer, Berlin, Heidelberg 2002) pp. 75–89

23.128 S. Thevuthasan, G.S. Herman, Y.J. Kim, S.A. Chambers, C.H.F. Peden, Z. Wang, R.X. Ynzunza, E.D. Tober, J. Morais, C.S. Fadley: The structure of formate on TiO_2(110) by scanned-energy and scanned-angle photoelectron diffraction, Surf. Sci. **401**, 261–268 (1998)

23.129 H. Uetsuka, A. Sasahara, A. Yamakata, H. Onishi: Microscopic identification of a bimolecular reaction intermediate, J. Phys. Chem. B **106**, 11549–11552 (2002)

23.130 D.R. Lide: *Handbook of Chemistry and Physics*, 81st edn. (CRC, Boca Raton 2000)

23.131 K. Kobayashi, H. Yamada, K. Matsushige: Dynamic force microscopy using FM detection in various environments, Appl. Surf. Sci. **188**, 430–434 (2002)

24. Low-Temperature Scanning Probe Microscopy

Markus Morgenstern, Alexander Schwarz, Udo D. Schwarz

This chapter is dedicated to scanning probe microscopy (SPM) operated at cryogenic temperatures, where the more fundamental aspects of phenomena important in the field of nanotechnology can be investigated with high sensitivity under well-defined conditions. In general, scanning probe techniques allow the measurement of physical properties down to the nanometer scale. Some techniques, such as scanning tunneling microscopy and scanning force microscopy, even go down to the atomic scale. Various properties are accessible. Most importantly, one can image the arrangement of atoms on conducting surfaces by scanning tunneling microscopy and on insulating substrates by scanning force microscopy. However, the arrangement of electrons (scanning tunneling spectroscopy), the force interaction between different atoms (scanning force spectroscopy), magnetic domains (magnetic force microscopy), the local capacitance (scanning capacitance microscopy), the local temperature (scanning thermo microscopy), and local light-induced excitations (scanning near-field microscopy) can also be measured with high spatial resolution. In addition, some techniques even allow the manipulation of atomic configurations.

Probably the most important advantage of the low-temperature operation of scanning probe techniques is that they lead to a significantly better signal-to-noise ratio than measuring at room temperature. This is why many researchers work below 100 K. However, there are also physical reasons to use low-temperature equipment. For example, the manipulation of atoms or scanning tunneling spectroscopy with high energy resolution can only be realized at low temperatures. Moreover, some physical effects such as superconductivity or the Kondo effect are restricted to low temperatures. Here, we describe the design criteria of low-temperature scanning probe equipment and summarize some of the most spectacular results achieved since the invention of the method about 30 years ago. We first focus on the scanning tunneling microscope, giving examples of atomic manipulation and the analysis of electronic properties in different material arrangements. Afterwards, we describe results obtained by scanning force microscopy, showing atomic-scale imaging on insulators, as well as force spectroscopy analysis. Finally, the magnetic force microscope, which images domain patterns in ferromagnets and vortex patterns in superconductors, is discussed. Although this list is far from complete, we feel that it gives an adequate impression of the fascinating possibilities of low-temperature scanning probe instruments.

In this chapter low temperatures are defined as lower than about 100 K and are normally achieved by cooling with liquid nitrogen or liquid helium. Applications in which SPMs are operated close to 0 °C are not covered in this chapter.

24.1	**Microscope Operation at Low Temperatures**	664
	24.1.1 Drift	664
	24.1.2 Noise	665
	24.1.3 Stability	665
	24.1.4 Piezo Relaxation and Hysteresis ...	665
24.2	**Instrumentation**	666
	24.2.1 A Simple Design for a Variable-Temperature STM ...	666
	24.2.2 A Low-Temperature SFM Based on a Bath Cryostat	668
24.3	**Scanning Tunneling Microscopy and Spectroscopy**	669
	24.3.1 Atomic Manipulation	669
	24.3.2 Imaging Atomic Motion	671
	24.3.3 Detecting Light from Single Atoms and Molecules.	672
	24.3.4 High-Resolution Spectroscopy	673
	24.3.5 Imaging Electronic Wavefunctions	679
	24.3.6 Imaging Spin Polarization: Nanomagnetism	686

24.4 **Scanning Force Microscopy and Spectroscopy**................................. 688
    24.4.1 Atomic-Scale Imaging................. 689
    24.4.2 Force Spectroscopy 692
    24.4.3 Atomic Manipulation 695
    24.4.4 Electrostatic Force Microscopy 695
    24.4.5 Magnetic Force Microscopy 696
    24.4.6 Magnetic Exchange Force Microscopy 698

References 700

Nearly three decades ago, the first design of an experimental setup was presented where a sharp tip was systematically scanned over a sample surface in order to obtain local information on the tip–sample interaction down to the atomic scale. This original instrument used the tunneling current between a conducting tip and a conducting sample as a feedback signal and was thus named the *scanning tunneling microscope* [24.1]. Soon after this historic breakthrough, it became widely recognized that virtually any type of tip–sample interaction could be used to obtain local information on the sample by applying the same general principle, provided that the selected interaction was reasonably short-ranged. Thus, a whole variety of new methods has been introduced, which are denoted collectively as *scanning probe methods*. An overview is given, e.g., by *Wiesendanger* [24.2].

The various methods, especially the above mentioned scanning tunneling microscopy (STM) and scanning force microscopy (SFM) – which is often further classified into subdisciplines such as topography-reflecting atomic force microscopy (AFM), magnetic force microscopy (MFM) or electrostatic force microscopy (EFM) – have been established as standard methods for surface characterization on the nanometer scale. The reason is that they feature extremely high resolution (often down to the atomic scale for STM and AFM), despite a principally simple, compact, and comparatively inexpensive design.

A side-effect of the simple working principle and the compact design of many scanning probe microscopes (SPMs) is that they can be adapted to different environments such as air, all kinds of gaseous atmospheres, liquids or vacuum with reasonable effort.

Another advantage is their ability to work within a wide temperature range. A microscope operation at higher temperatures is chosen to study surface diffusion, surface reactivity, surface reconstructions that only manifest at elevated temperatures, high-temperature phase transitions, or to simulate conditions as they occur, e.g., in engines, catalytic converters or reactors. Ultimately, the upper limit for the operation of an SPM is determined by the stability of the sample, but thermal drift, which limits the ability to move the tip in a controlled manner over the sample, as well as the depolarization temperature of the piezoelectric positioning elements might further restrict successful measurements.

On the other hand, low-temperature (LT) application of SPMs is much more widespread than operation at high temperatures. Essentially five reasons make researchers adapt their experimental setups to low-temperature compatibility. These are: (1) the reduced thermal drift, (2) lower noise levels, (3) enhanced stability of tip and sample, (4) the reduction in piezo hysteresis/creep, and (5) probably the most obvious, the fact that many physical effects are restricted to low temperature. Reasons 1–4 only apply unconditionally if the whole microscope body is kept at low temperature (typically in or attached to a bath cryostat, see Sect. 24.2). Setups in which only the sample is cooled may show considerably less favorable operating characteristics. As a result of 1–4, ultrahigh resolution and long-term stability can be achieved on a level that significantly exceeds what can be accomplished at room temperature even under the most favorable circumstances. Typical examples of effect 5 are superconductivity [24.3] and the Kondo effect [24.4].

24.1 Microscope Operation at Low Temperatures

Nevertheless, before we devote ourselves to a short overview of experimental LT-SPM work, we will take a closer look at the specifics of microscope operation at low temperatures, including a discussion of the corresponding instrumentation.

24.1.1 Drift

Thermal drift originates from thermally activated movements of the individual atoms, which are reflected by the thermal expansion coefficient. At room temperature, typical values for solids are on the order of $(1-50) \times 10^{-6}\,\mathrm{K}^{-1}$. If the temperature could be kept precisely constant, any thermal drift would vanish, regardless of the absolute temperature of the system. The close coupling of the microscope to a large temperature bath that maintains a constant temperature ensures a significant reduction in thermal drift and allows for distortion-free long-term measurements. Microscopes that are efficiently attached to sufficiently large bath cryostats, therefore, show a one- to two-order-of-magnitude increase in thermal stability compared with nonstabilized setups operated at room temperature.

A second effect also helps suppress thermally induced drift of the probing tip relative to a specific location on the sample surface. The thermal expansion coefficients at liquid-helium temperatures are two or more orders of magnitude smaller than at room temperature. Consequently, the thermal drift during low-temperature operation decreases accordingly.

For some specific scanning probe methods, there may be additional ways in which a change in temperature can affect the quality of the data. In *frequency-modulation SFM* (FM-SFM), for example, the measurement principle relies on the accurate determination of the eigenfrequency of the cantilever, which is determined by its spring constant and its effective mass. However, the spring constant changes with temperature due to both thermal expansion (i. e., the resulting change in the cantilever dimensions) and the variation of the Young's modulus with temperature. Assuming drift rates of about $2\,\mathrm{mK/min}$, as is typical for room-temperature measurements, this effect might have a significant influence on the obtained data.

24.1.2 Noise

The theoretically achievable resolution in SPM often increases with decreasing temperature due to a decrease in thermally induced noise. An example is the thermal noise in SFM, which is proportional to the square root of the temperature [24.5, 6]. Lowering the temperature from $T=300\,\mathrm{K}$ to $T=10\,\mathrm{K}$ thus results in a reduction of the thermal frequency noise by more than a factor of five. Graphite, e.g., has been imaged with atomic resolution only at low temperatures due to its extremely low corrugation, which was below the room-temperature noise level [24.7, 8].

Another, even more striking, example is the spectroscopic resolution in *scanning tunneling spectroscopy* (STS). This depends linearly on the temperature [24.2] and is consequently reduced even more at LT than the thermal noise in AFM. This provides the opportunity to study structures or physical effects not accessible at room temperature such as spin and Landau levels in semiconductors [24.9].

Finally, it might be worth mentioning that the enhanced stiffness of most materials at low temperatures (increased Young's modulus) leads to a reduced coupling to external noise. Even though this effect is considered small [24.6], it should not be ignored.

24.1.3 Stability

There are two major stability issues that considerably improve at low temperature. First, low temperatures close to the temperature of liquid helium inhibit most of the thermally activated diffusion processes. As a consequence, the sample surfaces show a significantly increased long-term stability, since defect motion or adatom diffusion is massively suppressed. Most strikingly, even single xenon atoms deposited on suitable substrates can be successfully imaged [24.10, 11] or even manipulated [24.12]. In the same way, low temperatures also stabilize the atomic configuration at the tip end by preventing sudden jumps of the most loosely bound, foremost tip atom(s). Secondly, the large cryostat that usually surrounds the microscope acts as an effective cryo-pump. Thus samples can be kept clean for several weeks, which is a multiple of the corresponding time at room temperature (about $3-4\,\mathrm{h}$).

24.1.4 Piezo Relaxation and Hysteresis

The last important benefit from low-temperature operation of SPMs is that artifacts from the response of the piezoelectric scanners are substantially reduced. After applying a voltage ramp to one electrode of a piezoelectric scanner, its immediate initial deflection, l_0, is followed by a much slower relaxation, Δl, with a logarithmic time dependence. This effect, known as piezo relaxation or *creep*, diminishes substantially at low temperatures, typically by a factor of ten or more. As a consequence, piezo nonlinearities and piezo hysteresis decrease accordingly. Additional information is given by *Hug* et al. [24.13].

24.2 Instrumentation

The two main design criteria for all vacuum-based scanning probe microscope systems are: (1) to provide an efficient decoupling of the microscope from the vacuum system and other sources of external vibrations, and (2) to avoid most internal noise sources through the high mechanical rigidity of the microscope body itself. In vacuum systems designed for low-temperature applications, a significant degree of complexity is added, since, on the one hand, close thermal contact of the SPM and cryogen is necessary to ensure the (approximately) drift-free conditions described above, while, on the other hand, good vibration isolation (both from the outside world, as well as from the boiling or flowing cryogen) has to be maintained.

Plenty of microscope designs have been presented in the last 10–15 years, predominantly in the field of STM. Due to the variety of the different approaches, we will, somewhat arbitrarily, give two examples at different levels of complexity that might serve as illustrative model designs.

24.2.1 A Simple Design for a Variable-Temperature STM

A simple design for a variable-temperature STM system is presented in Fig. 24.1; similar systems are also offered by Omicron (Germany) or Jeol (Japan). It should give an impression of what the minimum requirements are, if samples are to be investigated successfully at low temperatures. It features a single ultrahigh-vacuum (UHV) chamber that houses the microscope in its center. The general idea to keep the setup simple is that only the sample is cooled, by means of a flow cryostat that ends in the small liquid-nitrogen (LN) reservoir.

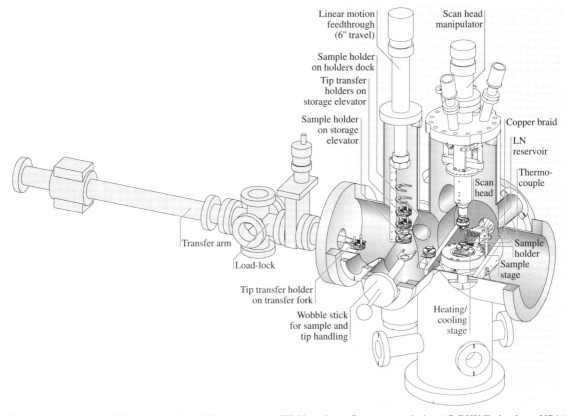

Fig. 24.1 One-chamber UHV system with variable-temperature STM based on a flow cryostat design. (© RHK Technology, USA)

This reservoir is connected to the sample holder with copper braids. The role of the copper braids is to attach the LN reservoir thermally to the sample located on the sample holder in an effective manner, while vibrations due to the flow of the cryogen should be blocked as much as possible. In this way, a sample temperature of about 100 K is reached. Alternatively, with liquid-helium operation, a base temperature of below 30 K can be achieved, while a heater that is integrated into the sample stage enables high-temperature operation up to 1000 K.

A typical experiment would run as follows. First, the sample is brought into the system by placing it in the so-called *load-lock*. This small part of the chamber can be separated from the rest of the system by a valve, so that the main part of the system can remain under vacuum at all times (i.e., even if the load-lock is opened to introduce the sample). After vacuum is reestablished, the sample is transferred to the main chamber using the transfer arm. A linear-motion feedthrough enables the storage of sample holders or, alternatively, specialized holders that carry replacement tips for the STM. Extending the transfer arm further, the sample can be placed on the sample stage and subsequently cooled down to the desired temperature. The scan head, which carries the STM tip, is then lowered with the scan-head manipulator onto the sample holder (see Fig. 24.2). The special design of the scan head (see [24.14] for details) allows not only flexible positioning of the tip on any desired location on the sample surface but also compensates to a certain degree for the thermal drift that inevitably occurs in such a design due to temperature gradients.

In fact, thermal drift is often much more prominent in LT-SPM designs, where only the sample is cooled, than in room-temperature designs. Therefore, to benefit fully from the high-stability conditions described in the introduction, it is mandatory to keep the whole microscope at the exact same temperature. This is mostly realized by using bath cryostats, which add a certain degree of complexity.

Fig. 24.2 Photograph of the STM located inside the system sketched in Fig. 24.1. After the scan head has been lowered onto the sample holder, it is fully decoupled from the scan head manipulator and can be moved laterally using the three piezo legs on which it stands. (© RHK Technology, USA)

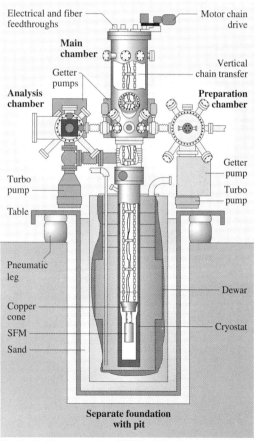

Fig. 24.3 Three-chamber UHV and bath cryostat system for scanning force microscopy, front view

24.2.2 A Low-Temperature SFM Based on a Bath Cryostat

As an example of an LT-SPM setup based on a bath cryostat, let us take a closer look at the LT-SFM system sketched in Fig. 24.3, which has been used to acquire the images on graphite, xenon, NiO, and InAs presented in Sect. 24.4. The force microscope is built into a UHV system that comprises three vacuum chambers: one for cantilever and sample preparation, which also serves as a transfer chamber, one for analysis purposes, and a main chamber that houses the microscope. A specially designed vertical transfer mechanism based on a double chain allows the lowering of the microscope into a UHV-compatible bath cryostat attached underneath the main chamber. To damp the system, it is mounted on a table carried by pneumatic damping legs, which, in turn, stand on a separate foundation to decouple it from building vibrations. The cryostat and dewar are separated from the rest of the UHV system by a bellow. In addition, the dewar is surrounded by sand for acoustic isolation.

In this design, tip and sample are exchanged at room temperature in the main chamber. After the transfer into the cryostat, the SFM can be cooled by either liquid nitrogen or liquid helium, reaching temperatures down to 10 K. An all-fiber interferometer as the detection mechanism for the cantilever deflection ensures high resolution, while simultaneously allowing the construction of a comparatively small, rigid, and symmetric microscope.

Figure 24.4 highlights the layout of the SFM body itself. Along with the careful choice of materials, the symmetric design eliminates most of the problems with drift inside the microscope encountered when cooling or warming it up. The microscope body has an overall cylindrical shape with a height of 13 cm and a diameter of 6 cm and exact mirror symmetry along the cantilever axis. The main body is made of a single block of macor, a machinable glass ceramic, which ensures a rigid and stable design. For most of the metallic parts titanium was used, which has a temperature coefficient similar to macor. The controlled but stable accomplishment of movements, such as coarse approach and lateral posi-

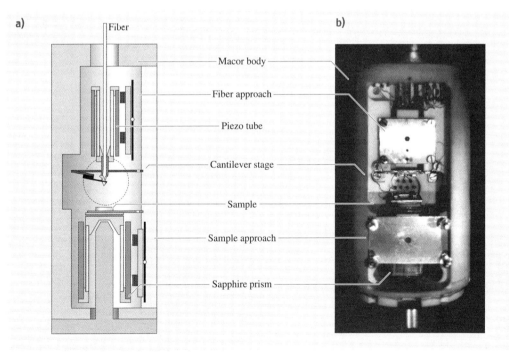

Fig. 24.4a,b The scanning force microscope incorporated into the system presented in Fig. 24.3. (**a**) Section along plane of symmetry. (**b**) Photo from the front

tioning in other microscope designs, is a difficult task at low temperatures. The present design uses a special type of piezo motor that moves a sapphire prism (see the *fiber approach* and the *sample approach* labels in Fig. 24.4); it is described in detail in [24.15]. More information regarding this design is given in [24.16].

24.3 Scanning Tunneling Microscopy and Spectroscopy

In this section, we review some of the most important results achieved by LT-STM. After summarizing the results, placing emphasis on the necessity for LT equipment, we turn to the details of the different experiments and the physical meaning of the results obtained.

As described in Sect. 24.1, the LT equipment has basically three advantages for scanning tunneling microscopy (STM) and spectroscopy (STS): First, the instruments are much more stable with respect to thermal drift and coupling to external noise, allowing the establishment of new functionalities of the instrument. In particular, the LT-STM has been used to move atoms on a surface [24.12], cut molecules into pieces [24.17], reform bonds [24.18], charge individual atoms [24.19], and, consequently, establish new structures on the nanometer scale. Also, the detection of light resulting from tunneling into a particular molecule [24.20, 21], the visualization of thermally induced atomic movements [24.22], and the detection of hysteresis curves of individual atoms [24.23] require LT instrumentation.

Second, the spectroscopic resolution in STS depends linearly on temperature and is, therefore, considerably reduced at LT. This provides the opportunity to study physical effects inaccessible at room temperature. Examples are the resolution of spin and Landau levels in semiconductors [24.9], or the investigation of lifetime-broadening effects on the nanometer scale [24.24]. Also the imaging of distinct electronic wavefunctions in real space requires LT-STM [24.25]. More recently, vibrational levels, spin-flip excitations, and phonons have been detected with high spatial resolution at LT using the additional inelastic tunneling channel [24.26–28].

Third, many physical effects, in particular, effects guided by electronic correlations, are restricted to low temperature. Typical examples are superconductivity [24.3], the Kondo effect [24.4], and many of the electron phases found in semiconductors [24.29]. Here, LT-STM provides the possibility to study electronic effects on a local scale, and intensive work has been done in this field, the most elaborate with respect to high-temperature superconductivity [24.30–32].

24.3.1 Atomic Manipulation

Although manipulation of surfaces on the atomic scale can be achieved at room temperature [24.33, 34], only the use of LT-STM allows the placement of individual atoms at desired atomic positions [24.35]. The main reason is that rotation, diffusion or charge transfer of entities could be excited at higher temperature, making the intentionally produced configurations unstable.

The usual technique to manipulate atoms is to increase the current above a certain atom, which reduces the tip–atom distance, then to move the tip with the atom to a desired position, and finally to reduce the current again in order to decouple the atom and tip. The first demonstration of this technique was performed by *Eigler* and *Schweizer* [24.12], who used Xe atoms on a Ni(110) surface to write the three letters "IBM" (their employer) on the atomic scale (Fig. 24.5a). Nowadays, many laboratories are able to move different kinds of atoms and molecules on different surfaces with high precision. An example featuring CO molecules on Cu(110) is shown in Fig. 24.5b–g. Even more complex structures than the "2000", such as cascades of CO molecules that by mutual repulsive interaction mimic different kinds of logic gates, have been assembled and their functionality tested [24.36]. Although these devices are slow and restricted to low temperature, they nicely demonstrate the high degree of control achieved on the atomic scale.

The basic modes of controlled motion of atoms and molecules by the tip are pushing, pulling, and sliding. The selection of the particular mode depends on the tunneling current, i.e., the distance between tip and molecule, as well as on the particular molecule–substrate combination [24.37]. It has been shown experimentally that the potential landscape for the adsorbate movement is modified by the presence of the tip [24.38, 39] and that excitations induced by the tunneling current can trigger atomic or molecular motion [24.40, 41]. Other sources of motion are the electric field between tip and molecule or electromigration caused by the high current density [24.35]. The required lateral tip force for atomic motion has been

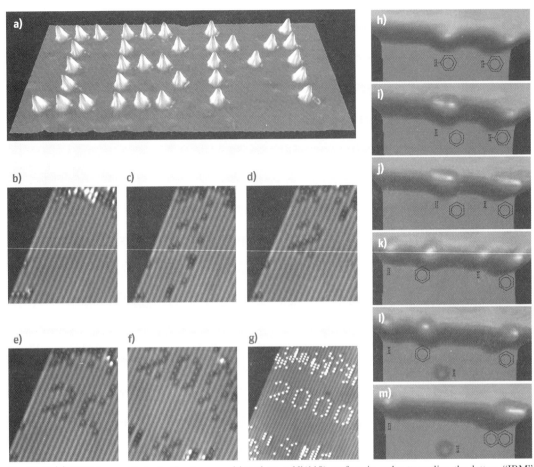

Fig. 24.5 (a) STM image of single Xe atoms positioned on a Ni(110) surface in order to realize the letters "IBM" on the atomic scale (© D. Eigler, Almaden); (b–f) STM images recorded after different positioning processes of CO molecules on a Cu(110) surface; (g) final artwork greeting the new millennium on the atomic scale ((b–g) © G. Meyer, Zürich). (h–m) Synthesis of biphenyl from two iodobenzene molecules on Cu(111): First, iodine is abstracted from both molecules (i,j); then the iodine between the two phenyl groups is removed from the step (k), and finally one of the phenyls is slid along the Cu step (l) until it reacts with the other phenyl (m); the line drawings symbolize the actual status of the molecules ((h–m) © S. W. Hla and K. H. Rieder, Berlin)

measured for typical adsorbate–substrate combinations to be ≈ 0.1 nN [24.42]. Other types of manipulation on the atomic scale are feasible. Some of them require a selective inelastic tunneling into vibrational or rotational modes of the molecules [24.43]. This leads to controlled desorption [24.44], diffusion [24.45], molecular rotation [24.46, 47], conformational change [24.48] or even controlled pick-up of molecules by the tip [24.18]. Dissociation can be achieved by voltage pulses [24.17] inducing local heating, even if the pulse is applied at distances of 100 nm away from the molecule [24.49]. Also, association of individual molecules [24.18, 50–52] can require voltage pulses in order to overcome local energy barriers. The process of controlled bond formation can even be used for doping of single C_{60} molecules by up to four potassium atoms [24.53]. As an example of controlled manipulation, Fig. 24.5h–m shows the production of biphenyl from two iodobenzene molecules [24.54]. The iodine is abstracted by voltage pulses (Fig. 24.5i,j), then the iodine is moved to the terrace by the pulling mode (Fig. 24.5k,l), and finally the two phenyl parts are slid along the step edge until they are close enough to react (Fig. 24.5m). The chemical identification of the product is not deduced straightforwardly and partly requires detailed vibrational STM spectroscopy (see below and [24.55]).

Finally, also the charge state of a single atom or molecule can be manipulated, tested, and read out. A Au atom has been switched reversibly between two charge states using an insulating thin film as the substrate [24.19]. In addition, the carrier capture rate of a single impurity level within the bandgap of a semiconductor has been quantified [24.56], and the point conductance of a single atom has been measured and turned out to be a reproducible quantity [24.57]. These promising results might trigger a novel electronic field of manipulation of matter on the atomic scale, which is tightly related to the currently very popular field of molecular electronics.

24.3.2 Imaging Atomic Motion

Since individual manipulation processes last seconds to minutes, they probably cannot be used to manufacture large and repetitive structures. A possibility to construct such structures is self-assembled growth [24.58]. This partly relies on the temperature dependence of different diffusion processes on the surface. Detailed knowledge of the diffusion parameters is required, which can be deduced from sequences of STM images measured at temperatures close to the onset of the process [24.59]. Since many diffusion processes have their onset at LT, LT are partly required [24.22]. Consecutive images of so-called hexa-*tert*-butyl-decacyclene (HtBDC) molecules on Cu(110) recorded at $T = 194$ K are shown in Fig. 24.6a–c [24.60]. As indicated by the arrows, the positions of the molecules change with time, implying diffusion. Diffusion parameters are obtained from Arrhenius plots of the determined hopping rate h, as shown in Fig. 24.6d. Of course, one must make sure that the diffusion process is not influenced by the presence of the tip, since it is known from manipulation experiments that the presence of the tip can move a molecule. However, particularly at low tunneling voltages, these conditions can be fulfilled.

Besides the determination of diffusion parameters, studies of the diffusion of individual molecules showed the importance of mutual interactions in diffusion, which can lead to concerted motion of several molecules [24.22], directional motion where smaller molecules carry larger ones [24.61] or, very interestingly, the influence of quantum tunneling [24.62]. The latter is deduced from the Arrhenius plot of hopping rates of H and D on Cu(001), as shown in Fig. 24.6e.

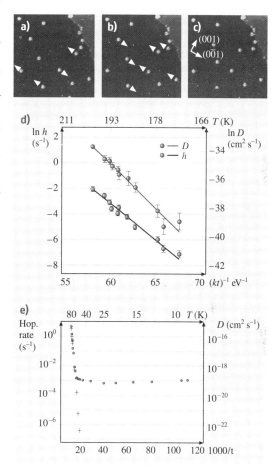

Fig. 24.6 (a–c) Consecutive STM images of hexa-*tert*-butyl decacyclene molecules on Cu(110) imaged at $T = 194$ K; *arrows* indicate the direction of motion of the molecules between two images. **(d)** Arrhenius plot of the hopping rate h determined from images such as **(a–c)** as a function of inverse temperature (*grey symbols*); the *brown symbols* show the corresponding diffusion constant D; *lines* are fit results revealing an energy barrier of 570 meV for molecular diffusion ((**a–d**) © M. Schunack and F. Besenbacher, Aarhus). **(e)** Arrhenius plot for D (*crosses*) and H (*circles*) on Cu(001). The constant hopping rate of H below 65 K indicates a nonthermal diffusion process, probably tunneling (© W. Ho, Irvine)

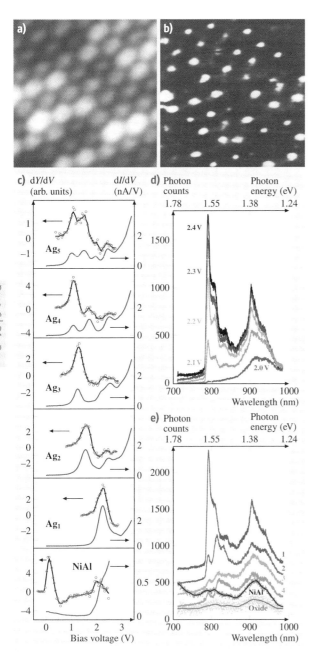

Fig. 24.7 (a) STM image of C_{60} molecules on Au(110) imaged at $T = 50$ K. (b) STM-induced photon intensity map of the same area; all photons from 1.5 to 2.8 eV contribute to the image, tunneling voltage $V = -2.8$ V (© R. Berndt, Kiel (a,b)). (c) Photon yield spectroscopy $dY/dV(V)$ obtained above Ag chains (Ag_n) of different length consisting of n atoms. For comparison, the differential conductivity $dI/dV(V)$ is also shown. The Ag chains are deposited on NiAl(110). The photon yield Y is integrated over the spectral range from 750 to 775 nm. (d) Photon yield spectra $Y(E)$ measured at different tip voltages as indicated. The tip is positioned above a ZnEtiol molecule deposited on Al_2O_3/NiAl(110). Note that the *peaks* in $Y(E)$ do not shift with applied tip voltage; (e) $Y(E)$ spectra determined at different positions above the ZnEtiol molecule, $V = 2.35$ V, $I = 0.5$ nA. ((c-e) © W. Ho, Irvine) ◀

found for vertical Sn displacements within a Sn adsorbate layer on Si(111) [24.63].

Other diffusion processes investigated by LT-STM include the movement of surface vacancies [24.64] or bulk interstitials close to the surface [24.65], the Brownian motion of vacancy islands [24.66] as well as laser-induced diffusion distinct from thermally excited diffusion [24.67].

24.3.3 Detecting Light from Single Atoms and Molecules

It had already been realized in 1988 that STM experiments are accompanied by light emission [24.68]. The fact that molecular resolution in the light intensity was achieved at LT (Fig. 24.7a,b) [24.20] raised the hope of performing quasi-optical experiments on the molecular scale. Meanwhile, it is clear that the basic emission process observed on metals is the decay of a local plasmon induced in the area around the tip by inelastic tunneling processes [24.69, 70]. Thus, the molecular resolution is basically a change in the plasmon environment, largely given by the increased height of the tip with respect to the surface above the molecule [24.71]. However, the electron can, in principle, also decay via single-particle excitations. Indeed, signatures of single-particle levels have been observed for a Na monolayer on Cu(111) [24.72] as well as for Ag adatom chains on NiAl(110) [24.21]. As shown in Fig. 24.7c, the peaks of differential photon yield dY/dV as a function of applied bias V are at identical voltages to the peaks in dI/dV intensity. This is evidence that the density of states of the Ag adsorbates is responsible for the radiative decay. Photon emission spectra displaying much

The hopping rate of H levels off at about 65 K, while the hopping rate of the heavier D atom goes down to nearly zero, as expected from thermally induced hopping. Quantum tunneling has surprisingly also been

more details could be detected by depositing the adsorbates of interest on a thin insulating film [24.73, 74]. Figure 24.7d shows spectra of ZnEtiol deposited on a 0.5 nm-thick Al_2O_3 layer on NiAl(110). Importantly, the peaks within the light spectra do not shift with applied voltage, ruling out that they are due to a plasmon mode induced by the tip. As shown in Fig. 24.7e, the photon spectra show distinct variations by changing the position within the molecule, demonstrating that atomically resolved maps of the excitation probability can be measured by STM.

Meanwhile, external laser light has also been coupled to the tunneling contact between the STM tip and a molecule deposited on an insulating film. A magnesium porphine molecule positioned below the tip could be charged reversibly either by increasing the voltage of the tip or by increasing the photon energy of the laser. This indicates selective absorption of light energy by the molecule leading to population of a novel charge level by tunneling electrons [24.75], a result that raises the hope that STM can probe photochemistry on the atomic scale. STM-induced light has also been detected from semiconductors [24.76], including heterostructures [24.77]. This light is again caused by single-particle relaxation of injected electrons, but without contrast on the atomic scale.

24.3.4 High-Resolution Spectroscopy

One of the most important modes of LT-STM is STS, which detects the differential conductivity dI/dV as a function of the applied voltage V and the position (x, y). The dI/dV signal is basically proportional to the local density of states (LDOS) of the sample, the sum over squared single-particle wavefunctions Ψ_i [24.2]

$$\frac{dI}{dV}(V, x, y) \propto LDOS(E, x, y)$$
$$= \sum_{\Delta E} |\Psi_i(E, x, y)|^2 , \quad (24.1)$$

where ΔE is the energy resolution of the experiment. In simple terms, each state corresponds to a tunneling channel, if it is located between the Fermi levels (E_F) of the tip and the sample. Thus, all states located in this energy interval contribute to I, while $dI/dV(V)$ detects only the states at the energy E corresponding to V. The local intensity of each channel depends further on the LDOS of the state at the corresponding surface position and its decay length into vacuum. For s-like tip states, *Tersoff* and *Hamann* have shown

that it is simply proportional to the LDOS at the position of the tip [24.78]. Therefore, as long as the decay length is spatially constant, one measures the LDOS at the surface (24.1). Note that the contributing states are not only surface states, but also bulk states. However, surface states usually dominate if present. *Chen* has shown that higher orbital tip states lead to the so-called derivation rule [24.79]: p_z-type tip states detect $d(LDOS)/dz$, d_{z^2}-states detect $d^2(LDOS)/dz^2$, and so on. As long as the decay into vacuum is exponential and spatially constant, this leads only to an additional, spatially constant factor in dI/dV. Thus, it is still the LDOS that is measured (24.1). The requirement of a spatially constant decay is usually fulfilled on larger length scales, but not on the atomic scale [24.79]. There, states located close to the atoms show a stronger decay into vacuum than the less localized states in the interstitial region. This effect can lead to STS corrugations that are larger than the real LDOS corrugations [24.80].

The voltage dependence of dI/dV is sensitive to a changing decay length with V, which increases with V. This influence can be reduced at higher V by displaying $dI/dV/(I/V)$ [24.81]. Additionally,

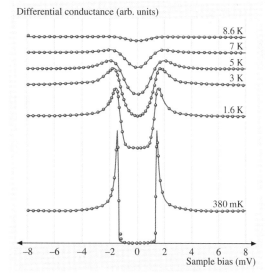

Fig. 24.8 Differential conductivity curve $dI/dV(V)$ measured on a Au surface by a Nb tip (*circles*). Different temperatures are indicated; the *lines* are fits according to the superconducting gap of Nb folded with the temperature-broadened Fermi distribution of the Au (© S.H. Pan, Houston)

Fig. 24.9 (a,b) Spatially averaged $dI/dV(V)$ curves of Ag(111) and Cu(111); both surfaces exhibit a surface state with parabolic dispersion, starting at -65 and -430 meV, respectively. The *lines* are drawn to determine the energetic width of the onset of these surface bands ((**a,b**) © R. Berndt, Kiel); (**c**) dI/dV intensity as a function of position away from a step edge of Cu(111) measured at the voltages ($E - E_F$), as indicated (*points*); the *lines* are fits assuming standing electron waves with a phase coherence length L_Φ as marked; (**d**) resulting phase coherence time as a function of energy for Ag(111) and Cu(111). *Inset* shows the same data on a double-logarithmic scale, evidencing the E^{-2} dependence (*line*) ((**c,d**) © H. Brune, Lausanne) ▶

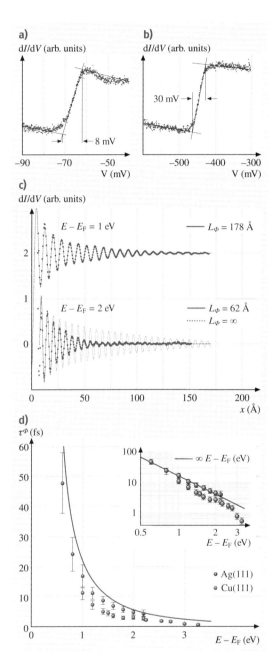

$dI/dV(V)$ curves might be influenced by possible structures in the DOS of the tip, which also contributes to the number of tunneling channels [24.82]. However, these structures can usually be identified, and only tips free of characteristic DOS structures are used for quantitative experiments.

Importantly, the energy resolution ΔE is largely determined by temperature. It is defined as the smallest energy distance of two δ-peaks in the LDOS that can still be resolved as two individual peaks in $dI/dV(V)$ curves and is $\Delta E = 3.3 k_B T$ [24.2]. The temperature dependence is nicely demonstrated in Fig. 24.8, where the tunneling gap of the superconductor Nb is measured at different temperatures [24.83]. The peaks at the rim of the gap get wider at temperatures well below the critical temperature of the superconductor ($T_c = 9.2$ K).

Lifetime Broadening

Besides ΔE, intrinsic properties of the sample lead to a broadening of spectroscopic features. Basically, the finite lifetime of the electron or hole in the corresponding state broadens its energetic width. Any kind of interaction such as electron–electron interaction can be responsible. Lifetime broadening has usually been measured by photoemission spectroscopy (PES), but it turned out that lifetimes of surface states on noble-metal surfaces determined by STS (Fig. 24.9a,b) are up to a factor of three larger than those measured by PES [24.84]. The reason is probably that defects broaden the PES spectrum. Defects are unavoidable in a spatially integrating technique such as PES, thus STS has the advantage of choosing a particularly clean area for lifetime measurements. The STS results can be successfully compared with theory, highlighting the dominating influence of intraband transitions for the surface-state lifetime on Au(111) and Cu(111), at least close to the onset of the surface band [24.24].

With respect to band electrons, the analysis of the width of the band onset on $dI/dV(V)$ curves has the disadvantage of being restricted to the onset energy. Another method circumvents this problem by measuring the decay of standing electron waves scattered from a step edge as a function of energy [24.85]. Figure 24.9c,d shows the resulting oscillating dI/dV signal measured for two different energies. To deduce the coherence length L_Φ, which is inversely proportional to the lifetime τ_Φ, one has to consider that the finite energy resolution ΔE in the experiment also leads to a decay of the standing wave away from the step edge. The dotted fit line using $L_\Phi = \infty$ indicates this effect and, more importantly, shows a discrepancy from the measured curve. Only including a finite coherence length of 6.2 nm results in good agreement, which in turn determines L_Φ and thus τ_Φ, as displayed in Fig. 24.9c. The found $1/E^2$ dependence of τ_Φ points to a dominating influence of electron–electron interactions at higher energies in the surface band.

Landau and Spin Levels

Moreover, the increased energy resolution at LT allows the resolution of electronic states that are not resolvable at room temperature (RT); for example, Landau and spin quantization appearing in a magnetic field B have been probed on InAs(110) [24.9, 86]. The corresponding quantization energies are given by $E_\text{Landau} = \hbar eB/m_\text{eff}$ and $E_\text{spin} = g\mu B$. Thus InAs is a good choice, since it exhibits a low effective mass $m_\text{eff}/m_\text{e} = 0.023$ and a high g-factor of 14 in the bulk conduction band. The values in metals are $m_\text{eff}/m_\text{e} \approx 1$ and $g \approx 2$, resulting in energy splittings of only 1.25 and 1.2 meV at $B = 10$ T. This is obviously lower than the typical lifetime broadenings discussed in the previous section and also close to $\Delta E = 1.1$ meV achievable at $T = 4$ K.

Fortunately, the electron density in doped semiconductors is much lower, and thus the lifetime increases significantly. Figure 24.10a shows a set of spectroscopy curves obtained on InAs(110) in different magnetic fields [24.9]. Above E_F, oscillations with increasing intensity and energy distance are observed. They show the separation expected from Landau quantization. In turn, they can be used to deduce m_eff from the peak separation (Fig. 24.10b). An increase of m_eff with increasing E has been found, as expected from theory. Also, at high fields, spin quantization is observed (Fig. 24.10c). It is larger than expected from the bare g-factor due to contributions from exchange enhancement [24.87].

Atomic Energy Levels

Another opportunity at LT is to study electronic states and resonances of single adatoms. A complicated resonance is the Kondo resonance described below.

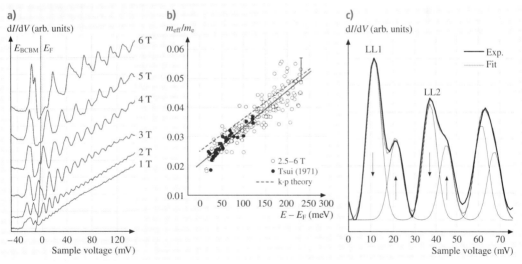

Fig. 24.10 (a) dI/dV curves of n-InAs(110) at different magnetic fields, as indicated; E_BCBM marks the bulk conduction band minimum; oscillations above E_BCBM are caused by Landau quantization; the *double peaks* at $B = 6$ T are caused by spin quantization. (b) Effective-mass data deduced from the distance of adjacent Landau peaks ΔE according to $\Delta E = \hbar eB/m_\text{eff}$ (*open symbols*); *filled symbols* are data from planar tunnel junctions (Tsui), the *solid line* is a mean-square fit of the data and the *dashed line* is the expected effective mass of InAs according to $k \cdot p$ theory. (c) Magnification of a dI/dV curve at $B = 6$ T, exhibiting spin splitting; the Gaussian curves marked by *arrows* are the fitted spin levels

A simpler resonance is a surface state bound at the adatom potential. It appears as a spatially localized peak below the onset of the extended surface state (Fig. 24.9a) [24.88, 89]. A similar resonance caused by a mixing of bulk states of the NiAl(110) substrate with atomic Au levels has been used to detect exchange splitting in Au dimers as a function of interatomic distance [24.90]. Single magnetic adatoms on the same surface also exhibit a double-peak resonance, but here due to the influence of spin-split d-levels of the adsorbate [24.91]. Atomic and molecular states decoupled from the substrate have finally been observed, if the atoms or molecules are deposited on an insulating thin film [24.19, 51].

Vibrational Levels

As discussed with respect to light emission in STM, inelastic tunneling processes contribute to the tunneling current. The coupling of electronic states to vibrational levels is one source of inelastic tunneling [24.26]. It provides additional channels contributing to $dI/dV(V)$ with final states at energies different from V. The final energy is simply shifted by the energy of the vibrational level. If only discrete vibrational energy levels couple to a smooth electronic DOS, one expects a peak in d^2I/dV^2 at the vibrational energy. This situation appears for molecules on noble-metal surfaces. As usual, the isotope effect can be used to verify the vibrational origin of the peak. First indications of vibrational levels have been found for H_2O and D_2O on TiO_2 [24.92], and completely convincing work has been performed for C_2H_2 and C_2D_2 on Cu(001) [24.26] (Fig. 24.11a). The technique has been used to identify individual molecules on the surface by their characteristic vibrational levels [24.55]. Moreover, the orientation of complexes with respect to the surface can be determined to a certain extent, since the vibrational excitation depends on the position of the tunneling current within the molecule. Finally, the excitation of certain molecular levels can induce such corresponding motions as hopping [24.45], rotation [24.47] (Fig. 24.11b–e) or desorption [24.44], leading to additional possibilities for manipulation on the atomic scale.

In turn, the manipulation efficiency as a function of applied voltage can be used to identify vibrational energies within the molecule, even if they are not detectable directly by d^2I/dV^2 spectroscopy [24.93]. Multiple vibronic excitations are found by positioning the molecule on an insulating film, leading to the observation of equidistant peaks in $d^2I/dV^2(V)$ [24.94].

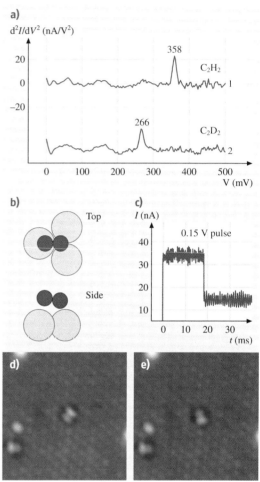

Fig. 24.11 (a) d^2I/dV^2 curves taken above a C_2H_2 and a C_2D_2 molecule on Cu(100); the *peaks* correspond to the C–H or C–D stretch-mode energy of the molecule, respectively. (b) Sketch of O_2 molecule on Pt(111). (c) Tunneling current above an O_2 molecule on Pt(111) during a voltage pulse of 0.15 V; the jump in current indicates rotation of the molecule. (d,e) STM images of an O_2 molecule on Pt(111) ($V = 0.05$ V), prior to and after rotation induced by a voltage pulse to 0.15 V ((a–e) © W. Ho, Irvine)

Other Inelastic Excitations

The tunneling current can not only couple to vibrational modes of molecules, but also to other degrees of freedom. It has been shown that phonon modes can be observed in carbon nanotubes [24.95, 96], on graphite [24.97], and on metal surfaces [24.28]. One finds distinct dependencies of excitation probability on the position of the STM tip with respect to the investigated structure.

First indications for the $d^2I/dV^2(V)$-based detection of extended magnons [24.98] and plasmons [24.97] have also been published.

The inelastic tunneling current has, moreover, been used to study single spin-flip excitations in magnetic field for atoms and atomic assemblies deposited on a thin insulator. The excitation probability was high enough to observe the spin flip even as a step in dI/dV instead of as a peak in d^2I/dV^2. Figure 24.12a shows the dI/dV curves recorded above a single Mn atom on Al_2O_3/NiAl(110) at different B fields. The linear shift of the step with B field is obvious, and the step voltage can be fitted by $eV = g\mu_B B$ with μ_B being the Bohr magneton and a reasonable g-factor of $g \approx 2$ [24.27]. Figure 24.12b shows the dI/dV spectra obtained on a Mn dimer embedded in CuN/Cu(100). A step is already visible at $B = 0$ T, splitting into three steps at higher field. This result can be explained straightforwardly, as sketched in the inset, by a singlet–triplet transition of the combined two spins (coupled by an exchange energy of $J \approx 6$ meV). Investigating longer chains revealed an even–odd asymmetry, i.e., chains consisting of $2, 4, 6, \ldots$ atoms exhibit a singlet–triplet transition, while chains of $1, 3, 5, \ldots$ atoms exhibit a transition from $S = 5/2$ to $S = 3/2$. This indicates antiferromagnetic coupling within the chain [24.99]. Figure 24.12c shows spectra of a single Fe atom embedded within CuN. The spectrum reveals several steps already at $B = 0$ T, showing that different spin orientations S_z must exhibit different energies due to magnetic anisotropy. In order to determine the anisotropy, the step energies and intensities were measured at different magnetic fields applied in three different directions. Amazingly, the results could be fitted completely by a single model with an out-of-plane anisotropy of $D = -1.55$ meV and an in-plane anisotropy of $E = 0.31$ meV. Therefore, one has to assume five different spin states of the Fe being mixtures of the five S_z states of a total Fe spin of $|S| = 2$. The excellent fit is shown for energy and intensity of a particular B-field direction in Fig. 24.12d,e [24.100].

The different experiments of inelastic tunneling demonstrate that details of atomic excitations in a solid environment can be probed by LT-STM, even if they are not of primary electronic origin. This might be a highly productive method in the near future. A com-plementary novel approach to inelastic effects might be the recently developed radiofrequency STM [24.101], which could give access to low-energy excitations, such as GHz spin-wave modes in nanostructures, which are not resolvable by d^2I/dV^2 at LT.

Kondo Resonance

A rather intricate interaction effect is the Kondo effect. It results from a second-order scattering process between itinerate states and a localized state [24.102]. The two states exchange some degree of freedom back and forth, leading to a divergence of the scattering probability at the Fermi level of the itinerate state. Due to the divergence, the effect strongly modifies sample properties. For example, it leads to an unexpected increase in resistance with decreasing temperature for metals containing magnetic impurities [24.4]. Here, the exchanged degree of freedom is the spin. A spectroscopic signature of the Kondo effect is a narrow peak in the DOS at the Fermi level, continuously disappearing above a characteristic temperature (the Kondo temperature). STS provides the opportunity to study this effect on the local scale [24.103, 104].

Figure 24.13a–d shows an example of Co clusters deposited on a carbon nanotube [24.105]. While only a small dip at the Fermi level, probably caused by curvature influences on the π-orbitals, is observed without Co (Fig. 24.13b) [24.106], a strong peak is found around a Co cluster deposited on top of the tube (Co cluster is marked in Fig. 24.13a). The peak is slightly shifted with respect to $V = 0$ mV due to the so-called Fano resonance [24.107], which results from interference of the tunneling processes into the localized Co level and the itinerant nanotube levels. The resonance disappears within several nanometers of the cluster, as shown in Fig. 24.13d.

The Kondo effect has also been detected for different magnetic atoms deposited on noble-metal surfaces [24.103, 104]. There, it disappears at about 1 nm from the magnetic impurity, and the effect of the Fano resonance is more pronounced, contributing to dips in $dI/dV(V)$ curves instead of peaks. Detailed investigations show that the d-level occupation of the adsorbate [24.108] as well as the surface charge density [24.109, 110] matter for the Kondo temperature. Exchange interaction between adsorbates tunable by their mutual distance can be used to tune the Kondo temperature [24.111] or even to destroy the Kondo resonance completely [24.112]. Meanwhile, magnetic molecules have also been shown to exhibit Kondo resonances. This increases the tunability of the Kondo

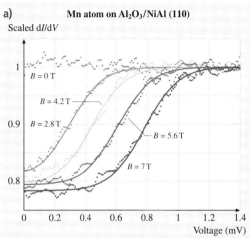
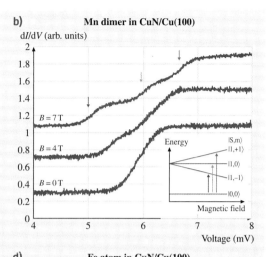
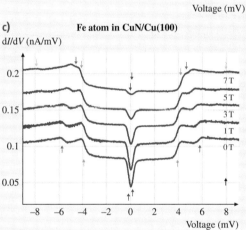
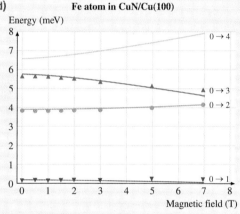
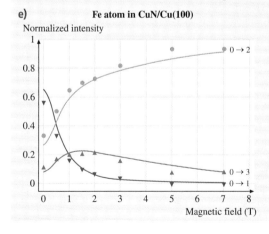

Fig. 24.12 (a) dI/dV curves taken above a single Mn atom deposited on Al$_2$O$_3$/NiAl(110) at different magnetic fields as indicated; (b) dI/dV curves taken above a Mn dimer deposited onto CuN/Cu(100) at different magnetic fields as indicated; *inset* shows the three possible spin-flip transitions between singlet and triplet; (c) dI/dV curves taken above a single Fe atom deposited onto CuN/Cu(100) at different magnetic fields as indicated; transitions are marked by *arrows*; (d,e) energy and intensity of the steps in dI/dV measured with magnetic field along the direction of the N rows of the CuN surface (*symbols*) in comparison with calculated results (*lines*) (© A. Heinrich, C. F. Hirjibehedin, Almaden) ◀

effect, e.g., by the selection of adequate ligands surrounding the localized spins [24.113, 114], by distant association of other molecules [24.115] or by conformational changes within the molecule [24.116].

A fascinating experiment has been performed by *Manoharan* et al. [24.117], who used manipulation to form an elliptic cage for the surface states of Cu(111) (Fig. 24.13e, bottom). This cage was constructed to have a quantized level at E_F. Then, a cobalt atom was placed in one focus of the elliptic cage, producing a Kondo resonance. Surprisingly, the same resonance reappeared in the opposite focus, but not away from the focus (Fig. 24.13e, top). This shows amazingly that complex local effects such as the Kondo resonance can be wave-guided to remote points.

24.3.5 Imaging Electronic Wavefunctions

Bloch Waves

Since STS measures the sum of squared wavefunctions (24.1), it is an obvious task to measure the local appearance of the most simple wavefunctions in solids, namely Bloch waves. The atomically periodic part of the Bloch wave is always measured if atomic resolution is achieved (inset of Fig. 24.15a). However, the long-range wavy part requires the presence of scatterers. The electron wave impinges on the scatterer and is reflected, leading to self-interference. In other words, the phase of the Bloch wave becomes fixed by the scatterer.

Such self-interference patterns were first found on graphite(0001) [24.118] and later on noble-metal surfaces, where adsorbates or step edges scatter the surface states (Fig. 24.14a) [24.25]. Fourier transforms of the real-space images reveal the k-space distribution of the corresponding states [24.119], which may include additional contributions besides the surface state [24.120]. Using particular geometries such as so-called quantum corrals, the Bloch waves can be confined (Fig. 24.14b). Depending on the geometry of the corral, the result state looks rather complex, but it can usually be reproduced

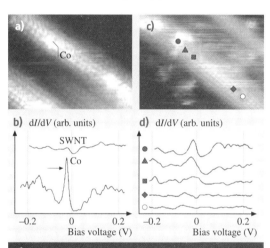

Fig. 24.13 (a) STM image of a Co cluster on a single-wall carbon nanotube (SWNT). (b) dI/dV curves taken directly above the Co cluster (Co) and far away from the Co cluster (SWNT); the *arrow* marks the Kondo peak. (c) STM image of another Co cluster on a SWNT with *symbols* marking the positions where the dI/dV curves displayed in (d) are taken. (d) dI/dV curves taken at the positions marked in (c) ((a–d) © C. Lieber, Cambridge). (e) *Lower part:* STM image of a quantum corral of elliptic shape made from Co atoms on Cu(111); one Co atom is placed at one of the foci of the ellipse. *Upper part:* map of the strength of the Kondo signal in the corral; note that there is also a Kondo signal at the focus that is not covered by a Co atom ((e) © D. Eigler, Almaden) ◀

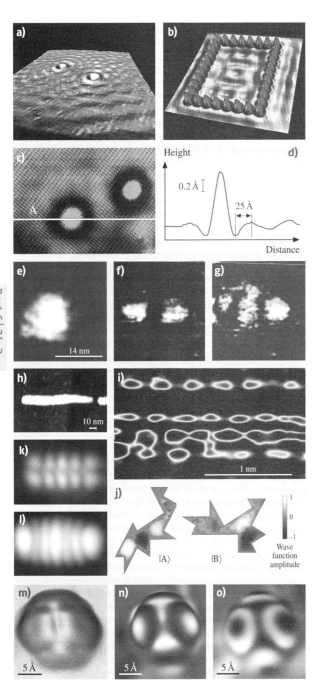

Fig. 24.14 (a) Low-voltage STM image of Cu(111) including two defect atoms; the waves are electronic Bloch waves scattered at the defects; (b) low-voltage STM image of a rectangular quantum corral made from single atoms on Cu(111); the pattern inside the corral is the confined state of the corral close to E_F; (© D. Eigler, Almaden (**a,b**)); (**c**) STM image of GaAs(110) around a Si donor, $V = -2.5$ V; the line scan along A, shown in (**d**), exhibits an additional oscillation around the donor caused by a standing Bloch wave; the grid-like pattern corresponds to the atomic corrugation of the Bloch wave (© H. van Kempen, Nijmegen (**c,d**)); (**e–g**) dI/dV images of a self-assembled InAs quantum dot deposited on GaAs(100) and measured at different V ((**e**) 1.05 V, (**f**) 1.39 V, (**g**) 1.60 V). The *images* show the squared wavefunctions confined within the quantum dot, which exhibit zero, one, and two nodal lines with increasing energy. (**h**) STM image of a short-cut carbon nanotube; (**i**) greyscale plot of the dI/dV intensity inside the short-cut nanotube as a function of position (x-axis) and tunneling voltage (y-axis); four wavy patterns of different wavelength are visible in the voltage range from -0.1 to 0.15 V (© C. Dekker, Delft (**h,i**)); (**j**) two reconstructed wavefunctions confined in so-called isospectral corrals made of CO molecules on Cu(111). Note that $\Psi(x)$ instead of $|\Psi(x)|^2$ is displayed, exhibiting positive and negative values. This is possible since the transplantation matrix transforming one isospectral wavefunction into another is known (© H. Manoharan, Stanford (**j**)); (**k,l**) STM images of a pentacene molecule deposited on NaCl/Cu(100) and measured with a pentacene molecule at the apex of the tip at $V = -2.5$ V ((**k**), HOMO = highest occupied molecular orbital) and $V = 2.5$ V ((**l**), LUMO = lowest unoccupied molecular orbital) (© J. Repp, Regensburg (**k,l**)) ; (**m**) STM image of a C_{60} molecule deposited on Ag(100), $V = 2.0$ V; (**n,o**) dI/dV images of the same molecule at $V = 0.4$ V (**n**), 1.6 V (**n**) ((**m–o**)) © M. Crommie, Berkeley ◄

Meanwhile, Bloch waves in semiconductors scattered at charged dopants (Fig. 24.14c,d) [24.122], Bloch states confined in semiconducting or organic quantum dots (Fig. 24.14e–g) [24.123–125], and quantum wells [24.126], as well as Bloch waves confined in short-cut carbon nanotubes (Fig. 24.14h,i) [24.127, 128] have been visualized. In special nanostructures, it was even possible to extract the phase of the wavefunction by using the mathematically known transformation matrices of so-called isospectral structures, i.e., geometrically different structures exhibiting exactly the same spatially averaged density of states. The resulting wavefunctions $\Psi(x)$ are shown in Fig. 24.14j [24.129] by simple calculations involving single-particle states only [24.121].

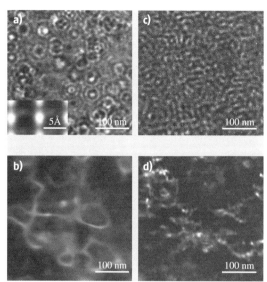

Fig. 24.15 (a) dI/dV image of InAs(110) at $V = 50\,\mathrm{mV}$, $B = 0\,\mathrm{T}$; circular wave patterns corresponding to standing Bloch waves around each sulphur donor are visible; *inset* shows a magnification revealing the atomically periodic part of the Bloch wave. (b) Same as (a), but at $B = 6\,\mathrm{T}$; the stripe structures are drift states. (c) dI/dV image of a 2-D electron system on InAs(110) induced by the deposition of Fe, $B = 0\,\mathrm{T}$. (d) Same as (c) but at $B = 6\,\mathrm{T}$; note that the contrast in (a) is increased by a factor of ten with respect to (b–d)

More localized structures, where a Bloch wave description is not appropriate, have been imaged, too. Examples are the highest occupied molecular orbital (HOMO) and lowest unoccupied molecular orbital (LUMO) of pentacene molecules deposited on NaCl/Cu(100) (Fig. 24.14k,l) [24.51], the different molecular states of C_{60} on Ag(110) (Fig. 24.3m–o) [24.130], the anisotropic states of Mn acceptors in a semiconducting host [24.131, 132], and the hybridized states developing within short monoatomic Au chains, which develop particular states at the end of the chains [24.133, 134]. Using pairs of remote Mn acceptors, even symmetric and antisymmetric pair wavefunctions have been imaged in real space [24.135].

The central requirements for a detailed imaging of wavefunctions are LT for an appropriate energetic distinction of an individual state, adequate decoupling of the state from the substrate in order to decrease lifetime-induced broadening effects, and, partly, the selection of a system with an increased Bohr radius in order to increase the spatial extension of details above the lateral resolution of STM, thereby improving, e.g., the visibility of bonding and antibonding pair states within a dimer [24.135].

Wavefunctions in Disordered Systems

More complex wavefunctions result from interactions. A nice playground to study such interactions is doped semiconductors. The reduced electron density with respect to metals increases the importance of electron interactions with potential disorder and other electrons. Applying a magnetic field quenches the kinetic energy, thus enhancing the importance of interactions. A dramatic effect can be observed on InAs(110), where three-dimensional (3-D) bulk states are measured. While the usual scattering states around individual dopants are observed at $B = 0\,\mathrm{T}$ (Fig. 24.15a) [24.136], stripe structures are found at high magnetic field (Fig. 24.15b) [24.137]. They run along equipotential lines of the disorder potential. This can be understood by recalling that the electron tries to move in a cyclotron circle, which becomes a cycloid path along an equipotential line within an inhomogeneous electrostatic potential [24.138].

The same effect has been found in two-dimensional (2-D) electron systems (2-DES) of InAs at the same large B-field (Fig. 24.15d) [24.139]. However the scattering states at $B = 0\,\mathrm{T}$ are much more complex in 2-D (Fig. 24.15c) [24.140]. The reason is the tendency of a 2-DES to exhibit closed scattering paths [24.141]. Consequently, the self-interference does not result from scattering at individual scatterers, but from complicated self-interference paths involving many scatterers. Nevertheless, the wavefunction pattern can be reproduced by including these effects within the calculations.

Reducing the dimensionality to one dimension (1-D) leads again to complicated self-interference patterns due to the interaction of the electrons with several impurities [24.142, 143]. For InAs, they can be reproduced by single-particle calculations. However, experiments imaging self-interference patterns close to the end of a C-nanotube are interpreted as indications of spin-charge separation, a genuine property of 1-D electrons not feasible within the single-particle description [24.144].

Charge Density Waves, Jahn–Teller Distortion

Another interaction modifying the LDOS is the electron–phonon interaction. Phonons scatter electrons between different Fermi points. If the wavevectors connecting Fermi points exhibit a preferential orien-

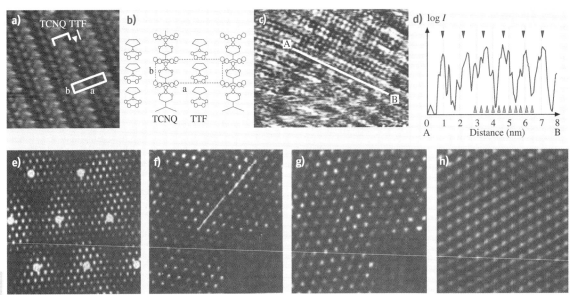

Fig. 24.16 (a) STM image of the *ab*-plane of the organic quasi-1-D conductor tetrathiafulvalene tetracyanoquinodimethane (TTF-TCNQ), $T = 300$ K; while the TCNQ chains are conducting, the TTF chains are insulating. (b) Stick-and-ball model of the *ab*-plane of TTF-TCNQ. (c) STM image taken at $T = 61$ K; the additional modulation due to the Peierls transition is visible in the profile along line AB shown in (d); the *brown triangles* mark the atomic periodicity and the *black triangles* the expected CDW periodicity ((a–d) © M. Kageshima, Kanagawa). (e–h) Low-voltage STM images of the two-dimensional CDW system 1 T-TaS$_2$ at $T = 242$ K (e), 298 K (f), 349 K (g), and 357 K (h). A long-range, hexagonal modulation is visible besides the atomic spots; its periodicity is highlighted by *large white dots* in (e); the additional modulation obviously weakens with increasing T, but is still apparent in (f) and (g), as evidenced in the lower-magnification images in the *insets* ((e–h) © C. Lieber, Cambridge)

tation, a so-called Peierls instability occurs [24.145]. The corresponding phonon energy goes to zero, the atoms are slightly displaced with the periodicity of the corresponding wavevector, and a charge density wave (CDW) with the same periodicity appears. Essentially, the CDW increases the overlap of the electronic states with the phonon by phase-fixing with respect to the atomic lattice. The Peierls transition naturally occurs in one-dimensional (1-D) systems, where only two Fermi points are present and hence preferential orientation is pathological. It can also occur in 2-D systems if large parts of the Fermi line run in parallel.

STS studies of CDWs are numerous (e.g., [24.146, 147]). Examples of a 1-D CDW on a quasi-1-D bulk material and of a 2-D CDW are shown in Fig. 24.16a–d and Fig. 24.16e–h, respectively [24.148, 149]. In contrast to usual scattering states, where LDOS corrugations are only found close to the scatterer, the corrugations of CDWs are continuous across the surface. Heating the substrate toward the transition temperature leads to a melting of the CDW lattice, as shown in Fig. 24.16f–h.

CDWs have also been found on monolayers of adsorbates such as a monolayer of Pb on Ge(111) [24.150]. These authors performed a nice temperature-dependent study revealing that the CDW is nucleated by scattering states around defects, as one might expect [24.151]. Some of the transitions have been interpreted as more complex Mott–Hubbard transitions caused primarily by electron–electron interactions [24.152]. One-dimensional systems have also been prepared on surfaces showing Peierls transitions [24.153, 154]. Finally, the energy gap occurring at the transition has been studied by measuring $dI/dV(V)$ curves [24.155].

A more local crystallographic distortion due to electron–lattice interactions is the Jahn–Teller effect. Here, symmetry breaking by elastic deformation can lead to the lifting of degeneracies close to the Fermi level. This results in an energy gain due to the lowering of the energy of the occupied levels. By tuning the Fermi level of an adsorbate layer to a degeneracy via doping, such a Jahn–Teller deformation has been induced on a surface and visualized by STM [24.156].

Superconductors

An intriguing effect resulting from electron–phonon interaction is superconductivity. Here, the attractive part of the electron–phonon interaction leads to the coupling of electronic states with opposite wavevector and mostly opposite spin [24.157]. Since the resulting Cooper pairs are bosons, they can condense at LT, forming a coherent many-particle phase, which can carry current without resistance. Interestingly, defect scattering does not influence the condensate if the coupling along the Fermi surface is homogeneous (s-wave superconductor). The reason is that the symmetry of the scattering of the two components of a Cooper pair effectively leads to a scattering from one Cooper pair state to another without affecting the condensate. This is different if the scatterer is magnetic, since the different spin components of the pair are scattered differently, leading to an effective pair breaking, which is visible as a single-particle excitation within the superconducting gap. On a local scale, this effect was first demonstrated by putting Mn, Gd, and Ag atoms on a Nb(110) surface [24.158]. While the nonmagnetic Ag does not modify the gap shown in Fig. 24.17a, it is modified in an asymmetric fashion close to Mn or Gd adsorbates, as shown in Fig. 24.17b. The asymmetry of the additional intensity is caused by the breaking of the particle–hole symmetry due to the exchange interaction between the localized Mn state and the itinerate Nb states.

Another important local effect is caused by the relatively large coherence length of the condensate. At a material interface, the condensate wavefunction cannot stop abruptly, but overlaps into the surrounding material (proximity effect). Consequently, a superconducting gap can be measured in areas of nonsuperconducting material. Several studies have shown this effect on the local scale using metals and doped semiconductors as surrounding materials [24.159, 160].

While the classical type I superconductors are ideal diamagnets, the so-called type II superconductors can contain magnetic flux. The flux forms vortices, each containing one flux quantum. These vortices are accompanied by the disappearance of the superconducting gap and, therefore, can be probed by STS [24.161]. LDOS maps measured inside the gap lead to bright features in the area of the vortex core. Importantly, the length scale of these features is different from the length scale of the magnetic flux due to the difference between the London penetration depth and the electronic coherence length. Thus, STS probes a different property of the vortex than the usual magnetic imaging techniques (see Sect. 24.4.4). Surprisingly, first measurements of the vortices on $NbSe_2$ revealed vortices shaped as a sixfold star [24.162] (Fig. 24.17c). With increasing voltage inside the gap, the orientation of the star rotates by 30° (Fig. 24.17d,e). The shape of these stars could finally be reproduced by theory, assuming an anisotropic pairing of electrons in the superconductor (Fig. 24.17f–h) [24.163]. Additionally, bound states inside the vortex core, which result from confinement by the surrounding superconducting material, are found [24.162]. Further experiments investigated the arrangement of the vortex lattice, including transitions between hexagonal and quadratic lattices [24.164], the influence of pinning centers [24.165], and the vortex motion induced by current [24.166].

A central topic is still the understanding of high-temperature superconductors (HTCS). An almost accepted property of HTCS is their d-wave pairing symmetry, which is partly combined with other contributions [24.167]. The corresponding k-dependent gap (where k is the reciprocal lattice vector) can be measured indirectly by STS using a Fourier transformation of the $LDOS(x, y)$ determined at different energies [24.168]. This shows that LDOS modulations in HTCS are dominated by simple self-interference patterns of the Bloch-like quasiparticles [24.169]. However, scattering can also lead to pair breaking (in contrast to s-wave superconductors), since the Cooper-pair density vanishes in certain directions. Indeed, scattering states (bound states in the gap) around nonmagnetic Zn impurities have been observed in $Bi_2Sr_2CaCu_2O_{8+\delta}$ (BSCCO) (Fig. 24.17i,j) [24.170]. They reveal a d-like symmetry, but not the one expected from simple Cooper-pair scattering. Other effects such as magnetic polarization in the environment probably have to be taken into account [24.171]. An interesting topic is the importance of inhomogeneities in HTCS materials. Evidence for inhomogeneities has indeed been found in underdoped materials, where puddles of the superconducting phase identified by the coherence peaks around the gap are shown to be embedded in nonsuperconducting areas [24.30].

In addition, temperature-dependent measurements of the gap size development at each spatial position exhibit a percolation-type behavior above T_c [24.32]. This stresses the importance of inhomogeneities, but the observed percolation temperature being higher than T_c shows that T_c is not caused by percolation of superconducting puddles only. On the other hand, it

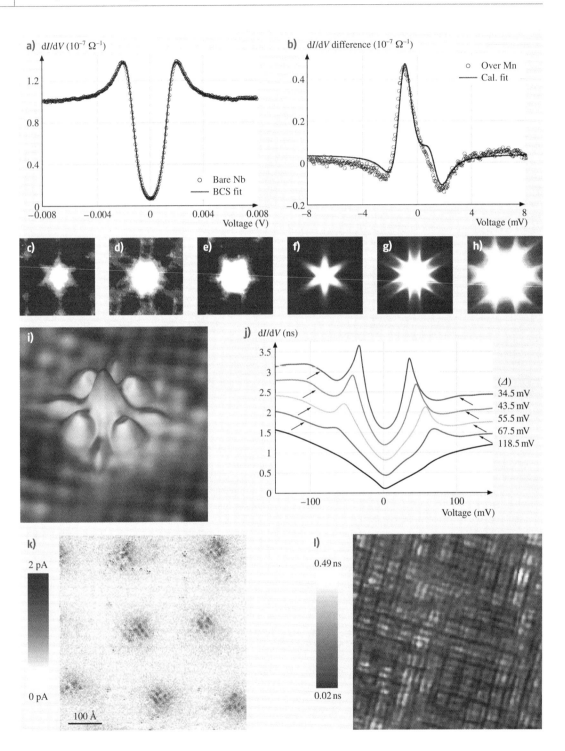

Fig. 24.17 (a) dI/dV curve of Nb(110) at $T = 3.8$ K (*symbols*) in comparison with a BCS fit of the superconducting gap of Nb (*line*). (b) Difference between the dI/dV curve taken directly above a Mn atom on Nb(110) and the dI/dV curve taken above clean Nb(110) (*symbols*) in comparison with a fit using the Bogulubov–de Gennes equations (*line*) (© D. Eigler, Almaden (**a,b**)). (**c–e**) dI/dV images of a vortex core in the type II superconductor 2H-NbSe$_2$ at 0 mV (**c**), 0.24 mV (**d**), and 0.48 mV (**e**) ((**c–e**) © H. F. Hess). (**f–h**) Corresponding calculated LDOS images within the Eilenberger framework ((**f–h**) © K. Machida, Okayama). (**i**) Overlap of an STM image at $V = -100$ mV (background 2-D image) and a dI/dV image at $V = 0$ mV (overlapped 3-D image) of optimally doped Bi$_2$Sr$_2$CaCu$_2$O$_{8+\delta}$ containing 0.6% Zn impurities. The STM image shows the atomic structure of the cleavage plane, while the dI/dV image shows a bound state within the superconducting gap, which is located around a single Zn impurity. The fourfold symmetry of the bound state reflects the d-like symmetry of the superconducting pairing function; (**j**) dI/dV spectra of Bi$_2$Sr$_2$CaCu$_2$O$_{8+\delta}$ measured at different positions of the surface at $T = 4.2$ K; the phonon peaks are marked by *arrows*, and the determined local gap size Δ is indicated; note that the strength of the phonon peak increases with the strength of the coherence peaks surrounding the gap; (**k**) LDOS in the vortex core of slightly overdoped Bi$_2$Sr$_2$CaCu$_2$O$_{8+\delta}$, $B = 5$ T; the dI/dV image taken at $B = 5$ T is integrated over $V = 1-12$ mV, and the corresponding dI/dV image at $B = 0$ T is subtracted to highlight the LDOS induced by the magnetic field. The checkerboard pattern within the seven vortex cores exhibits a periodicity, which is fourfold with respect to the atomic lattice shown in (**i**) and is thus assumed to be a CDW; (**l**) STM image of cleaved Ca$_{1.9}$Na$_{0.1}$CuO$_2$Cl$_2$ at $T = 0.1$ K, i. e., within the superconducting phase of the material; a checkerboard pattern with fourfold periodicity is visible on top of the atomic resolution (© S. Davis, Cornell and S. Uchida, Tokyo (**i–l**)) ◀

was found that for overdoped and optimally doped samples the gap develops continuously across T_c, showing a universal relation between the local gap size $\Delta(T=0)$ (measured at low temperature) and the local critical temperature T_p (at which the gap completely disappears): $2\Delta(T=0)/k_B T_p \approx 8$. The latter result is evidence that the so-called pseudogap phase is a phase with incoherent Cooper pairs. The results are less clear in the underdoped region, where probably two gaps complicate the analysis. Below T_c, it turns out that the strength of the coherence peak is anticorrelated to the local oxygen acceptor density [24.169] and, in addition, correlated to the energy of an inelastic phonon excitation peak in dI/dV spectra [24.31]. Figure 24.17j shows corresponding spectra taken at different positions, where the coherence peaks and the nearby phonon peaks marked by arrows are clearly visible. The phonon origin of the peak has been proven by the isotope effect, similar to Fig. 24.11a. The strong intensity of the phonon side-peak as well as the correlation of its strength with the coherence peak intensity points towards an important role of electron–phonon coupling for the pairing mechanism. However, since the gap size does not scale with the strength of the phonon peak [24.172], other contributions must be involved too.

Of course, vortices have also been investigated for HTCS [24.173]. Bound states are found, but at energies that are in disagreement with simple models, assuming a Bardeen–Cooper–Schrieffer (BCS)-like d-wave superconductor [24.174, 175]. Theory predicts, instead, that the bound states are magnetic-field-induced spin density waves, stressing the competition between antiferromagnetic order and superconductivity in HTCS materials [24.176]. Since the spin density wave is accompanied by a charge density wave of half wavelength, it can be probed by STS [24.177]. Indeed, a checkerboard pattern of the right periodicity has been found in and around vortex cores in BSCCO (Fig. 24.17k). Similar checkerboards, which do not show any $E(k)$ dispersion, have also been found in the underdoped pseudogap phase at temperatures higher than the superconducting transition temperature [24.178] or at dopant densities lower than the critical doping [24.179]. Depending on the sample, the patterns can be either homogeneous or inhomogeneous and exhibit slightly different periodicities. However, the fact that the pattern persists within the superconducting phase as shown in Fig. 24.17l, at least for Na-CCOC, indicates that the corresponding phase can coexist with superconductivity. This raises the question of whether spin density waves are the central opponent to HTCS. Interestingly, a checkerboard pattern of similar periodicity, but without long-range order, is also found, if one displays the particle–hole asymmetry of $dI/dV(V)$ intensity in underdoped samples at low temperature [24.180]. Since the observed asymmetry is known to be caused by the lifting of the correlation gap with doping, the checkerboard pattern might be directly linked to the corresponding localized holes in the CuO planes appearing at low doping. Although a comprehensive model for HTCS materials is still lacking, STS contributes significantly to disentangling this puzzle.

Even more complex superconductors are based on heavy fermions, where superconductivity is known to coexist with ferromagnetism. First attempts to obtain information about these materials by STM have been made using very low temperature (190 mK). They exhibit indeed spatial fluctuations of the superconducting gap [24.181, 182]. However, the key issue for these materials is still the preparation of high-quality surfaces similar to the HTCS materials, where cleavage was extremely advantageous to obtain high-quality data.

Notice that all the measurements described above have probed the superconducting phase only indirectly by measuring the quasiparticle LDOS. The superconducting condensate itself could principally also be probed directly using Cooper-pair tunneling between a superconducting tip and a superconducting sample. A proof of principle of this detection scheme has indeed been given at low tunneling resistance ($R \approx 50\,\mathrm{k\Omega}$) [24.183], but meaningful spatially resolved data are still lacking.

Complex Systems (Manganites)

Complex phase diagrams are not restricted to HTCS materials (cuprates). They exist with similar complexity for other doped oxides such as manganites. Only a few studies of these materials have been performed by STS. Some of them show the inhomogeneous evolution of metallic and insulating phases across a metal–insulator transition [24.184, 185]. Within layered materials, such a phase separation has been found to be absent [24.186]. This experiment performed on LaSrMnO revealed, in addition, a peculiar atomic structure, which appears only locally. It has been attributed to the observation of a local polaron bound to a defect. Since inhomogeneities seem to be crucial also in these materials, a local method such as STS might continue to be important for the understanding of their complex properties.

24.3.6 Imaging Spin Polarization: Nanomagnetism

Conventional STS couples to the LDOS, i.e., the charge distribution of the electronic states. Since electrons also have spin, it is desirable to also probe the spin distribution of the states. This can be achieved by spin-polarized STM (SP-STM) using a tunneling tip covered by a ferromagnetic material [24.187]. The coating acts as a spin filter or, more precisely, the tunneling current depends on the relative angle α_{ij} between the spins of the tip and the sample according to $\cos(\alpha_{ij})$. Consequently, a particular tip is not sensitive to spin orientations of the sample perpendicular to the spin orientation of the tip.

Different tips have to be prepared to detect different spin orientations. Moreover, the stray magnetic field of the tip can perturb the spin orientation of the sample. To avoid this, a technique using antiferromagnetic Cr as a tip coating material has been developed [24.188]. This avoids stray fields, but still provides a preferential spin orientation of the few atoms at the tip apex that dominate the tunneling current. Depending on the thickness of the Cr coating, spin orientations perpendicular or parallel to the sample surface, implying corresponding sensitivities to the spin directions of the sample, are achieved.

SP-STM has been used to image the evolution of magnetic domains with increasing B field (Fig. 24.18a–d) [24.189], the antiferromagnetic order of a Mn monolayer on W(110) [24.190], as well as of a Fe monolayer on W(100) (Fig. 24.18e) [24.191], and the out-of-plane orientation of a magnetic vortex core in the center of a nanomagnet exhibiting the flux closure configuration [24.192].

In addition, more complex atomic spin structures showing chiral or noncollinear arrangements have been identified [24.193–195]. Even the spin orientation of a single adatom could be detected, if the adatom is placed either directly on a ferromagnetic island [24.196] or close to a ferromagnetic stripe [24.23]. In the latter case, hysteresis curves of the ferromagnetic adatoms could be measured, as shown in Fig. 24.18f–h. It was found that the adatoms couple either ferromagnetically (Fig. 24.18g) or antiferromagnetically (Fig. 24.18h) to the close-by magnetic stripe; i.e., the hysteresis is either in phase or out of phase with the hysteresis of the stripe. This behavior, depending on adatom–stripe distance in an oscillating fashion, directly visualizes the famous Ruderman–Kittel–Kasuya–Yoshida (RKKY) interaction [24.23].

An interesting possibility of SP-STM is the observation of magnetodynamics on the nanoscale. Nanoscale ferromagnetic islands become unstable at a certain temperature, the so-called superparamagnetic transition temperature. Above this temperature, the direction of magnetization switches back and forth due to thermal excitations. This switching results in a stripe-like contrast in SP-STM images, as visible in the inset of Fig. 24.18i. The island appears dark during the time when the orientation of the island spin is opposite to the orientation of the tip spin, and switches to bright when the island spin orientation changes. By observing the switching as a function of time on different islands at different temperatures the energy barriers of individual islands can be determined [24.197]. Even more

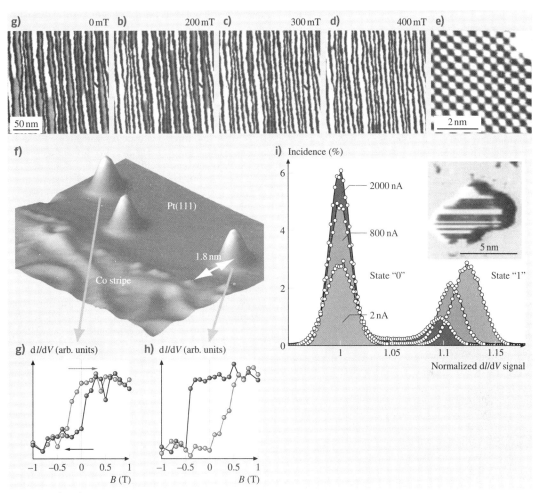

Fig. 24.18 (a–d) Spin-polarized STM images of 1.65 monolayers of Fe deposited on a stepped W(110) surface measured at different B fields, as indicated. Double-layer and monolayer Fe stripes are formed on the W substrate; only the double-layer stripes exhibit magnetic contrast with an out-of-plane sensitive tip, as used here. *White* and *grey areas* correspond to different domains. Note that more white areas appear with increasing field (© M. Bode, Argonne (a–d)). (e) STM image of an antiferromagnetic Fe monolayer on W(001) exhibiting a checkerboard pattern of spin-down (*dark*) and spin-up (*bright*) atoms (© A. Kubetzka, Hamburg); (f) STM image of a Pt(111) surface with a Co stripe deposited at the Pt edge as marked. Single Co atoms, visible as *three hills*, are deposited subsequently on the surface at $T = 25$ K; (g,h) $dI/dV(B)$ curves obtained above the Co atoms marked in (f) using a spin-polarized tip at $V = 0.3$ V. The *colors* mark the sweeping direction of the B field. Obviously the resulting contrast is hysteretic with B and opposite for the two Co atoms. This indicates a different sign of ferromagnetic coupling to the Co stripe. (© J. Wiebe, Hamburg (f–h)); (i) observed incidences of differential conductivities above a single monolayer Fe island on W(110) with a spin-polarized tip. The *three curves* are recorded at different tunneling currents and the increasing asymmetry shows a preferential spin direction with increasing spin-polarized current. *Inset*: dI/dV image of the Fe island at $T = 56$ K showing the irregular change of dI/dV intensity (© S. Krause, Hamburg)

importantly, the preferential orientation during switching can be tuned by the tunneling current. This is visible in Fig. 24.18i, which shows the measured orientational probability at different tunneling currents [24.198]. The observed asymmetry in the peak intensity increases with current, providing evidence that current-induced magnetization switching is possible even on the atomic scale.

24.4 Scanning Force Microscopy and Spectroscopy

The examples discussed in the previous section show the wide variety of physical questions that have been tackled with the help of LT-STM. Here, we turn to the other prominent scanning probe method that is applied at low temperatures, namely SFM, which gives complementary information on sample properties on the atomic scale.

The ability to detect *forces* sensitively with spatial resolution down to the atomic scale is of great interest, since force is one of the most fundamental quantities in physics. Mechanical force probes usually consist of a cantilever with a tip at its free end that is brought close to the sample surface. The cantilever can be mounted parallel or perpendicular to the surface (general aspects of force probe designs are described in Chap. 22). Basically, two methods exist to detect forces with cantilever-based probes: the *static* and the *dynamic* mode (see Chap. 21). They can be used to generate a laterally resolved image (*microscopy* mode) or determine its distance dependence (*spectroscopy* mode). One can argue about this terminology, since spectroscopy is usually related to energies and not to distance dependencies. Nevertheless, we will use it throughout the text, because it avoids lengthy paraphrases and is established in this sense throughout the literature.

In the static mode, a force that acts on the tip bends the cantilever. By measuring its deflection Δz the tip–sample force F_{ts} can be directly calculated from Hooke's law: $F_{ts} = c_z \Delta z$, where c_z denotes the spring constant of the cantilever. In the various dynamic modes, the cantilever is oscillated with amplitude A at or near its eigenfrequency f_0, but in some applications also off-resonance. At ambient pressures or in liquids, amplitude modulation (AM-SFM) is used to detect amplitude changes or the phase shift between the driving force and cantilever oscillation. In vacuum, the frequency shift Δf of the cantilever due to a tip–sample interaction is measured by the frequency-modulation technique (FM-SFM). The nomenclature is not standardized. Terms such as tapping mode or intermittent contact mode are used instead of AM-SFM, and NC-AFM (noncontact atomic force microscopy) or DFM (dynamic force microscopy) instead of FM-SFM or FM-AFM. However, all these modes are *dynamic*, i.e., they involve an oscillating cantilever and can be used in the noncontact, as well as in the contact, regime. Therefore, we believe that the best and most consistent way is to distinguish them by their different detection schemes. Converting the measured quantity (amplitude, phase or frequency shift) into a physically meaningful quantity, e.g., the tip–sample interaction force F_{ts} or the force gradient $\partial F_{ts}/\partial z$, is not always straightforward and requires an analysis of the equation of motion of the oscillating tip (see Chaps. 23 and 26).

Whatever method is used, the resolution of a cantilever-based force detection is fundamentally limited by its intrinsic *thermomechanical* noise. If the cantilever is in thermal equilibrium at a temperature T, the equipartition theorem predicts a thermally induced *root-mean-square* (RMS) motion of the cantilever in the z direction of $z_{RMS} = (k_B T/c_{eff})^{1/2}$, where k_B is the Boltzmann constant and $c_{eff} = c_z + \partial F_{ts}/\partial z$. Note that usually $dF_{ts}/dz \gg c_z$ in the contact mode and $dF_{ts}/dz < c_z$ in the noncontact mode. Evidently, this fundamentally limits the force resolution in the static mode, particularly if operated in the noncontact mode. Of course, the same is true for the different dynamic modes, because the thermal energy $k_B T$ excites the eigenfrequency f_0 of the cantilever. Thermal noise is *white* noise, i.e., its spectral density is flat. However, if the cantilever transfer function is taken into account, one can see that the thermal energy mainly excites f_0. This explains the term "thermo" in thermomechanical noise, but what is the "mechanical" part?

A more detailed analysis reveals that the thermally induced cantilever motion is given by

$$z_{RMS} = \sqrt{\frac{2k_B T B}{\pi c_z f_0 Q}}, \qquad (24.2)$$

where B is the measurement bandwidth and Q is the quality factor of the cantilever. Analogous expressions can be obtained for all quantities measured in dynamic modes, because the deflection noise translates, e.g., into frequency noise [24.5]. Note that f_0 and c_z are correlated with each other via $2\pi f_0 = (c_z/m_{\text{eff}})^{1/2}$, where the effective mass m_{eff} depends on the geometry, density, and elasticity of the material. The Q-factor of the cantilever is related to the external damping of the cantilever motion in a medium and to the intrinsic damping within the material. This is the "mechanical" part of the fundamental cantilever noise.

It is possible to operate a low-temperature force microscope directly immersed in the cryogen [24.199, 200] or in the cooling gas [24.201], whereby the cooling is simple and very effective. However, it is evident from (24.2) that the smallest fundamental noise is achievable in vacuum, where the Q-factors are more than 100 times larger than in air, and at low temperatures.

The best force resolution up to now, which is better than 1×10^{-18} N/Hz$^{1/2}$, has been achieved by *Mamin* et al. [24.202] in vacuum at a temperature below 300 mK. Due to the reduced thermal noise and the lower thermal drift, which results in a higher stability of the tip–sample gap and a better signal-to-noise ratio, the highest resolution is possible at low temperatures in ultrahigh vacuum with FM-SFM. A vertical RMS noise below 2 pm [24.203, 204] and a force resolution below 1 aN [24.202] have been reported.

Besides the reduced noise, the application of force detection at low temperatures is motivated by the increased stability and the possibility to observe phenomena that appear below a certain critical temperature T_c, as outlined on page 664. The experiments, which have been performed at low temperatures until now, were motivated by at least one of these reasons and can be roughly divided into four groups:

(i) Atomic-scale imaging
(ii) Force spectroscopy
(iii) Investigation of quantum phenomena by measuring electrostatic forces
(iv) Utilizing magnetic probes to study ferromagnets, superconductors, and single spins

In the following, we describe some exemplary results.

24.4.1 Atomic-Scale Imaging

In a simplified picture, the dimensions of the tip end and its distance to the surface limit the lateral resolution of force microscopy, since it is a near-field technique. Consequently, atomic resolution requires a stable single atom at the tip apex that has to be brought within a distance of some tenths of a nanometer of an atomically flat surface. The latter condition can only be fulfilled in the dynamic mode, where the additional restoring force $c_z A$ at the lower turnaround point prevents the jump-to-contact. As described in Chap. 23, by preventing the so-called jump-to-contact, *true* atomic resolution is nowadays routinely obtained in vacuum by FM-AFM. The nature of the short-range tip–sample interaction during imaging with atomic resolution has been studied experimentally as well as theoretically. Si(111)-(7×7) was the first surface on which true atomic resolution was achieved [24.205], and several studies have been performed at low temperatures on this well-known material [24.206–208]. First-principles simulations performed on semiconductors with a silicon tip revealed that *chemical* interactions, i.e., a significant charge redistribution between the dangling bonds of the tip and sample, dominate the atomic-scale contrast [24.209–211]. On V–III semiconductors, it was found that only one atomic species, the group V atoms, is imaged as protrusions with a silicon tip [24.210, 211]. Furthermore, these simulations revealed that the sample, as well as the tip atoms, are noticeably displaced from their equilibrium position due to the interaction forces. At low temperatures, both aspects could be observed with silicon tips on indium arsenide [24.203, 212]. On weakly interacting surfaces the short-range interatomic van der Waals force has been believed responsible for the atomic-scale contrast [24.213–215].

Chemical Sensitivity of Force Microscopy

The (110) surface of the III–V semiconductor indium arsenide exhibits both atomic species in the top layer (see Fig. 24.19a). Therefore, this sample is well suited to study the chemical sensitivity of force microscopy [24.203]. In Fig. 24.19b, the usually observed atomic-scale contrast on InAs(110) is displayed. As predicted, the arsenic atoms, which are shifted by 80 pm above the indium layer due to the (1×1) relaxation, are imaged as protrusions. While this general appearance was similar for most tips, two other distinctively different contrasts were also observed: a second protrusion (Fig. 24.19c) and a sharp depression (Fig. 24.19d). The arrangement of these two features corresponds well to the zigzag configuration of the indium and arsenic atoms along the [1$\bar{1}$0] direction. A sound explanation would be as follows: the contrast usually obtained with one feature per surface unit cell corresponds to a silicon-terminated tip, as predicted by simulations. A different atomic species at the tip apex, however, can result in a very different charge redistribution. Since the atomic-scale contrast is due to a chemical interaction, the two

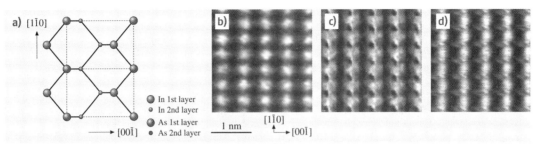

Fig. 24.19a–d The structure of InAs(110) as seen from above (**a**) and three FM-AFM images of this surface obtained with different tips at 14 K (**b–d**). In (**b**), only the arsenic atoms are imaged as protrusions, as predicted for a silicon tip. The two features in (**c**) and (**d**) corresponds to the zigzag arrangement of the indium and arsenic atoms. Since force microscopy is sensitive to short-range chemical forces, the appearance of the indium atoms can be associated with a chemically different tip apex

other contrasts would then correspond to a tip that has been accidentally contaminated with sample material (an arsenic- or indium-terminated tip apex). Nevertheless, this explanation has not yet been verified by simulations for this material.

Tip-Induced Atomic Relaxation

Schwarz et al. [24.203] were able to visualize directly the predicted tip-induced relaxation during atomic-scale imaging near a point defect. Figure 24.20 shows two FM-AFM images of the same point defect recorded with different constant frequency shifts on InAs(110), i.e., the tip was closer to the surface in Fig. 24.20b compared with Fig. 24.20a. The arsenic atoms are imaged as protrusions with the silicon tip used. From the symmetry of the defect, an indium-site defect can be inferred, since the distance-dependent contrast is consistent with what is expected for an indium vacancy. This expectation is based on calculations performed for the similar III–V semiconductor GaP(110), where the two surface gallium atoms around a P-vacancy were found to relax downward [24.216]. This corresponds to the situation in Fig. 24.20a, where the tip is relatively far away and an inward relaxation of the two arsenic atoms is observed. The considerably larger attractive force in Fig. 24.20b, however, pulls the two arsenic atoms toward the tip. All

other arsenic atoms are also pulled, but they are less displaced, because they have three bonds to the bulk, while the two arsenic atoms in the neighborhood of an indium vacancy have only two bonds. This direct experimental proof of the presence of tip-induced relaxations is

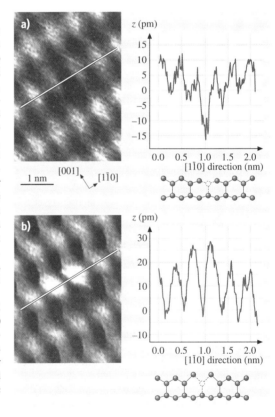

Fig. 24.20a,b Two FM-AFM images of the identical indium-site point defect (presumably an indium vacancy) recorded at 14 K. If the tip is relatively far away, the theoretically predicted inward relaxation of two arsenic atoms adjacent to an indium vacancy is visible (**a**). At a closer tip–sample distance (**b**), the two arsenic atoms are pulled farther toward the tip compared with the other arsenic atoms, since they have only two instead of three bonds ▶

also relevant for STM measurements, because the tip–sample distances are similar during atomic-resolution imaging. Moreover, the result demonstrates that FM-AFM can probe elastic properties on an atomic level.

Imaging of Weakly Interacting van der Waals Surfaces

For weakly interacting van der Waals surfaces, much smaller atomic corrugation amplitudes are expected compared with strongly interacting surfaces of semiconductors. A typical example is graphite, a layered material, where the carbon atoms are covalently bonded and arranged in a honeycomb structure within the (0001) plane. Individual graphene layers stick together by van der Waals forces. Due to the *ABA* stacking, three distinctive sites exist on the (0001) surface: carbon atoms with (*A*-type) and without (*B*-type) neighbor in the next graphite layer and the *hollow site* (*H*-site) in the hexagon center. In static contact force microscopy as well as in STM the contrast usually exhibits a trigonal symmetry with a periodicity of 246 pm, where *A*- and *B*-site carbon atoms could not be distinguished. However, in high-resolution FM-AFM images acquired at low temperatures, a large maximum and two different minima have been resolved, as demonstrated by the profiles along the three equivalent [1-100] directions in Fig. 24.21a. A simulation using the Lennard-Jones (LJ) potential, given by the short-range interatomic van der Waals force, reproduced these three features very well (dotted line). Therefore, the large maximum could be assigned to the *H*-site, while the two different minima represent *A*- and *B*-type carbon atoms [24.214].

Compared with graphite, the carbon atoms in a single-walled carbon nanotube (SWNT), which consists of a single rolled-up graphene layer, are indistinguishable. For the first time *Ashino* et al. [24.215] successfully imaged the curved surface of a SWNT with atomic resolution. Note that, for geometric reasons, atomic resolution is only achieved on the top (see Fig. 24.21b). Indeed, as shown in Fig. 24.21b, all profiles between two hollow sites across two neighboring carbon atoms are symmetric [24.217]. Particularly, curves 1 and 2 exhibit two minima of equal depth, as predicted by theory (cf., dotted line). The assumption used in the simulation (dotted lines in the profiles of Fig. 24.21) that interatomic van der Waals forces are responsible for the atomic-scale contrast has been supported by a quantitative evaluation of force spectroscopy data obtained on SWNTs [24.215].

Interestingly, the image contrast on graphite and SWNTs is inverted with respect to the arrangement of the atoms, i.e., the minima correspond to the positions of the carbon atoms. This can be related to the

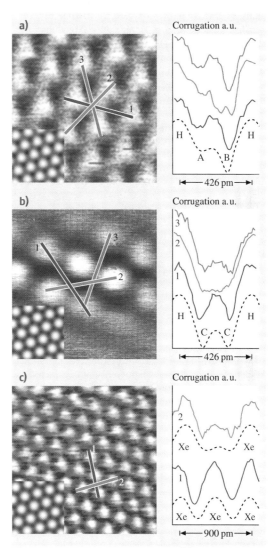

Fig. 24.21a–c FM-AFM images of (**a**) graphite(0001), (**b**) a single-walled carbon nanotube (SWNT), and (**c**) Xe(111) recorded at 22 K. On the right side, line sections taken from the experimental data (*solid lines*) are compared with simulations (*dotted lines*). *A*- and *B*-type carbon atoms, as well as the hollow site (*H*-site) on graphite can be distinguished, but are imaged with inverted contrast, i.e., the carbon sites are displayed as minima. Such an inversion does not occur on Xe(111)

small carbon–carbon distance of only 142 pm, which is in fact the smallest interatomic distance that has been resolved with FM-AFM so far. The van der Waals radius of the front tip atom, (e.g., 210 pm for silicon) has a radius that is significantly larger than the intercarbon distance. Therefore, next-nearest-neighbor interactions become important and result in contrast inversion [24.217].

While experiments on graphite and SWNTs basically take advantage of the increased stability and signal-to-noise ratio at low temperatures, solid xenon (melting temperature $T_m = 161$ K) can only be observed at sufficient low temperatures [24.8]. In addition, xenon is a pure van der Waals crystal and, since it is an insulator, FM-AFM is the only real-space method available today that allows the study of solid xenon on the atomic scale.

Allers et al. [24.8] adsorbed a well-ordered xenon film on cold graphite(0001) ($T < 55$ K) and studied it subsequently at 22 K by FM-AFM (Fig. 24.21c). The sixfold symmetry and the distance between the protrusions corresponds well with the nearest-neighbor distance in the close-packed (111) plane of bulk xenon, which crystallizes in a face-centered cubic structure. A comparison between experiment and simulation confirmed that the protrusions correspond to the position of the xenon atoms [24.214]. However, the simulated corrugation amplitudes do not fit as well as for graphite (see sections in Fig. 24.21c). A possible reason is that tip-induced relaxations, which were not considered in the simulations, are more important for this pure van der Waals crystal xenon than they are for graphite, because in-plane graphite exhibits strong covalent bonds. Nevertheless, the results demonstrated for the first time that a weakly bonded van der Waals crystal could be imaged nondestructively on the atomic scale. Note that on Xe(111) no contrast inversion exists, presumably because the separation between Xe sites is about 450 pm, i.e., twice as large as the van der Waals radius of a silicon atom at the tip end.

Atomic Resolution
Using Small Oscillation Amplitudes

All the examples above described used spring constants and amplitudes on the order of 40 N/m and 10 nm, respectively, to obtain atomic resolution. However, *Giessibl* et al. [24.218] pointed out that the optimal amplitude should be on the order of the characteristic decay length λ of the relevant tip–sample interaction. For short-range interactions, which are responsible for the atomic-scale contrast, λ is on the order of 0.1 nm. On the other hand, stable imaging without a jump-to-contact is only possible as long as the restoring force $c_z A$ at the lower turnaround point of each cycle is larger than the maximal attractive tip–sample force. Therefore, reducing the desired amplitude by a factor of 100 requires a 100 times larger spring constant. Indeed, *Hembacher* et al. [24.219] could demonstrate atomic resolution with small amplitudes (about 0.25 nm) and large spring constants (about 1800 N/m) utilizing a qPlus sensor [24.220]. Figure 24.22 shows a constant-height image of graphite recorded at 4.9 K within the repulsive regime. Note that, compared with Fig. 24.21a,b, the contrast is inverted, i.e., the carbon atoms appear as maxima. This is expected, because the imaging interaction is switched from attractive to repulsive regime [24.213, 217].

24.4.2 Force Spectroscopy

A wealth of information about the nature of the tip–sample interaction can be obtained by measuring its distance dependence. This is usually done by recording the measured quantity (deflection, frequency shift, amplitude change, phase shift) and applying an appropriate voltage ramp to the z-electrode of the scanner piezo, while the z-feedback is switched off. According to (24.2), low temperatures and high Q-factors (vacuum) considerably increase the force resolution. In the static mode, long-range forces and contact forces can be examined. Force measurements at small tip–sample distances are inhibited by the *jump-to-contact* phenomenon: If the force gradient $\partial F_{ts}/\partial z$ becomes larger than the spring constant c_z, the cantilever cannot resist the attractive tip–sample forces and the tip snaps onto the surface. Sufficiently large spring constants prevent this effect, but reduce the force resolution. In the dynamic modes, the jump-to-contact can be avoided due to the additional restoring force ($c_z A$) at the lower turnaround point. The highest sensitivity can be achieved in vacuum by using the FM technique, i.e., by recording $\Delta f(z)$ curves. An alternative FM spectroscopy method, the recording of $\Delta f(A)$ curves, has been suggested by *Hölscher* et al. [24.221]. Note that, if the amplitude is much larger than the characteristic decay length of the tip–sample force, the frequency shift cannot simply be converted into force gradients by using $\partial F_{ts}/\partial z = 2c_z \Delta f / f_0$ [24.222]. Several methods have been published to convert $\Delta f(z)$ data into the tip–sample potential $V_{ts}(z)$ and tip–sample force $F_{ts}(z)$ [24.223–226].

Measurement of Interatomic Forces at Specific Atomic Sites

FM force spectroscopy has been successfully used to measure and determine quantitatively the short-range chemical force between the foremost tip atom and specific surface atoms [24.177, 227, 228]. Figure 24.23 displays an example for the quantitative determination of the short-range force. Figure 24.23a shows two $\Delta f(z)$ curves measured with a silicon tip above a corner hole and above an adatom. Their position is indicated by arrows in the inset, which displays the atomically resolved Si(111)-(7×7) surface. The two curves differ from each other only for small tip–sample distances, because the long-range forces do not contribute to the atomic-scale contrast. The low, thermally induced lateral drift and the high stability at low temperatures were required to precisely address the two specific sites. To extract the short-range force, the long-range van der Waals and/or electrostatic forces can be subtracted from the total force. The black curve in Fig. 24.23b has been reconstructed from the $\Delta f(z)$ curve recorded above an adatom and represents the total force. After removing the long-range contribution from the data, the much steeper brown line is obtained, which corresponds to the short-range force between the adatom and the atom at the tip apex. The measured maximum attractive force (-2.1 nN) agrees well with that obtained from first-principles calculations (-2.25 nN).

By determining the maximal attractive short-range force between tip apex atom and surface atom as a fingerprint *Sugimoto* et al. [24.229] were able to utilize force spectroscopy data for chemical identification. They demonstrated this concept using a Si tip and a surface with Sn, Pb, and Si adatoms located at equivalent lattice sites on a Si(111) substrate. Since the experiment was performed at room temperature, the signal-to-noise-ratio had to be increased by averaging about 100 curves at every atom species, which required an appropriate atom tracking scheme [24.230].

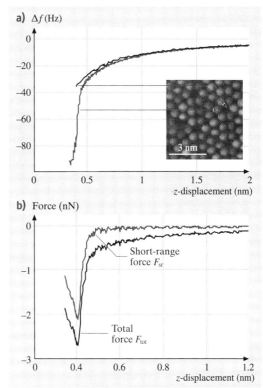

Fig. 24.23a,b FM force spectroscopy on specific atomic sites at 7.2 K. In (**a**), an FM-SFM image of the Si(111)-(7×7) surface is displayed together with two $\Delta f(z)$ curves, which have been recorded at the positions indicated by the *arrows*, i.e., above the corner hole (*black*) and above an adatom (*brown*). In (**b**), the total force above an adatom (*black line*) has been recovered from the $\Delta f(z)$ curve. After subtraction of the long-range part, the short-range force can be determined (*brown line*) (courtesy of H. J. Hug; cf. [24.227])

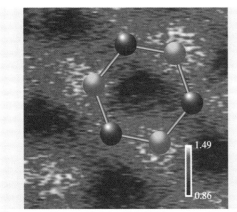

Fig. 24.22 Constant-height FM-AFM image of graphite (0001) recorded at 4.9 K using a small amplitude ($A = 0.25$ nm) and a large spring constant ($c_z = 1800$ N/m). As in Fig. 24.20a, A- and B-site carbon atoms can be distinguished. However, they appear as maxima, because imaging has been performed in the repulsive regime (© F. J. Giessibl [24.219])

Three-Dimensional Force Field Spectroscopy

Further progress with the FM technique has been made by *Hölscher* et al. [24.231]. They acquired a complete 3-D force field on NiO(001) with atomic resolution (*3-D force field spectroscopy*). In Fig. 24.24, the atomically resolved FM-AFM image of NiO(001) is shown together with the coordinate system used and the tip to illustrate the measurement principle. NiO(001) crystallizes in the rock-salt structure. The distance between the protrusions corresponds to the lattice constant of 417 pm, i.e., only one type of atom (most likely the oxygen) is imaged as a protrusion. In an area of 1 nm × 1 nm, 32 × 32 individual $\Delta f(z)$ curves have been recorded at every (x, y) image point and converted into $F_{ts}(z)$ curves. The $\Delta f(x, y, z)$ data set is thereby converted into the 3-D force field $F_{ts}(x, y, z)$. Figure 24.24, where a specific x–z-plane is displayed, demonstrates that atomic resolution is achieved. It represents a 2-D cut $F_{ts}(x, y = \text{const}, z)$ along the [100] direction (corresponding to the shaded slice marked in Fig. 24.24). Since a large number of curves have been recorded, *Langkat* et al. [24.228] could evaluate the whole data set by standard statistical means to extract the long- and short-range forces. A possible future application of 3-D force field spectroscopy could be to map the short-range forces of complex molecules with functionalized tips in order to resolve locally their chemical reactivity. A first step in this direction has been accomplished on SWNTs. Its structural unit, a hexagonal carbon ring, is common to all aromatic molecules. Like the constant frequency-shift image of an SWNT shown in Fig. 24.21b the force map shows clear differences between hollow sites and carbon sites [24.215]. Analyzing site-specific individual force curves extracted from the 3-D data revealed a maximum attractive force of ≈ -0.106 nN above H-sites and ≈ -0.075 nN above carbon sites. Since the attraction is one order of magnitude weaker than on Si(111)-(7×7) (Fig. 24.23b), it has been inferred that the short-range interatomic van der Waals force and not a chemical force is responsible for atomic-scale contrast formation on such nonreactive surfaces. It is worth mentioning that 3-D force field spectroscopy data have been acquired at room temperature as well [24.232, 233].

Apart from calculating the vertical tip–sample force *Schwarz* et al. [24.234] demonstrated that it is also possible to obtain the lateral tip–sample force from 3-D data sets. First, the tip–sample potential $V_{ts}(x, y, z)$ has to be determined. Then the lateral force components can be calculated by taking the derivative with respect to the x- and y-coordinate, respectively. This technique has been employed to determine the lateral force needed to move an atom sideways by *Ternes* et al. [24.42].

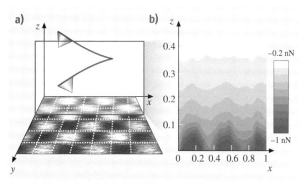

Fig. 24.24a,b Principle of the 3-D force field spectroscopy method (**a**) and a 2-D cut through the 3-D force field $F_{ts}(x, y, z)$ recorded at 14 K (**b**). At all 32 × 32 image points of the 1 nm × 1 nm scan area on NiO(001), a $\Delta f(z)$ curve has been recorded. The $\Delta f(x, y, z)$ data set obtained is then converted into the 3-D tip–sample force field $F_{ts}(x, y, z)$. The *shaded slice* $F_{ts}(x, y = \text{const}, z)$ in (**a**) corresponds to a cut along the [100] direction and demonstrates that atomic resolution has been obtained, because the distance between the protrusions corresponds well to the lattice constant of nickel oxide

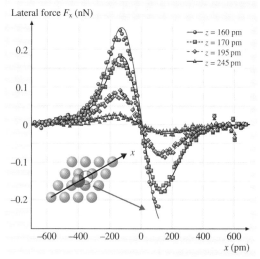

Fig. 24.25 Lateral force curves recorded at constant tip–sample separation z. At the lowest separation a discontinuity appears, which marks the jump of the Co atom form one site to the next as indicated in the *inset* (©. M. Ternes [24.42]).

They recorded constant-height data at different tip–sample distances to obtain first $V_{ts}(x, z)$ and then the lateral force $F_x(x, z) = d/dx V_{ts}(x, z)$. Four curves of the whole data set are displayed in Fig. 24.25. The discontinuity of the lateral force at the lowest adjusted tip–sample distance ($z = 160$ pm) indicates the jump of the Co atom from one hollow site to the next on Pt(111), cf., inset. It takes place at about 210 pN.

24.4.3 Atomic Manipulation

Nowadays, atomic-scale manipulation is routinely performed using an STM tip (see Sect. 24.3.1). In most of these experiments an adsorbate is dragged with the tip by using an attractive force between the foremost tip apex atoms and the adsorbate. By adjusting a large or a small tip–surface distance via the tunneling resistance, it is possible to switch between imaging and manipulation. Recently, it has been demonstrated that controlled manipulation of individual atoms is also possible in the dynamic mode of atomic force microscopy, i.e., FM-AFM. Vertical manipulation was demonstrated by pressing the tip in a controlled manner into a Si(111)-(7×7) surface [24.236]. The strong repulsion leads to the removal of the selected silicon atom. The process could be traced by recording the frequency shift and the damping signal during the approach. For lateral manipulation a *rubbing* technique has been utilized [24.235], where the slow scan axis is halted above a selected atom, while the tip–surface distance is gradually reduced until the selected atom hops to a new stable position. Figure 24.26 shows a Ge adatom on Ge(111)-c(2×8) that was moved during scanning in two steps from its original position (Fig. 24.26a) to its final position (Fig. 24.26c). In fact, manipulation by FM-AFM is reproducible and fast enough to write nanostructures in a bottom-up process with single atoms [24.237].

24.4.4 Electrostatic Force Microscopy

Electrostatic forces are readily detectable by a force microscope, because the tip and sample can be regarded as two electrodes of a capacitor. If they are electrically connected via their back sides and have different work functions, electrons will flow between the tip and sample until their Fermi levels are equalized. As a result, an electric field, and consequently an attractive electrostatic force, exists between them at zero bias. This *contact potential difference* can be balanced by applying an appropriate bias voltage. It has been demonstrated that individual doping atoms in semiconducting materials can be detected by electrostatic interactions due to the local variation of the surface potential around them [24.238, 239].

Detection of Edge Channels in the Quantum Hall Regime

At low temperatures, electrostatic force microscopy has been used to measure the electrostatic potential in the quantum Hall regime of a *two-dimensional electron gas* (2-DEG) buried in epitaxially grown GaAs/AlGaAs heterostructures [24.240–243]. In the 2-DEG, electrons can move freely in the x–y-plane, but they cannot move in z-direction. Electrical transport properties of a 2-DEG are very different compared with normal metallic conduction. Particularly, the Hall resistance $R_H = h/ne^2$ (where h represents Planck's constant, e is the electron charge, and $n = 1, 2, \ldots$) is quantized in the quantum Hall regime, i.e., at sufficiently low temperatures ($T < 4$ K) and high magnetic fields (up to 20 T). Under these conditions, theoretical calculations

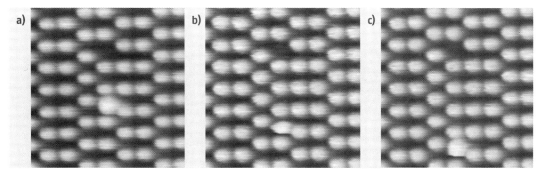

Fig. 24.26a–c Consecutively recorded FM-AFM images showing the tip-induced manipulation of a Ge adatom on Ge(111)-c(2×8) at 80 K. Scanning was performed from bottom to top (© N. Oyabu [24.235])

predict the existence of *edge channels* in a Hall bar. A Hall bar is a strip conductor that is contacted in a specific way to allow longitudinal and transversal transport measurements in a perpendicular magnetic field. The current is not evenly distributed over the cross-section of the bar, but passes mainly along rather thin paths close to the edges. This prediction has been verified by measuring profiles of the electrostatic potential across a Hall bar in different perpendicular external magnetic fields [24.240–242].

Figure 24.27a shows the experimental setup used to observe these edge channels on top of a Hall bar with a force microscope. The tip is positioned above the surface of a Hall bar under which the 2-DEG is buried. The direction of the magnetic field is oriented perpendicular to the 2-DEG. Note that, although the 2-DEG is located several tens of nanometers below the surface, its influence on the electrostatic surface potential can be detected. In Fig. 24.27b, the results of scans perpendicular to the Hall bar are plotted against the magnitude of the external magnetic field. The value of the electrostatic potential is grey-coded in arbitrary units. In certain field ranges, the potential changes linearly across the Hall bar, while in other field ranges the potential drop is confined to the edges of the Hall bar. The predicted edge channels can explain this behavior. The periodicity of the phenomenon is related to the filling factor ν, i.e., the number of Landau levels that are filled with electrons (Sect. 24.3.4). Its value depends on $1/B$ and is proportional to the electron concentration n_e in the 2-DEG ($\nu = n_e h / eB$, where h represents Planck's constant and e the electron charge).

24.4.5 Magnetic Force Microscopy

To detect magnetostatic tip–sample interactions with magnetic force microscopy (MFM), a ferromagnetic probe has to be used. Such probes are readily prepared by evaporating a thin magnetic layer, e.g., 10 nm iron, onto the tip. Due to the in-plane shape anisotropy of thin films, the magnetization of such tips lies predominantly along the tip axis, i.e., perpendicular to the surface. Since magnetostatic interactions are long range, they can be separated from the topography by scanning at a certain constant height (typically around 20 nm) above the surface, where the z-component of the sample stray field is probed (Fig. 24.28a). Therefore, MFM is always operated in noncontact mode. The signal from the cantilever is directly recorded while the z-feedback is switched off. MFM can be operated in the static mode or in the dynamic modes (AM-MFM at ambient pressures and FM-MFM in vacuum). A lateral resolution below 50 nm can be routinely obtained.

Observation of Domain Patterns

MFM is widely used to visualize domain patterns of ferromagnetic materials. At low temperatures, *Moloni* et al. [24.244] observed the domain structure of magnetite below its Verwey transition temperature ($T_V = 122$ K), but most of the work has concentrated on thin films of $La_{1-x}Ca_xMnO_3$ [24.245–247]. Below

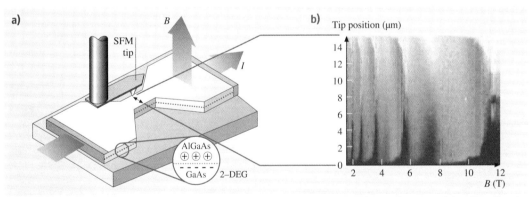

Fig. 24.27a,b Configuration of the Hall bar within a low-temperature ($T < 1$ K) force microscope (**a**) and profiles (y-axis) at different magnetic field (x-axis) of the electrostatic potential across a 14-μm-wide Hall bar in the quantum Hall regime (**b**). The external magnetic field is oriented perpendicular to the 2-DEG, which is buried below the surface. *Bright* and *dark regions* reflect the characteristic changes of the electrostatic potential across the Hall bar at different magnetic fields and can be explained by the existence of the theoretically predicted edge channels (© E. Ahlswede [24.242])

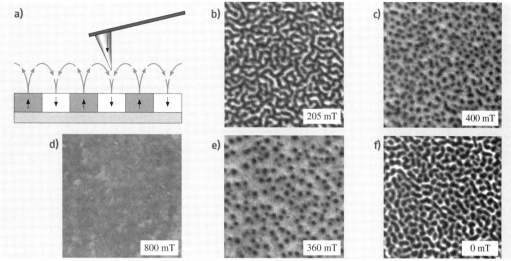

Fig. 24.28a–f Principle of MFM operation (**a**) and field-dependent domain structure of a ferromagnetic thin film (**b–f**) recorded at 5.2 K with FM-MFM. All images were recorded on the same 4 μm × 4 μm scan area. The $La_{0.7}Ca_{0.3}MnO_3/LaAlO_3$ system exhibits a substrate-induced out-of-plane anisotropy. *Bright* and *dark areas* are visible and correspond to attractive and repulsive magnetostatic interactions, respectively. The series shows how the domain pattern evolves along the major hysteresis loop, i. e., from zero field to saturation at 600 mT and back to zero field

T_V, the conductivity decreases by two orders of magnitude and a small structural distortion is observed. The domain structure of this mixed-valence manganite is of great interest, because its resistivity strongly depends on the external magnetic field, i. e., it exhibits a large colossal-magnetoresistive effect. To investigate the field dependence of the domain patterns under ambient conditions, electromagnets have to be used. They can cause severe thermal drift problems due to Joule heating of the coils by large currents. Flux densities on the order of 100 mT can be achieved. In contrast, much larger flux densities (more than 10 T) can be rather easily produced by implementing a superconducting magnet in low-temperature setups. Using such a design, *Liebmann* et al. [24.247] recorded the domain structure along the major hysteresis loop of $La_{0.7}Ca_{0.3}MnO_3$ epitaxially grown on $LaAlO_3$ (Fig. 24.28b–f). The film geometry (with thickness of 100 nm) favors an in-plane magnetization, but the lattice mismatch with the substrate induces an out-of-plane anisotropy. Thereby, an irregular pattern of strip domains appears at zero field. If the external magnetic field is increased, the domains with antiparallel orientation shrink and finally disappear in saturation (Fig. 24.28b,c). The residual contrast in saturation (Fig. 24.28d) reflects topographic features. If the field is decreased after saturation (Fig. 24.28e,f), cylindrical domains first nucleate and then start to grow. At zero field, the maze-type domain pattern has evolved again. Such data sets can be used to analyze domain nucleation and the domain growth mode. Moreover, due to the negligible drift, domain structure and surface morphology can be directly compared, because every MFM can be used as a regular topography-imaging force microscope.

Detection of Individual Vortices in Superconductors

Numerous low-temperature MFM experiments have been performed on superconductors [24.248–255]. Some basic features of superconductors have been mentioned already in Sect. 24.3.5. The main difference of STM/STS compared to MFM is its high sensitivity to the electronic properties of the surface. Therefore, careful sample preparation is a prerequisite. This is not so important for MFM experiments, since the tip is scanned at a certain distance above the surface.

Superconductors can be divided into two classes with respect to their behavior in an external magnetic field. For type I superconductors, all magnetic flux is entirely excluded below their critical temperature T_c (Meissner effect), while for type II superconductors, cylindrical inclusions (*vortices*) of normal material exist in a superconducting matrix (*vortex* state). The radius of the vortex *core*, where the Cooper-pair density decreases to zero, is on the order of the coherence length ξ. Since the superconducting gap vanishes in the core, they can be detected by STS (see Sect. 24.3.5). Additionally, each vortex contains one magnetic quantum flux $\Phi = h/2e$ (where h represents Planck's constant and e the electron charge). Circular supercurrents around the core screen the magnetic field associated with a vortex; their radius is given by the London penetration depth λ of the material. This magnetic field of the vortices can be detected by MFM. Investigations have been performed on the two most popular copper oxide high-T_c superconductors, YBa$_2$Cu$_3$O$_7$ [24.248, 249, 251] and Bi$_2$Sr$_2$CaCu$_2$O$_8$ [24.249, 255], on the only elemental conventional type II superconductor Nb [24.252, 253] and on the layered compound crystal NbSe$_2$ [24.250, 252].

Most often, vortices have been generated by cooling the sample from the normal state to below T_c in an external magnetic field. After such a *field-cooling* procedure, the most energetically favorable vortex arrangement is a regular triangular Abrikosov lattice. *Volodin* et al. [24.250] were able to observe such an Abrikosov lattice on NbSe$_2$. The intervortex distance d is related to the external field during B cool down via $d = (4/3)^{1/4}(\Phi/B)^{1/2}$. Another way to introduce vortices into a type II superconductor is vortex penetration from the edge by applying a magnetic field at temperatures below T_c. According to the Bean model, a vortex density gradient exists under such conditions within the superconducting material. *Pi* et al. [24.255] slowly increased the external magnetic field until the vortex front approaching from the edge reached the scanning area.

If the vortex configuration is dominated by the *pinning* of vortices at randomly distributed structural defects, no Abrikosov lattice emerges. The influence of pinning centers can be studied easily by MFM, because every MFM can be used to scan the topography in its AFM mode. This has been done for natural growth defects by *Moser* et al. [24.251] on YBa$_2$Cu$_3$O$_7$ and for YBa$_2$Cu$_3$O$_7$ and niobium thin films, respectively, by *Volodin* et al. [24.254]. *Roseman* and *Grütter* [24.256] investigated the formation of vortices in the presence of an artificial structure on niobium films, while *Pi* et al. [24.255] produced columnar defects by heavy-ion bombardment in a Bi$_2$Sr$_2$CaCu$_2$O$_8$ single crystal to study the strong pinning at these defects.

Figure 24.29 demonstrates that MFM is sensitive to the polarity of vortices. In Fig. 24.29a, six vortices have been produced in a niobium film by field cooling in $+0.5$ mT. The external magnetic field and tip magnetization are parallel, and therefore the tip–vortex interaction is attractive (bright contrast). To remove the vortices, the niobium was heated above T_c (≈ 9 K). Thereafter, vortices of opposite polarity were produced by field-cooling in -0.5 mT, which appear dark in Fig. 24.29b. The vortices are probably bound to strong pinning sites, because the vortex positions are identical in both images of Fig. 24.29. By imaging the vortices at different scanning heights, *Roseman* et al. [24.253] tried to extract values for the London penetration depth from the scan-height dependence of their profiles. While good qualitative agreement with theoretical predictions has been found, the absolute values do not agree with published literature values. The disagreement was attributed to the convolution between the tip and vortex stray fields. Better values might be obtained with calibrated tips.

24.4.6 Magnetic Exchange Force Microscopy

The resolution of MFM is limited to the nanometer range, because the long-range magnetostatic tip–sample interaction is not localized between individual surface atoms and the foremost tip apex atom [24.257]. As early as 1991 *Wiesendanger* et al. [24.258] proposed that the short-range magnetic exchange interaction could be utilized to image the configuration of magnetic moments with atomic resolution. For the suggested test system NiO(001), an antiferromagnetic insulator, *Momida* and *Oguchi* [24.259] provided density-functional calcula-

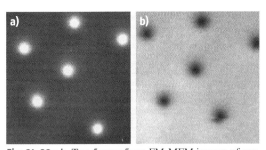

Fig. 24.29a,b Two 5 μm × 5 μm FM-MFM images of vortices in a niobium thin film after field-cooling at 0.5 mT (**a**) and −0.5 mT (**b**), respectively. Since the external magnetic field was parallel in (**a**) and antiparallel in (**b**) with respect to the tip magnetization, the vortices exhibit opposite contrast. Strong pinning dominates the position of the vortices, since they appear at identical locations in (**a**) and (**b**) and are not arranged in a regular Abrikosov lattice (© P. Grütter [24.253])

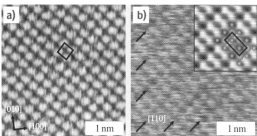

Fig. 24.30 (a) Pure chemical contrast on NiO(001) obtained with AFM using a nonmagnetic tip. Oxygen and nickel atoms are represented as maxima and minima, respectively, forming the (1×1) surface unit cell (*black square*). Arrows indicate the main crystallographic directions. (b) Additional modulation on neighboring nickel rows along the [110] direction (see *arrows*) due to the magnetic exchange interaction obtained with MExFM using a magnetic tip. The (2×1) structure (*black rectangle in the inset*) represents the magnetic surface unit cell. The *inset* is tiled together from the averaged magnetic unit cell calculated from the raw data, whereby the signal-to-noise ratio is significantly increased

tions. They found a magnetic exchange force between the magnetic moments of a single iron atom (the tip) and nickel surface atoms of more than 0.1 nN at tip–sample distances below 0.5 nm. Recently, *Kaiser* et al. [24.260] were able to prove the feasibility of magnetic exchange force microscopy (MExFM) on NiO(001). The superexchange between neighboring {111} planes via bridging oxygen atoms results in a row-wise antiferromagnetic configuration of magnetic moments on the (001) surface. Hence the magnetic surface unit cell is twice as large as the chemical surface unit cell. Figure 24.30a shows the atomic-scale contrast due to a pure chemical interaction. Maxima and minima correspond to the oxygen and nickel atoms, respectively. Their arrangement represents the (1×1) surface unit cell. Figure 24.30b exhibits an additional modulation on chemically and structurally equivalent rows of nickel atoms (the minima). The structure corresponds to the (2×1) magnetic surface unit cell. Since the spin-carrying nickel 3d states are highly localized, the magnetic contrast only becomes significant at very small tip–sample distances. More recently *Schmidt* et al. [24.261] were able to perform MExFM with much better signal-to-noise ratio on an itinerant metallic system: the antiferromagnetic iron monolayer on W(001). Density-functional theory performed with a realistic tip model indicated significant relaxations of tip and sample atoms during imaging.

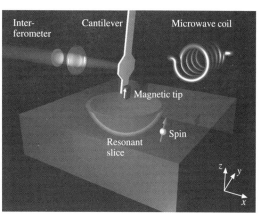

Fig. 24.31 MRFM setup. The cantilever with the magnetic tip oscillates parallel to the surface. Only electron spins within a hemispherical slice, where the stray field of the tip plus the external field matches the condition for magnetic resonance, can contribute to the MRFM signal due to cyclic spin inversion (© D. Rugar [24.262])

Moreover, a comparison between simulation and experimental data revealed complex interplay between chemical and magnetic interaction, which results in the observed atomic-scale contrast.

Even more ambitious is the proposed detection of individual nuclear spins by magnetic resonance force microscopy (MRFM) using a magnetic tip [24.263, 264]. Conventionally, nuclear spins are investigated by nuclear magnetic resonance (NMR), a spectroscopic technique to obtain microscopic chemical and physical information about molecules. An important application of NMR for medical diagnostics of the inside of humans is magnetic resonance imaging (MRI). This tomographic imaging technique uses the NMR signal from thin slices through the body to reconstruct its three-dimensional structure. Currently, at least 10^{12} nuclear spins must be present in a given volume to obtain a significant MRI signal. The ultimate goal of MRFM is to combine aspects of force microscopy with MRI to achieve true 3-D imaging with atomic resolution and elemental selectivity.

The experimental setup is sketched in Fig. 24.31. An oscillating cantilever with a magnetic tip at its end points toward the surface. The spherical resonant slice within the sample represents those points where the stray field from the tip and the external field match the condition for magnetic resonance. The cyclic spin flip causes a slight shift of the cantilever frequency due to the magnetic force exerted by the spin on the tip. Since the forces are extremely small, very low temperatures are required.

To date, no individual nuclear spins have been detected by MRFM. However, the design of ultrasensitive cantilevers has made considerable progress, and the detection of forces below 1×10^{-18} N has been achieved [24.202]. Therefore, it has become possible to perform nuclear magnetic resonance [24.265], and ferromagnetic resonance [24.266] experiments of spin ensembles with micrometer resolution. Moreover, in SiO_2 the magnetic moment of a single electron, which is three orders of magnitude larger than the nuclear magnetic moment, could be detected [24.262] using the setup shown in Fig. 24.31 at 1.6 K. This major breakthrough demonstrates the capability of force microscopy to detect single spins.

References

24.1 G. Binnig, H. Rohrer, C. Gerber, E. Weibel: Surface studies by scanning tunneling microscopy, Phys. Rev. Lett. **49**, 57–61 (1982)

24.2 R. Wiesendanger: *Scanning Probe Microscopy and Spectroscopy* (Cambridge Univ. Press, Cambridge 1994)

24.3 M. Tinkham: *Introduction to Superconductivity* (McGraw–Hill, New York 1996)

24.4 J. Kondo: Theory of dilute magnetic alloys, Solid State Phys. **23**, 183–281 (1969)

24.5 T.R. Albrecht, P. Grütter, H.K. Horne, D. Rugar: Frequency modulation detection using high-Q cantilevers for enhanced force microscope sensitivity, J. Appl. Phys. **69**, 668–673 (1991)

24.6 F.J. Giessibl, H. Bielefeld, S. Hembacher, J. Mannhart: Calculation of the optimal imaging parameters for frequency modulation atomic force microscopy, Appl. Surf. Sci. **140**, 352–357 (1999)

24.7 W. Allers, A. Schwarz, U.D. Schwarz, R. Wiesendanger: Dynamic scanning force microscopy at low temperatures on a van der Waals surface: Graphite(0001), Appl. Surf. Sci. **140**, 247–252 (1999)

24.8 W. Allers, A. Schwarz, U.D. Schwarz, R. Wiesendanger: Dynamic scanning force microscopy at low temperatures on a noble-gas crystal: Atomic resolution on the xenon(111) surface, Europhys. Lett. **48**, 276–279 (1999)

24.9 M. Morgenstern, D. Haude, V. Gudmundsson, C. Wittneven, R. Dombrowski, R. Wiesendanger: Origin of Landau oscillations observed in scanning tunneling spectroscopy on n-InAs(110), Phys. Rev. B **62**, 7257–7263 (2000)

24.10 D.M. Eigler, P.S. Weiss, E.K. Schweizer, N.D. Lang: Imaging Xe with a low-temperature scanning tunneling microscope, Phys. Rev. Lett. **66**, 1189–1192 (1991)

24.11 P.S. Weiss, D.M. Eigler: Site dependence of the apparent shape of a molecule in scanning tunneling micoscope images: Benzene on Pt(111), Phys. Rev. Lett. **71**, 3139–3142 (1992)

24.12 D.M. Eigler, E.K. Schweizer: Positioning single atoms with a scanning tunneling microscope, Nature **344**, 524–526 (1990)

24.13 H. Hug, B. Stiefel, P.J.A. van Schendel, A. Moser, S. Martin, H.-J. Güntherodt: A low temperature ultrahigh vacuum scanning force microscope, Rev. Sci. Instrum. **70**, 3627–3640 (1999)

24.14 S. Behler, M.K. Rose, D.F. Ogletree, F. Salmeron: Method to characterize the vibrational response of a beetle type scanning tunneling microscope, Rev. Sci. Instrum. **68**, 124–128 (1997)

24.15 C. Wittneven, R. Dombrowski, S.H. Pan, R. Wiesendanger: A low-temperature ultrahigh-vacuum scanning tunneling microscope with rotatable magnetic field, Rev. Sci. Instrum. **68**, 3806–3810 (1997)

24.16 W. Allers, A. Schwarz, U.D. Schwarz, R. Wiesendanger: A scanning force microscope with atomic resolution in ultrahigh vacuum and at low temperatures, Rev. Sci. Instrum. **69**, 221–225 (1998)

24.17 G. Dujardin, R.E. Walkup, P. Avouris: Dissociation of individual molecules with electrons from the tip of a scanning tunneling microscope, Science **255**, 1232–1235 (1992)

24.18 H.J. Lee, W. Ho: Single-bond formation and characterization with a scanning tunneling microscope, Science **286**, 1719–1722 (1999)

24.19 J. Repp, G. Meyer, F.E. Olsson, M. Persson: Controlling the charge state of individual gold adatoms, Science **305**, 493–495 (2004)

24.20 R. Berndt, R. Gaisch, J.K. Gimzewski, B. Reihl, R.R. Schlittler, W.D. Schneider, M. Tschudy: Photon emission at molecular resolution induced by a scanning tunneling microscope, Science **262**, 1425–1427 (1993)

24.21 G.V. Nazin, X.H. Qui, W. Ho: Atomic engineering of photon emission with a scanning tunneling microscopy, Phys. Rev. Lett. **90**, 216110-1–216110-4 (2003)

24.22 B.G. Briner, M. Doering, H.P. Rust, A.M. Bradshaw: Microscopic diffusion enhanced by adsorbate interaction, Science **278**, 257–260 (1997)

24.23 F. Meier, L. Zhou, J. Wiebe, R. Wiesendanger: Revealing magnetic interactions from single-atom magnetization curves, Science **320**, 82–86 (2008)

24.24 J. Kliewer, R. Berndt, E.V. Chulkov, V.M. Silkin, P.M. Echenique, S. Crampin: Dimensionality effects in the lifetime of surface states, Science **288**, 1399–1401 (2000)

24.25 M.F. Crommie, C.P. Lutz, D.M. Eigler: Imaging standing waves in a two-dimensional electron gas, Nature **363**, 524–527 (1993)

24.26 B.C. Stipe, M.A. Rezaei, W. Ho: Single-molecule vibrational spectroscopy and microscopy, Science **280**, 1732–1735 (1998)

24.27 A.J. Heinrich, J.A. Gupta, C.P. Lutz, D. Eigler: Single-atom spin-flip spectroscopy, Science **306**, 466–469 (2004)

24.28 H. Gawronski, M. Mehlhorn, K. Morgenstern: Imaging phonon excitation with atomic resolution, Science **319**, 930–933 (2008)

24.29 C.W.J. Beenakker, H. van Houten: Quantum transport in semiconductor nanostructures, Solid State Phys. **44**, 1–228 (1991)

24.30 K.M. Lang, V. Madhavan, J.E. Hoffman, E.W. Hudson, H. Eisaki, S. Uchida, J.C. Davis: Imaging the granular structure of high-T_c superconductivity in underdoped $Bi_2Sr_2CaCu_2O_{8+\delta}$, Nature **415**, 412–416 (2002)

24.31 J. Lee, K. Fujita, K. McElroy, J.A. Slezak, M. Wang, Y. Aiura, H. Bando, M. Ishikado, T. Masui, J.X. Zhu, A.V. Balatsky, H. Eisaki, S. Uchida, J.C. Davis: Interplay of electron-lattice interactions and superconductivity in $Bi_2Sr_2CaCu_2O_{8+\delta}$, Nature **442**, 546–550 (2006)

24.32 K.K. Gomes, A.N. Pasupathy, A. Pushp, S. Ono, Y. Ando, A. Yazdani: Visualizing pair formation on the atomic scale in the high-T_c superconductor $Bi_2Sr_2CaCu_2O_{8+\delta}$, Nature **447**, 569–572 (2007)

24.33 R.S. Becker, J.A. Golovchenko, B.S. Swartzentruber: Atomic-scale surface modifications using a tunneling microscope, Nature **325**, 419–421 (1987)

24.34 P.G. Piva, G.A. DiLabio, J.L. Pitters, J. Zikovsky, M. Rezeq, S. Dogel, W.A. Hofer, R.A. Wolkow: Field regulation of single-molecule conductivity by a charged surface atom, Nature **435**, 658–661 (2005)

24.35 J.A. Stroscio, D.M. Eigler: Atomic and molecular manipulation with the scanning tunneling microscope, Science **254**, 1319–1326 (1991)

24.36 A.J. Heinrich, J.A. Gupta, C.P. Lutz, D. Eigler: Molecule cascades, Science **298**, 1381–1387 (2002)

24.37 L. Bartels, G. Meyer, K.H. Rieder: Basic steps of lateral manipulation of single atoms and diatomic clusters with a scanning tunneling microscope, Phys. Rev. Lett. **79**, 697–700 (1997)

24.38 J.A. Stroscio, R.J. Celotta: Controlling the dynamics of a single atom in lateral atom manipulation, Science **306**, 242–247 (2004)

24.39 J.J. Schulz, R. Koch, K.H. Rieder: New mechanism for single atom manipulation, Phys. Rev. Lett. **84**, 4597–4600 (2000)

24.40 J.A. Stroscio, F. Tavazza, J.N. Crain, R.J. Celotta, A.M. Chaka: Electronically induced atom motion in engineered $CoCu_n$ nanostructures, Science **313**, 948–951 (2006)

24.41 M. Lastapis, M. Martin, D. Riedel, L. Hellner, G. Comtet, G. Dujardin: Picometer-scale electronic control of molecular dynamics inside a single molecule, Science **308**, 1000–1003 (2005)

24.42 M. Ternes, C.P. Lutz, C.F. Hirjibehedin, F.J. Giessibl, A.J. Heinrich: The force needed to move an atom on a surface, Science **319**, 1066–1069 (2008)

24.43 J.I. Pascual, N. Lorente, Z. Song, H. Conrad, H.P. Rust: Selectivity in vibrationally mediated single-molecule chemistry, Nature **423**, 525–528 (2003)

24.44 T.C. Shen, C. Wang, G.C. Abeln, J.R. Tucker, J.W. Lyding, P. Avouris, R.E. Walkup: Atomic-scale desorption through electronic and vibrational excitation mechanisms, Science **268**, 1590–1592 (1995)

24.45 T. Komeda, Y. Kim, M. Kawai, B.N.J. Persson, H. Ueba: Lateral hopping of molecules induced by excitations of internal vibration mode, Science **295**, 2055–2058 (2002)

24.46 Y.W. Mo: Reversible rotation of antimony dimers on the silicon(001) surface with a scanning tunneling microscope, Science **261**, 886–888 (1993)

24.47 B.C. Stipe, M.A. Rezaei, W. Ho: Inducing and viewing the rotational motion of a single molecule, Science **279**, 1907–1909 (1998)

24.48 P. Liljeroth, J. Repp, G. Meyer: Current-induced hydrogen tautomerization and conductance switching of napthalocyanine molecules, Science **317**, 1203–1206 (2007)

24.49 P. Maksymovych, D.B. Dougherty, X.-Y. Zhu, J.T. Yates Jr.: Nonlocal dissociative chemistry of adsorbed molecules induced by localized electron injection into metal surfaces, Phys. Rev. Lett. **99**, 016101-1–016101-4 (2007)

24.50 G.V. Nazin, X.H. Qiu, W. Ho: Visualization and spectroscopy of a metal-molecule-metal bridge, Science **302**, 77–81 (2003)

24.51 J. Repp, G. Meyer, S. Paavilainen, F.E. Olsson, M. Persson: Imaging bond formation between a gold atom and pentacene on an insulating surface, Science **312**, 1196–1199 (2006)

24.52 S. Katano, Y. Kim, M. Hori, M. Trenary, M. Kawai: Reversible control of hydrogenation of a single molecule, Science **316**, 1883–1886 (2007)

24.53 R. Yamachika, M. Grobis, A. Wachowiak, M.F. Crommie: Controlled atomic doping of a single C_{60} molecule, Science **304**, 281–284 (2004)

24.54 S.W. Hla, L. Bartels, G. Meyer, K.H. Rieder: Inducing all steps of a chemical reaction with the scanning tunneling microscope tip: Towards single molecule engineering, Phys. Rev. Lett. **85**, 2777–2780 (2000)

24.55 Y. Kim, T. Komeda, M. Kawai: Single-molecule reaction and characterization by vibrational ex-

24.56 M. Berthe, R. Stiufiuc, B. Grandidier, D. Deresmes, C. Delerue, D. Stievenard: Probing the carrier capture rate of a single quantum level, Science **319**, 436–438 (2008)

24.57 N. Neel, J. Kröger, L. Limot, K. Palotas, W.A. Hofer, R. Berndt: Conductance and Kondo effect in a controlled single-atom contact, Phys. Rev. Lett. **98**, 016801-1–016801-4 (2007)

24.58 F. Rosei, M. Schunack, P. Jiang, A. Gourdon, E. Laegsgaard, I. Stensgaard, C. Joachim, F. Besenbacher: Organic molecules acting as templates on metal surfaces, Science **296**, 328–331 (2002)

24.59 E. Ganz, S.K. Theiss, I.S. Hwang, J. Golovchenko: Direct measurement of diffusion by hot tunneling microscopy: Activations energy, anisotropy, and long jumps, Phys. Rev. Lett. **68**, 1567–1570 (1992)

24.60 M. Schunack, T.R. Linderoth, F. Rosei, E. Laegsgaard, I. Stensgaard, F. Besenbacher: Long jumps in the surface diffusion of large molecules, Phys. Rev. Lett. **88**, 156102-1–156102-4 (2002)

24.61 K.L. Wong, G. Pawin, K.Y. Kwon, X. Lin, T. Jiao, U. Solanki, R.H.J. Fawcett, L. Bartels, S. Stolbov, T.S. Rahman: A molecule carrier, Science **315**, 1391–1393 (2007)

24.62 L.J. Lauhon, W. Ho: Direct observation of the quantum tunneling of single hydrogen atoms with a scanning tunneling microscope, Phys. Rev. Lett. **85**, 4566–4569 (2000)

24.63 F. Ronci, S. Colonna, A. Cricenti, G. LeLay: Evidence of Sn adatoms quantum tunneling at the α-Sn/Si(111) surface, Phys. Rev. Lett. **99**, 166103-1–166103-4 (2007)

24.64 N. Kitamura, M. Lagally, M.B. Webb: Real-time observation of vacancy diffusion on Si(001)-(2×1) by scanning tunneling microscopy, Phys. Rev. Lett. **71**, 2082–2085 (1993)

24.65 M. Morgenstern, T. Michely, G. Comsa: Onset of interstitial diffusion determined by scanning tunneling microscopy, Phys. Rev. Lett. **79**, 1305–1308 (1997)

24.66 K. Morgenstern, G. Rosenfeld, B. Poelsema, G. Comsa: Brownian motion of vacancy islands on Ag(111), Phys. Rev. Lett. **74**, 2058–2061 (1995)

24.67 L. Bartels, F. Wang, D. Möller, E. Knoesel, T.F. Heinz: Real-space observation of molecular motion induced by femtosecond laser pulses, Science **305**, 648–651 (2004)

24.68 B. Reihl, J.H. Coombs, J.K. Gimzewski: Local inverse photoemission with the scanning tunneling microscope, Surf. Sci. **211/212**, 156–164 (1989)

24.69 R. Berndt, J.K. Gimzewski, P. Johansson: Inelastic tunneling excitation of tip-induced plasmon modes on noble-metal surfaces, Phys. Rev. Lett. **67**, 3796–3799 (1991)

24.70 P. Johansson, R. Monreal, P. Apell: Theory for light emission from a scanning tunneling microscope, Phys. Rev. B **42**, 9210–9213 (1990)

24.71 J. Aizpurua, G. Hoffmann, S.P. Apell, R. Berndt: Electromagnetic coupling on an atomic scale, Phys. Rev. Lett. **89**, 156803-1–156803-4 (2002)

24.72 G. Hoffmann, J. Kliewer, R. Berndt: Luminescence from metallic quantum wells in a scanning tunneling microscope, Phys. Rev. Lett. **78**, 176803-1–176803-4 (2001)

24.73 X.H. Qiu, G.V. Nazin, W. Ho: Vibrationally resolved fluorescence excited with submolecular precission, Science **299**, 542–546 (2003)

24.74 E. Cavar, M.C. Blüm, M. Pivetta, F. Patthey, M. Chergui, W.D. Schneider: Fluorescence and phosphorescence from individual C_{60} molecules excited by local electron tunneling, Phys. Rev. Lett. **95**, 196102-1–196102-4 (2005)

24.75 S.W. Wu, N. Ogawa, W. Ho: Atomic-scale coupling of photons to single-molecule junctions, Science **312**, 1362–1365 (2006)

24.76 A. Downes, M.E. Welland: Photon emission from Si(111)-(7×7) induced by scanning tunneling microscopy: atomic scale and material contrast, Phys. Rev. Lett. **81**, 1857–1860 (1998)

24.77 M. Kemerink, K. Sauthoff, P.M. Koenraad, J.W. Geritsen, H. van Kempen, J.H. Wolter: Optical detection of ballistic electrons injected by a scanning-tunneling microscope, Phys. Rev. Lett. **86**, 2404–2407 (2001)

24.78 J. Tersoff, D.R. Hamann: Theory and application for the scanning tunneling microscope, Phys. Rev. Lett. **50**, 1998–2001 (1983)

24.79 C.J. Chen: *Introduction to Scanning Tunneling Microscopy* (Oxford Univ. Press, Oxford 1993)

24.80 J. Winterlin, J. Wiechers, H. Brune, T. Gritsch, H. Hofer, R.J. Behm: Atomic-resolution imaging of close-packed metal surfaces by scanning tunneling microscopy, Phys. Rev. Lett. **62**, 59–62 (1989)

24.81 J.A. Stroscio, R.M. Feenstra, A.P. Fein: Electronic structure of the Si(111) 2×1 surface by scanning-tunneling microscopy, Phys. Rev. Lett. **57**, 2579–2582 (1986)

24.82 A.L. Vázquez de Parga, O.S. Hernan, R. Miranda, A. Levy Yeyati, N. Mingo, A. Martín-Rodero, F. Flores: Electron resonances in sharp tips and their role in tunneling spectroscopy, Phys. Rev. Lett. **80**, 357–360 (1998)

24.83 S.H. Pan, E.W. Hudson, J.C. Davis: Vacuum tunneling of superconducting quasiparticles from atomically sharp scanning tunneling microscope tips, Appl. Phys. Lett. **73**, 2992–2994 (1998)

24.84 J.T. Li, W.D. Schneider, R. Berndt, O.R. Bryant, S. Crampin: Surface-state lifetime measured by scanning tunneling spectroscopy, Phys. Rev. Lett. **81**, 4464–4467 (1998)

24.85 L. Bürgi, O. Jeandupeux, H. Brune, K. Kern: Probing hot-electron dynamics with a cold scanning tun-

24.86 J.W.G. Wildoer, C.J.P.M. Harmans, H. van Kempen: Observation of Landau levels at the InAs(110) surface by scanning tunneling spectroscopy, Phys. Rev. B **55**, R16013–R16016 (1997)

24.87 M. Morgenstern, V. Gudmundsson, C. Wittneven, R. Dombrowski, R. Wiesendanger: Nonlocality of the exchange interaction probed by scanning tunneling spectroscopy, Phys. Rev. B **63**, 201301(R)-1–201301(R)-4 (2001)

24.88 F.E. Olsson, M. Persson, A.G. Borisov, J.P. Gauyacq, J. Lagoute, S. Fölsch: Localization of the Cu(111) surface state by single Cu adatoms, Phys. Rev. Lett. **93**, 206803-1–206803-4 (2004)

24.89 L. Limot, E. Pehlke, J. Kröger, R. Berndt: Surface-state localization at adatoms, Phys. Rev. Lett. **94**, 036805-1–036805-4 (2005)

24.90 N. Nilius, T.M. Wallis, M. Persson, W. Ho: Distance dependence of the interaction between single atoms: Gold dimers on NiAl(110), Phys. Rev. Lett. **90**, 196103-1–196103-4 (2003)

24.91 H.J. Lee, W. Ho, M. Persson: Spin splitting of s and p states in single atoms and magnetic coupling in dimers on a surface, Phys. Rev. Lett. **92**, 186802-1–186802-4 (2004)

24.92 M.V. Grishin, F.I. Dalidchik, S.A. Kovalevskii, N.N. Kolchenko, B.R. Shub: Isotope effect in the vibrational spectra of water measured in experiments with a scanning tunneling microscope, JETP Lett. **66**, 37–40 (1997)

24.93 Y. Sainoo, Y. Kim, T. Okawa, T. Komeda, H. Shigekawa, M. Kawai: Excitation of molecular vibrational modes with inelastic scanning tunneling microscopy: Examination through action spectra of *cis*-2-butene on Pd(110), Phys. Rev. Lett. **95**, 246102-1–246102-4 (2005)

24.94 X.H. Qiu, G.V. Nazin, W. Ho: Vibronic states in single molecule electron transport, Phys. Rev. Lett. **92**, 206102-1–206102-4 (2004)

24.95 L. Vitali, M. Burghard, M.A. Schneider, L. Liu, S.Y. Wu, C.S. Jayanthi, K. Kern: Phonon spectromicroscopy of carbon nanostructures with atomic resolution, Phys. Rev. Lett. **93**, 136103-1–136103-4 (2004)

24.96 B.J. LeRoy, S.G. Lemay, J. Kong, C. Dekker: Electrical generation and absorption of phonons in carbon nanotubes, Nature **432**, 371–374 (2004)

24.97 L. Vitali, M.A. Schneider, K. Kern, L. Wirtz, A. Rubio: Phonon and plasmon excitation in inelastic scanning tunneling spectroscopy of graphite, Phys. Rev. B **69**, 121414-1–121414-4 (2004)

24.98 T. Balashov, A.F. Takacz, W. Wulfhekel, J. Kirschner: Magnon excitation with spin-polarized scanning tunneling microscopy, Phys. Rev. Lett. **97**, 187201-1–187201-4 (2006)

24.99 C.F. Hirjibehedin, C.P. Lutz, A.J. Heinrich: Spin coupling in engineered atomic structures, Science **312**, 1021–1024 (2006)

24.100 C.F. Hirjibehedin, C.Y. Lin, A.F. Otte, M. Ternes, C.P. Lutz, B.A. Jones, A.J. Heinrich: Large magnetic anisotropy of a single atomic spin embedded in a surface molecular network, Science **317**, 1199–1203 (2007)

24.101 U. Kemiktarak, T. Ndukum, K.C. Schwab, K.L. Ekinci: Radio-frequency scanning tunneling microscopy, Nature **450**, 85–89 (2007)

24.102 A. Hewson: *From the Kondo Effect to Heavy Fermions* (Cambridge Univ. Press, Cambridge 1993)

24.103 V. Madhavan, W. Chen, T. Jamneala, M.F. Crommie, N.S. Wingreen: Tunneling into a single magnetic atom: Spectroscopic evidence of the Kondo resonance, Science **280**, 567–569 (1998)

24.104 J. Li, W.D. Schneider, R. Berndt, B. Delley: Kondo scattering observed at a single magnetic impurity, Phys. Rev. Lett. **80**, 2893–2896 (1998)

24.105 T.W. Odom, J.L. Huang, C.L. Cheung, C.M. Lieber: Magnetic clusters on single-walled carbon nanotubes: the Kondo effect in a one-dimensional host, Science **290**, 1549–1552 (2000)

24.106 M. Ouyang, J.L. Huang, C.L. Cheung, C.M. Lieber: Energy gaps in metallic single-walled carbon nanotubes, Science **292**, 702–705 (2001)

24.107 U. Fano: Effects of configuration interaction on intensities and phase shifts, Phys. Rev. **124**, 1866–1878 (1961)

24.108 P. Wahl, L. Diekhöner, M.A. Schneider, L. Vitali, G. Wittich, K. Kern: Kondo temperature of magnetic impurities at surfaces, Phys. Rev. Lett. **93**, 176603-1–176603-4 (2004)

24.109 Y.S. Fu, S.H. Ji, X. Chen, X.C. Ma, R. Wu, C.C. Wang, W.H. Duan, X.H. Qiu, B. Sun, P. Zhang, J.F. Jia, Q.K. Xue: Manipulating the Kondo resonance through quantum size effects, Phys. Rev. Lett. **99**, 256601-1–256601-4 (2007)

24.110 J. Henzl, K. Morgenstern: Contribution of the surface state to the observation of the surface Kondo resonance, Phys. Rev. Lett. **98**, 266601-1–266601-4 (2007)

24.111 P. Wahl, P. Simon, L. Diekhöner, V.S. Stepanyuk, P. Bruno, M.A. Schneider, K. Kern: Exchange interaction between single magnetic adatoms, Phys. Rev. Lett. **98**, 056601-1–056601-4 (2007)

24.112 T. Jamneala, V. Madhavan, M.F. Crommie: Kondo response of a single antiferromagnetic chromium trimer, Phys. Rev. Lett. **87**, 256804-1–256804-4 (2001)

24.113 P. Wahl, L. Diekhöner, G. Wittich, L. Vitali, M.A. Schneider, K. Kern: Kondo effect of molecular complexes at surfaces: Ligand control of the local spin coupling, Phys. Rev. Lett. **95**, 166601-1–166601-4 (2005)

24.114 A. Zhao, Q. Li, L. Chen, H. Xiang, W. Wang, S. Pan, B. Wang, X. Xiao, J. Yang, J.G. Hou, Q. Zhu: Controlling the Kondo effect of an adsorbed magnetic ion through its chemical bonding, Science **309**, 1542–1544 (2005)

24.115 V. Iancu, A. Deshpande, S.W. Hla: Manipulation of the Kondo effect via two-dimensional molecular assembly, Phys. Rev. Lett. **97**, 266603-1–266603-4 (2006)

24.116 L. Gao, W. Ji, Y.B. Hu, Z.H. Cheng, Z.T. Deng, Q. Liu, N. Jiang, X. Lin, W. Guo, S.X. Du, W.A. Hofer, X.C. Xie, H.J. Gao: Site-specific Kondo effect at ambient temperatures in iron-based molecules, Phys. Rev. Lett. **99**, 106402-1–106402-4 (2007)

24.117 H.C. Manoharan, C.P. Lutz, D.M. Eigler: Quantum mirages formed by coherent projection of electronic structure, Nature **403**, 512–515 (2000)

24.118 H.A. Mizes, J.S. Foster: Long-range electronic perturbations caused by defects using scanning tunneling microscopy, Science **244**, 559–562 (1989)

24.119 P.T. Sprunger, L. Petersen, E.W. Plummer, E. Laegsgaard, F. Besenbacher: Giant Friedel oscillations on beryllium(0001) surface, Science **275**, 1764–1767 (1997)

24.120 P. Hofmann, B.G. Briner, M. Doering, H.P. Rust, E.W. Plummer, A.M. Bradshaw: Anisotropic two-dimensional Friedel oscillations, Phys. Rev. Lett. **79**, 265–268 (1997)

24.121 E.J. Heller, M.F. Crommie, C.P. Lutz, D.M. Eigler: Scattering and adsorption of surface electron waves in quantum corrals, Nature **369**, 464–466 (1994)

24.122 M.C.M.M. van der Wielen, A.J.A. van Roij, H. van Kempen: Direct observation of Friedel oscillations around incorporated Si_{Ga} dopants in GaAs by low-temperature scanning tunneling microscopy, Phys. Rev. Lett. **76**, 1075–1078 (1996)

24.123 O. Millo, D. Katz, Y.W. Cao, U. Banin: Imaging and spectroscopy of artificial-atom states in core/shell nanocrystal quantum dots, Phys. Rev. Lett. **86**, 5751–5754 (2001)

24.124 T. Maltezopoulos, A. Bolz, C. Meyer, C. Heyn, W. Hansen, M. Morgenstern, R. Wiesendanger: Wave-function mapping of InAs qunatum dots by scanning tunneling spectroscopy, Phys. Rev. Lett. **91**, 196804-1–196804-4 (2003)

24.125 R. Temirov, S. Soubatch, A. Luican, F.S. Tautz: Free-electron-like dispersion in an organic monolayer film on a metal substrate, Nature **444**, 350–353 (2006)

24.126 K. Suzuki, K. Kanisawa, C. Janer, S. Perraud, K. Takashina, T. Fujisawa, Y. Hirayama: Spatial imaging of two-dimensional electronic states in semiconductor quantum wells, Phys. Rev. Lett. **98**, 136802-1–136802-4 (2007)

24.127 L.C. Venema, J.W.G. Wildoer, J.W. Janssen, S.J. Tans, L.J.T. Tuinstra, L.P. Kouwenhoven, C. Dekker: Imaging electron wave functions of quantized energy levels in carbon nanotubes, Nature **283**, 52–55 (1999)

24.128 S.G. Lemay, J.W. Jannsen, M. van den Hout, M. Mooij, M.J. Bronikowski, P.A. Willis, R.E. Smalley, L.P. Kouwenhoven, C. Dekker: Two-dimensional imaging of electronic wavefunctions in carbon nanotubes, Nature **412**, 617–620 (2001)

24.129 C.R. Moon, L.S. Matos, B.K. Foster, G. Zeltzer, W. Ko, H.C. Manoharan: Quantum phase extraction in isospectral electronic nanostructures, Science **319**, 782–787 (2008)

24.130 X. Lu, M. Grobis, K.H. Khoo, S.G. Louie, M.F. Crommie: Spatially mapping the spectral density of a single C_{60} molecule, Phys. Rev. Lett. **90**, 096802-1–096802-4 (2003)

24.131 A.M. Yakunin, A.Y. Silov, P.M. Koenraad, J.H. Wolter, W. van Roy, J. de Boeck, J.M. Tang, M.E. Flatte: Spatial structure of an individual Mn acceptor in GaAs, Phys. Rev. Lett. **92**, 216806-1–216806-4 (200405)

24.132 F. Marczinowski, J. Wiebe, J.M. Tang, M.E. Flatte, F. Meier, M. Morgenstern, R. Wiesendanger: Local electronic structure near Mn acceptors in InAs: Surface-induced symmetry breaking and coupling to host states, Phys. Rev. Lett. **99**, 157202-1–157202-4 (2007)

24.133 N. Nilius, T.M. Wallis, W. Ho: Development of one-dimensional band structure in artificial gold chains, Science **297**, 1853–1856 (2002)

24.134 J.N. Crain, D.T. Pierce: End states in one-dimensional atom chains, Science **307**, 703–706 (2005)

24.135 D. Kitchen, A. Richardella, J.M. Tang, M.E. Flatte, A. Yazdani: Atom-by-atom substitution of Mn in GaAs and visualization of their hole-mediated interactions, Nature **442**, 436–439 (2006)

24.136 C. Wittneven, R. Dombrowski, M. Morgenstern, R. Wiesendanger: Scattering states of ionized dopants probed by low temperature scanning tunneling spectroscopy, Phys. Rev. Lett. **81**, 5616–5619 (1998)

24.137 D. Haude, M. Morgenstern, I. Meinel, R. Wiesendanger: Local density of states of a three-dimensional conductor in the extreme quantum limit, Phys. Rev. Lett. **86**, 1582–1585 (2001)

24.138 R. Joynt, R.E. Prange: Conditions for the quantum Hall effect, Phys. Rev. B **29**, 3303–3317 (1984)

24.139 M. Morgenstern, J. Klijn, R. Wiesendanger: Real space observation of drift states in a two-dimensional electron system at high magnetic fields, Phys. Rev. Lett. **90**, 056804-1–056804-4 (2003)

24.140 M. Morgenstern, J. Klijn, C. Meyer, M. Getzlaff, R. Adelung, R.A. Römer, K. Rossnagel, L. Kipp, M. Skibowski, R. Wiesendanger: Direct comparison between potential landscape and local density of states in a disordered two-dimensional electron system, Phys. Rev. Lett. **89**, 136806-1–136806-4 (2002)

24.141 E. Abrahams, P.W. Anderson, D.C. Licciardello, T.V. Ramakrishnan: Scaling theory of localization: absence of quantum diffusion in two dimensions, Phys. Rev. Lett. **42**, 673–676 (1979)

24.142 C. Meyer, J. Klijn, M. Morgenstern, R. Wiesendanger: Direct measurement of the local density of states of a disordered one-dimensional conductor, Phys. Rev. Lett. **91**, 076803-1–076803-4 (2003)

24.143 N. Oncel, A. van Houselt, J. Huijben, A.S. Hallbäck, O. Gurlu, H.J.W. Zandvliet, B. Poelsema: Quantum confinement between self-organized Pt nanowires on Ge(001), Phys. Rev. Lett. **95**, 116801-1–116801-4 (2005)

24.144 J. Lee, S. Eggert, H. Kim, S.J. Kahng, H. Shinohara, Y. Kuk: Real space imaging of one-dimensional standing waves: Direct evidence for a Luttinger liquid, Phys. Rev. Lett. **93**, 166403-1–166403-4 (2004)

24.145 R.E. Peierls: *Quantum Theory of Solids* (Clarendon, Oxford 1955)

24.146 C.G. Slough, W.W. McNairy, R.V. Coleman, B. Drake, P.K. Hansma: Charge-density waves studied with the use of a scanning tunneling microscope, Phys. Rev. B **34**, 994–1005 (1986)

24.147 X.L. Wu, C.M. Lieber: Hexagonal domain-like charge-density wave of TaS_2 determined by scanning tunneling microscopy, Science **243**, 1703–1705 (1989)

24.148 T. Nishiguchi, M. Kageshima, N. Ara-Kato, A. Kawazu: Behaviour of charge density waves in a one-dimensional organic conductor visualized by scanning tunneling microscopy, Phys. Rev. Lett. **81**, 3187–3190 (1998)

24.149 X.L. Wu, C.M. Lieber: Direct observation of growth and melting of the hexagonal-domain charge-density-wave phase in $1T-TaS_2$ by scanning tunneling microscopy, Phys. Rev. Lett. **64**, 1150–1153 (1990)

24.150 J.M. Carpinelli, H.H. Weitering, E.W. Plummer, R. Stumpf: Direct observation of a surface charge density wave, Nature **381**, 398–400 (1996)

24.151 H.H. Weitering, J.M. Carpinelli, A.V. Melechenko, J. Zhang, M. Bartkowiak, E.W. Plummer: Defect-mediated condensation of a charge density wave, Science **285**, 2107–2110 (1999)

24.152 S. Modesti, L. Petaccia, G. Ceballos, I. Vobornik, G. Panaccione, G. Rossi, L. Ottaviano, R. Larciprete, S. Lizzit, A. Goldoni: Insulating ground state of Sn/Si(111)-($\sqrt{3}\times\sqrt{3}$)R30°, Phys. Rev. Lett. **98**, 126401-1–126401-4 (2007)

24.153 H.W. Yeom, S. Takeda, E. Rotenberg, I. Matsuda, K. Horikoshi, J. Schäfer, C.M. Lee, S.D. Kevan, T. Ohta, T. Nagao, S. Hasegawa: Instability and charge density wave of metallic quantum chains on a silicon surface, Phys. Rev. Lett. **82**, 4898–4901 (1999)

24.154 K. Swamy, A. Menzel, R. Beer, E. Bertel: Charge-density waves in self-assembled halogen-bridged metal chains, Phys. Rev. Lett. **86**, 1299–1302 (2001)

24.155 J.J. Kim, W. Yamaguchi, T. Hasegawa, K. Kitazawa: Observation of Mott localization gap using low temperature scanning tunneling spectroscopy in commensurate $1T-TaSe_2$, Phys. Rev. Lett. **73**, 2103–2106 (1994)

24.156 A. Wachowiak, R. Yamachika, K.H. Khoo, Y. Wang, M. Grobis, D.H. Lee, S.G. Louie, M.F. Crommie: Visualization of the molecular Jahn–Teller effect in an insulating K_4C_{60} monolayer, Science **310**, 468–470 (2005)

24.157 J. Bardeen, L.N. Cooper, J.R. Schrieffer: Theory of superconductivity, Phys. Rev. **108**, 1175–1204 (1957)

24.158 A. Yazdani, B.A. Jones, C.P. Lutz, M.F. Crommie, D.M. Eigler: Probing the local effects of magnetic impurities on superconductivity, Science **275**, 1767–1770 (1997)

24.159 S.H. Tessmer, M.B. Tarlie, D.J. van Harlingen, D.L. Maslov, P.M. Goldbart: Probing the superconducting proximity effect in $NbSe_2$ by scanning tunneling micrsocopy, Phys. Rev. Lett. **77**, 924–927 (1996)

24.160 K. Inoue, H. Takayanagi: Local tunneling spectroscopy of Nb/InAs/Nb superconducting proximity system with a scanning tunneling microscope, Phys. Rev. B **43**, 6214–6215 (1991)

24.161 H.F. Hess, R.B. Robinson, R.C. Dynes, J.M. Valles, J.V. Waszczak: Scanning-tunneling-microscope observation of the Abrikosov flux lattice and the density of states near and inside a fluxoid, Phys. Rev. Lett. **62**, 214–217 (1989)

24.162 H.F. Hess, R.B. Robinson, J.V. Waszczak: Vortex-core structure observed with a scanning tunneling microscope, Phys. Rev. Lett. **64**, 2711–2714 (1990)

24.163 N. Hayashi, M. Ichioka, K. Machida: Star-shaped local density of states around vortices in a type-II superconductor, Phys. Rev. Lett. **77**, 4074–4077 (1996)

24.164 H. Sakata, M. Oosawa, K. Matsuba, N. Nishida: Imaging of vortex lattice transition in YNi_2B_2C by scanning tunneling spectroscopy, Phys. Rev. Lett. **84**, 1583–1586 (2000)

24.165 S. Behler, S.H. Pan, P. Jess, A. Baratoff, H.-J. Güntherodt, F. Levy, G. Wirth, J. Wiesner: Vortex pinning in ion-irradiated $NbSe_2$ studied by scanning tunneling microscopy, Phys. Rev. Lett. **72**, 1750–1753 (1994)

24.166 R. Berthe, U. Hartmann, C. Heiden: Influence of a transport current on the Abrikosov flux lattice observed with a low-temperature scanning tunneling microscope, Ultramicroscopy **42–44**, 696–698 (1992)

24.167 N.C. Yeh, C.T. Chen, G. Hammerl, J. Mannhart, A. Schmehl, C.W. Schneider, R.R. Schulz, S. Tajima, K. Yoshida, D. Garrigus, M. Strasik: Evidence of doping-dependent pairing symmetry in cuprate superconductors, Phys. Rev. Lett. **87**, 087003-1–087003-4 (2001)

24.168 K. McElroy, R.W. Simmonds, J.E. Hoffman, D.H. Lee, J. Orenstein, H. Eisaki, S. Uchida, J.C. Davis: Relating atomic-scale electronic phenomena to wave-like quasiparticle states in superconducting $Bi_2Sr_2CaCu_2O_{8+\delta}$, Nature **422**, 592–596 (2003)

24.169 K. McElroy, J. Lee, J.A. Slezak, D.H. Lee, H. Eisaki, S. Uchida, J.C. Davis: Atomic-scale sources and mechanism of nanoscale electronic disorder in $Bi_2Sr_2CaCu_2O_{8+\delta}$, Science **309**, 1048–1052 (2005)

24.170 S.H. Pan, E.W. Hudson, K.M. Lang, H. Eisaki, S. Uchida, J.C. Davis: Imaging the effects of individual zinc impurity atoms on superconductivity in $Bi_2Sr_2CaCu_2O_{8+\delta}$, Nature **403**, 746–750 (2000)

24.171 A. Polkovnikov, S. Sachdev, M. Vojta: Impurity in a d-wave superconductor: Kondo effect and STM spectra, Phys. Rev. Lett. **86**, 296–299 (2001)

24.172 A.N. Pasupathy, A. Pushp, K.K. Gomes, C.V. Parker, J. Wen, Z. Xu, G. Gu, S. Ono, Y. Ando, A. Yazdani: Electronic origin of the inhomogeneous pairing interaction in the high-T_c superconductor $Bi_2Sr_2CaCu_2O_{8+\delta}$, Science **320**, 196–201 (2008)

24.173 I. Maggio-Aprile, C. Renner, E. Erb, E. Walker, Ø. Fischer: Direct vortex lattice imaging and tunneling spectroscopy of flux lines on $YBa_2Cu_3O_{7-\delta}$, Phys. Rev. Lett. **75**, 2754–2757 (1995)

24.174 C. Renner, B. Revaz, K. Kadowaki, I. Maggio-Aprile, Ø. Fischer: Observation of the low temperature pseudogap in the vortex cores of $Bi_2Sr_2CaCu_2O_{8+\delta}$, Phys. Rev. Lett. **80**, 3606–3609 (1998)

24.175 S.H. Pan, E.W. Hudson, A.K. Gupta, K.W. Ng, H. Eisaki, S. Uchida, J.C. Davis: STM studies of the electronic structure of vortex cores in $Bi_2Sr_2CaCu_2O_{8+\delta}$, Phys. Rev. Lett. **85**, 1536–1539 (2000)

24.176 D.P. Arovas, A.J. Berlinsky, C. Kallin, S.C. Zhang: Superconducting vortex with antiferromagnetic core, Phys. Rev. Lett. **79**, 2871–2874 (1997)

24.177 J.E. Hoffmann, E.W. Hudson, K.M. Lang, V. Madhavan, H. Eisaki, S. Uchida, J.C. Davis: A four unit cell periodic pattern of quasi-particle states surrounding vortex cores in $Bi_2Sr_2CaCu_2O_{8+\delta}$, Science **295**, 466–469 (2002)

24.178 M. Vershinin, S. Misra, S. Ono, Y. Abe, Y. Ando, A. Yazdani: Local ordering in the pseudogap state of the high-T_c superconductor $Bi_2Sr_2CaCu_2O_{8+\delta}$, Science **303**, 1995–1998 (2004)

24.179 T. Hanaguri, C. Lupien, Y. Kohsaka, D.H. Lee, M. Azuma, M. Takano, H. Takagi, J.C. Davis: A 'checkerboard' electronic crystal state in lightly hole-doped $Ca_{2-x}Na_xCuO_2Cl_2$, Nature **430**, 1001–1005 (2004)

24.180 Y. Kohsaka, C. Taylor, K. Fujita, A. Schmidt, C. Lupien, T. Hanaguri, M. Azuma, M. Takano, H. Eisaki, H. Takagi, S. Uchida, J.C. Davis: An intrinsic bond-centered electronic glass with unidirectional domains in underdoped cuprates, Science **315**, 1380–1385 (2007)

24.181 M. Crespo, H. Suderow, S. Vieira, S. Bud'ko, P.C. Canfield: Local superconducting density of states of $ErNi_2B_2C$, Phys. Rev. Lett. **96**, 027003-1–027003-4 (2006)

24.182 H. Suderow, S. Vieira, J.D. Strand, S. Bud'ko, P.C. Canfield: Very-low-temperature tunneling spectroscopy in the heavy-fermion superconductor $PrOs_4Sb_{12}$, Phys. Rev. B **69**, 060504-1–060504-4 (2004)

24.183 O. Naaman, W. Teizer, R.C. Dynes: Fluctuation dominated Josephson tunneling with a scanning tunneling microscope, Phys. Rev. Lett. **87**, 097004-1–097004-4 (2001)

24.184 M. Fäth, S. Freisem, A.A. Menovsky, Y. Tomioka, J. Aarts, J.A. Mydosh: Spatially inhomogeneous metal-insulator transition in doped manganites, Science **285**, 1540–1542 (1999)

24.185 C. Renner, G. Aeppli, B.G. Kim, Y.A. Soh, S.W. Cheong: Atomic-scale images of charge ordering in a mixed-valence manganite, Nature **416**, 518–521 (2000)

24.186 H.M. Ronnov, C. Renner, G. Aeppli, T. Kimura, Y. Tokura: Polarons and confinement of electronic motion to two dimensions in a layered manganite, Nature **440**, 1025–1028 (2006)

24.187 M. Bode, M. Getzlaff, R. Wiesendanger: Spin-polarized vacuum tunneling into the exchange-split surface state of Gd(0001), Phys. Rev. Lett. **81**, 4256–4259 (1998)

24.188 A. Kubetzka, M. Bode, O. Pietzsch, R. Wiesendanger: Spin-polarized scanning tunneling microscopy with antiferromagnetic probe tips, Phys. Rev. Lett. **88**, 057201-1–057201-4 (2002)

24.189 O. Pietzsch, A. Kubetzka, M. Bode, R. Wiesendanger: Observation of magnetic hysteresis at the nanometer scale by spin-polarized scanning tunneling spectroscopy, Science **292**, 2053–2056 (2001)

24.190 S. Heinze, M. Bode, A. Kubetzka, O. Pietzsch, X. Xie, S. Blügel, R. Wiesendanger: Real-space imaging of two-dimensional antiferromagnetism on the atomic scale, Science **288**, 1805–1808 (2000)

24.191 A. Kubetzka, P. Ferriani, M. Bode, S. Heinze, G. Bihlmayer, K. von Bergmann, O. Pietzsch, S. Blügel, R. Wiesendanger: Revealing antiferromagnetic order of the Fe monolayer on W(001): Spin-polarized scanning tunneling microscopy and first-principles calculations, Phys. Rev. Lett. **94**, 087204-1–087204-4 (2005)

24.192 A. Wachowiak, J. Wiebe, M. Bode, O. Pietzsch, M. Morgenstern, R. Wiesendanger: Internal spin-structure of magnetic vortex cores observed by spin-polarized scanning tunneling microscopy, Science **298**, 577–580 (2002)

24.193 M. Bode, M. Heide, K. von Bergmann, P. Ferriani, S. Heinze, G. Bihlmeyer, A. Kubetzka, O. Pietzsch,

S. Blügel, R. Wiesendanger: Chiral magnetic order at surfaces driven by inversion asymmetry, Nature **447**, 190–193 (2007)

24.194 K. von Bergmann, S. Heinze, M. Bode, E.Y. Vedmedenko, G. Bihlmayer, S. Blügel, R. Wiesendanger: Observation of a complex nanoscale magnetic structure in a hexagonal Fe monolayer, Phys. Rev. Lett. **96**, 167203-1–167203-4 (2006)

24.195 C.L. Gao, U. Schlickum, W. Wulfhekel, J. Kirschner: Mapping the surface spin structure of large unit cells: Reconstructed Mn films on Fe(001), Phys. Rev. Lett. **98**, 107203-1–107203-4 (2007)

24.196 Y. Yayon, V.W. Brar, L. Senapati, S.C. Erwin, M.F. Crommie: Observing spin polarization of individual magnetic adatoms, Phys. Rev. Lett. **99**, 067202-1–067202-4 (2007)

24.197 M. Bode, O. Pietzsch, A. Kubetzka, R. Wiesendanger: Shape-dependent thermal switching behavior of superparamagnetic nanoislands, Phys. Rev. Lett. **92**, 067201-1–067201-4 (2004)

24.198 S. Krause, L. Berbil-Bautista, G. Herzog, M. Bode, R. Wiesendanger: Current-induced magnetization switching with a spin-polarized scanning tunneling microscope, Science **317**, 1537–1540 (2007)

24.199 M.D. Kirk, T.R. Albrecht, C.F. Quate: Low-temperature atomic force microscopy, Rev. Sci. Instrum. **59**, 833–835 (1988)

24.200 D. Pelekhov, J. Becker, J.G. Nunes: Atomic force microscope for operation in high magnetic fields at millikelvin temperatures, Rev. Sci. Instrum. **70**, 114–120 (1999)

24.201 J. Mou, Y. Jie, Z. Shao: An optical detection low temperature atomic force microscope at ambient pressure for biological research, Rev. Sci. Instrum. **64**, 1483–1488 (1993)

24.202 H.J. Mamin, D. Rugar: Sub-attonewton force detection at millikelvin temperatures, Appl. Phys. Lett. **79**, 3358–3360 (2001)

24.203 A. Schwarz, W. Allers, U.D. Schwarz, R. Wiesendanger: Dynamic mode scanning force microscopy of n-InAs(110)-(1×1) at low temperatures, Phys. Rev. B **61**, 2837–2845 (2000)

24.204 W. Allers, S. Langkat, R. Wiesendanger: Dynamic low-temperature scanning force microscopy on nickel oxide(001), Appl. Phys. A **72**, S27–S30 (2001)

24.205 F.J. Giessibl: Atomic resolution of the silicon(111)-(7×7) surface by atomic force microscopy, Science **267**, 68–71 (1995)

24.206 M.A. Lantz, H.J. Hug, P.J.A. van Schendel, R. Hoffmann, S. Martin, A. Baratoff, A. Abdurixit, H.-J. Güntherodt: Low temperature scanning force microscopy of the Si(111)-(7×7) surface, Phys. Rev. Lett. **84**, 2642–2645 (2000)

24.207 K. Suzuki, H. Iwatsuki, S. Kitamura, C.B. Mooney: Development of low temperature ultrahigh vacuum force microscope/scanning tunneling microscope, Jpn. J. Appl. Phys. **39**, 3750–3752 (2000)

24.208 N. Suehira, Y. Sugawara, S. Morita: Artifact and fact of Si(111)-(7×7) surface images observed with a low temperature noncontact atomic force microscope (LT-NC-AFM), Jpn. J. Appl. Phys. **40**, 292–294 (2001)

24.209 R. Pérez, M.C. Payne, I. Štich, K. Terakura: Role of covalent tip–surface interactions in noncontact atomic force microscopy on reactive surfaces, Phys. Rev. Lett. **78**, 678–681 (1997)

24.210 S.H. Ke, T. Uda, R. Pérez, I. Štich, K. Terakura: First principles investigation of tip–surface interaction on GaAs(110): Implication for atomic force and tunneling microscopies, Phys. Rev. B **60**, 11631–11638 (1999)

24.211 J. Tobik, I. Štich, R. Pérez, K. Terakura: Simulation of tip–surface interactions in atomic force microscopy of an InP(110) surface with a Si tip, Phys. Rev. B **60**, 11639–11644 (1999)

24.212 A. Schwarz, W. Allers, U.D. Schwarz, R. Wiesendanger: Simultaneous imaging of the In and As sublattice on InAs(110)-(1×1) with dynamic scanning force microscopy, Appl. Surf. Sci. **140**, 293–297 (1999)

24.213 H. Hölscher, W. Allers, U.D. Schwarz, A. Schwarz, R. Wiesendanger: Interpretation of 'true atomic resolution' images of graphite (0001) in noncontact atomic force microscopy, Phys. Rev. B **62**, 6967–6970 (2000)

24.214 H. Hölscher, W. Allers, U.D. Schwarz, A. Schwarz, R. Wiesendanger: Simulation of NC-AFM images of xenon(111), Appl. Phys. A **72**, S35–S38 (2001)

24.215 M. Ashino, A. Schwarz, T. Behnke, R. Wiesendanger: Atomic-resolution dynamic force microscopy and spectroscopy of a single-walled carbon nanotube: characterization of interatomic van der Waals forces, Phys. Rev. Lett. **93**, 136101-1–136101-4 (2004)

24.216 G. Schwarz, A. Kley, J. Neugebauer, M. Scheffler: Electronic and structural properties of vacancies on and below the GaP(110) surface, Phys. Rev. B **58**, 1392–1499 (1998)

24.217 M. Ashino, A. Schwarz, H. Hölscher, U.D. Schwarz, R. Wiesendanger: Interpretation of the atomic scale contrast obtained on graphite and single-walled carbon nanotubes in the dynamic mode of atomic force microscopy, Nanotechnology **16**, 134–137 (2005)

24.218 F.J. Giessibl, H. Bielefeldt, S. Hembacher, J. Mannhart: Calculation of the optimal imaging parameters for frequency modulation atomic force microscopy, Appl. Surf. Sci. **140**, 352–357 (1999)

24.219 S. Hembacher, F.J. Giessibl, J. Mannhart, C.F. Quate: Local spectroscopy and atomic imaging of tunneling current, forces, and dissipation on graphite, Phys. Rev. Lett. **94**, 056101-1–056101-4 (2005)

24.220 F.J. Giessibl: High-speed force sensor for force microscopy and profilometry utilizing a quartz tuning fork, Appl. Phys. Lett. **73**, 3956–3958 (1998)

24.221 H. Hölscher, W. Allers, U.D. Schwarz, A. Schwarz, R. Wiesendanger: Determination of tip–sample interaction potentials by dynamic force spectroscopy, Phys. Rev. Lett. **83**, 4780–4783 (1999)

24.222 H. Hölscher, U.D. Schwarz, R. Wiesendanger: Calculation of the frequency shift in dynamic force microscopy, Appl. Surf. Sci. **140**, 344–351 (1999)

24.223 B. Gotsman, B. Anczykowski, C. Seidel, H. Fuchs: Determination of tip–sample interaction forces from measured dynamic force spectroscopy curves, Appl. Surf. Sci. **140**, 314–319 (1999)

24.224 U. Dürig: Extracting interaction forces and complementary observables in dynamic probe microscopy, Appl. Phys. Lett. **76**, 1203–1205 (2000)

24.225 F.J. Giessibl: A direct method to calculate tip–sample forces from frequency shifts in frequency-modulation atomic force microscopy, Appl. Phys. Lett. **78**, 123–125 (2001)

24.226 J.E. Sader, S.P. Jarvis: Accurate formulas for interaction force and energy in frequency modulation force spectroscopy, Appl. Phys. Lett. **84**, 1801–1803 (2004)

24.227 M.A. Lantz, H.J. Hug, R. Hoffmann, P.J.A. van Schendel, P. Kappenberger, S. Martin, A. Baratoff, H.-J. Güntherodt: Quantitative measurement of short-range chemical bonding forces, Science **291**, 2580–2583 (2001)

24.228 S.M. Langkat, H. Hölscher, A. Schwarz, R. Wiesendanger: Determination of site specific forces between an iron coated tip and the NiO(001) surface by force field spectroscopy, Surf. Sci. **527**, 12–20 (2002)

24.229 Y. Sugimoto, P. Pou, M. Abe, P. Jelinek, R. Pérez, S. Morita, O. Custance: Chemical identification of individual surface atoms by atomic force microscopy, Nature **446**, 64–67 (2007)

24.230 M. Abe, Y. Sugimoto, O. Custance, S. Morita: Room-temperature reproducible spatial force spectroscopy using atom-tracking technique, Appl. Phys. Lett. **87**, 173503 (2005)

24.231 H. Hölscher, S.M. Langkat, A. Schwarz, R. Wiesendanger: Measurement of three-dimensional force fields with atomic resolution using dynamic force spectroscopy, Appl. Phys. Lett. **81**, 4428–4430 (2002)

24.232 A. Schirmeisen, D. Weiner, H. Fuchs: Single atom contact mechanics: From atomic scale energy barrier to mechanical relaxation hysteresis, Phys. Rev. Lett. **97**, 136101 (2006)

24.233 M. Abe, Y. Sugimoto, T. Namikawa, K. Morita, N. Oyabu, S. Morita: Drift-compensated data acquisition performed at room temperature with frequency modulation atomic force microscopy, Appl. Phys. Lett. **90**, 203103 (2007)

24.234 A. Schwarz, H. Hölscher, S.M. Langkat, R. Wiesendanger: Three-dimensional force field spectroscopy, AIP Conf. Proc. **696**, 68 (2003)

24.235 N. Oyabu, Y. Sugimoto, M. Abe, O. Custance, S. Morita: Lateral manipulation of single atoms at semiconductor surfaces using atomic force microscopy, Nanotechnology **16**, 112–117 (2005)

24.236 N. Oyabu, O. Custance, I. Yi, Y. Sugawara, S. Morita: Mechanical vertical manipulation of selected single atoms by soft nanoindentation using near contact atomic force microscopy, Phys. Rev. Lett. **90**, 176102-1–176102-4 (2004)

24.237 Y. Sugimoto, M. Abe, S. Hirayama, N. Oyabu, O. Custance, S. Morita: Atom inlays performed at room temperature using atomic force microscopy, Nat. Mater. **4**, 156–160 (2005)

24.238 C. Sommerhalter, T.W. Matthes, T. Glatzel, A. Jäger-Waldau, M.C. Lux-Steiner: High-sensitivity quantitative Kelvin probe microscopy by noncontact ultra-high-vacuum atomic force microscopy, Appl. Phys. Lett. **75**, 286–288 (1999)

24.239 A. Schwarz, W. Allers, U.D. Schwarz, R. Wiesendanger: Dynamic mode scanning force microscopy of n-InAs(110)-(1×1) at low temperatures, Phys. Rev. B **62**, 13617–13622 (2000)

24.240 K.L. McCormick, M.T. Woodside, M. Huang, M. Wu, P.L. McEuen, C. Duruoz, J.S. Harris: Scanned potential microscopy of edge and bulk currents in the quantum Hall regime, Phys. Rev. B **59**, 4656–4657 (1999)

24.241 P. Weitz, E. Ahlswede, J. Weis, K. von Klitzing, K. Eberl: Hall-potential investigations under quantum Hall conditions using scanning force microscopy, Physica E **6**, 247–250 (2000)

24.242 E. Ahlswede, P. Weitz, J. Weis, K. von Klitzing, K. Eberl: Hall potential profiles in the quantum Hall regime measured by a scanning force microscope, Physica B **298**, 562–566 (2001)

24.243 M.T. Woodside, C. Vale, P.L. McEuen, C. Kadow, K.D. Maranowski, A.C. Gossard: Imaging interedge-state scattering centers in the quantum Hall regime, Phys. Rev. B **64**, 041310-1–041310-4 (2001)

24.244 K. Moloni, B.M. Moskowitz, E.D. Dahlberg: Domain structures in single crystal magnetite below the Verwey transition as observed with a low-temperature magnetic force microscope, Geophys. Res. Lett. **23**, 2851–2854 (1996)

24.245 Q. Lu, C.C. Chen, A. de Lozanne: Observation of magnetic domain behavior in colossal magnetoresistive materials with a magnetic force microscope, Science **276**, 2006–2008 (1997)

24.246 G. Xiao, J.H. Ross, A. Parasiris, K.D.D. Rathnayaka, D.G. Naugle: Low-temperature MFM studies of CMR manganites, Physica C **341–348**, 769–770 (2000)

24.247 M. Liebmann, U. Kaiser, A. Schwarz, R. Wiesendanger, U.H. Pi, T.W. Noh, Z.G. Khim, D.W. Kim: Domain nucleation and growth of $La_{0.7}Ca_{0.3}MnO_{3-\delta}$/LaAlO$_3$ films studied by low temperature MFM, J. Appl. Phys. **93**, 8319–8321 (2003)

24.248 A. Moser, H.J. Hug, I. Parashikov, B. Stiefel, O. Fritz, H. Thomas, A. Baratoff, H.J. Güntherodt, P. Chaudhari: Observation of single vortices condensed into a vortex-glass phase by magnetic force microscopy, Phys. Rev. Lett. **74**, 1847–1850 (1995)

24.249 C.W. Yuan, Z. Zheng, A.L. de Lozanne, M. Tortonese, D.A. Rudman, J.N. Eckstein: Vortex images in thin films of $YBa_2Cu_3O_{7-x}$ and $Bi_2Sr_2Ca_1Cu_2O_{8-x}$ obtained by low-temperature magnetic force microscopy, J. Vac. Sci. Technol. B **14**, 1210–1213 (1996)

24.250 A. Volodin, K. Temst, C. van Haesendonck, Y. Bruynseraede: Observation of the Abrikosov vortex lattice in $NbSe_2$ with magnetic force microscopy, Appl. Phys. Lett. **73**, 1134–1136 (1998)

24.251 A. Moser, H.J. Hug, B. Stiefel, H.J. Güntherodt: Low temperature magnetic force microscopy on $YBa_2Cu_3O_{7-\delta}$ thin films, J. Magn. Magn. Mater. **190**, 114–123 (1998)

24.252 A. Volodin, K. Temst, C. van Haesendonck, Y. Bruynseraede: Imaging of vortices in conventional superconductors by magnetic force microscopy images, Physica C **332**, 156–159 (2000)

24.253 M. Roseman, P. Grütter: Estimating the magnetic penetration depth using constant-height magnetic force microscopy images of vortices, New J. Phys. **3**, 24.1–24.8 (2001)

24.254 A. Volodin, K. Temst, C. van Haesendonck, Y. Bruynseraede, M.I. Montero, I.K. Schuller: Magnetic force microscopy of vortices in thin niobium films: Correlation between the vortex distribution and the thickness-dependent film morphology, Europhys. Lett. **58**, 582–588 (2002)

24.255 U.H. Pi, T.W. Noh, Z.G. Khim, U. Kaiser, M. Liebmann, A. Schwarz, R. Wiesendanger: Vortex dynamics in $Bi_2Sr_2CaCu_2O_8$ single crystal with low density columnar defects studied by magnetic force microscopy, J. Low Temp. Phys. **131**, 993–1002 (2003)

24.256 M. Roseman, P. Grütter, A. Badia, V. Metlushko: Flux lattice imaging of a patterned niobium thin film, J. Appl. Phys. **89**, 6787–6789 (2001)

24.257 A. Schwarz, R. Wiesendanger: Magnetic sensitive force microscopy, Nano Today **3**, 28–39 (2008)

24.258 R. Wiesendanger, D. Bürgler, G. Tarrach, A. Wadas, D. Brodbeck, H.J. Güntherodt, G. Güntherodt, R.J. Gambio, R. Ruf: Vacuum tunneling of spin-polarized electrons detected by scanning tunneling microscopy, J. Vac. Sci. Technol. B **9**, 519–524 (1991)

24.259 H. Momida, T. Oguchi: First-principles study on exchange force image of NiO(001) surface using a ferromagnetic Fe probe, Surf. Sci. **590**, 42–50 (2005)

24.260 U. Kaiser, A. Schwarz, R. Wiesendanger: Magnetic exchange force microscopy with atomic resolution, Nature **446**, 522–525 (2007)

24.261 R. Schmidt, C. Lazo, H. Hölscher, U.H. Pi, V. Caciuc, A. Schwarz, R. Wiesendanger, S. Heinze: Probing the magnetic exchange forces of iron on the atomic scale, Nano Lett. **9**, 200–204 (2008)

24.262 D. Rugar, R. Budakian, H.J. Mamin, B.W. Chui: Single spin detection by magnetic resonance force microscopy, Nature **430**, 329–332 (2004)

24.263 J.A. Sidles, J.L. Garbini, G.P. Drobny: The theory of oscillator-coupled magnetic resonance with potential applications to molecular imaging, Rev. Sci. Instrum. **63**, 3881–3899 (1992)

24.264 J.A. Sidles, J.L. Garbini, K.J. Bruland, D. Rugar, O. Züger, S. Hoen, C.S. Yannoni: Magnetic resonance force microscopy, Rev. Mod. Phys. **67**, 249–265 (1995)

24.265 D. Rugar, O. Züger, S. Hoen, C.S. Yannoni, H.M. Vieth, R.D. Kendrick: Force detection of nuclear magnetic resonance, Science **264**, 1560–1563 (1994)

24.266 Z. Zhang, P.C. Hammel, P.E. Wigen: Observation of ferromagnetic resonance in a microscopic sample using magnetic resonance force microscopy, Appl. Phys. Lett. **68**, 2005–2007 (1996)

25. Higher Harmonics and Time-Varying Forces in Dynamic Force Microscopy

Ozgur Sahin, Calvin F. Quate, Olav Solgaard, Franz J. Giessibl

In atomic force microscopy, a force-sensing cantilever probes a sample and thereby creates a topographic image of its surface. The simplest implementation uses the static deflection of the cantilever to probe the forces. More recently, dynamic operation modes have been introduced, which either work at a constant oscillation frequency and sense the amplitude variations caused by tip–sample forces (amplitude modulation or tapping mode) or operate at constant amplitude and varying frequency (frequency modulation mode). Here, we report about new operational concepts capturing the higher harmonics in either amplitude modulation or frequency modulation mode. Higher-harmonic detection in atomic force microscopy allows us to measure time-varying tip–sample forces that contain detailed information about the material characteristics of the sample, while higher-harmonic detection in small-amplitude frequency modulation mode allows a significant improvement in spatial resolution, in particular when operating in vacuum at low temperatures. The most widely used mode of operation of atomic force microscopy (AFM) is tapping mode, because in this mode lateral tip–sample interaction forces are minimized. The gentle interaction between the AFM tip and the sample under test reduces wear on the sample and localizes the deformations to give nanometer, or even molecular, resolution [25.1, 2].

In tapping mode, the AFM cantilever is vibrated at resonance in the vicinity of the sample so that the tip makes contact with the sample once during each cycle. The tip–sample forces reduce the vibration amplitude of the cantilever. The vibrating cantilever is scanned across the surface while a feedback mechanism adjusts the height of the cantilever base to maintain the vibration amplitude at a constant setpoint value. The topography of the surface is then obtained by recording the feedback signal.

Tapping-mode AFM has the potential to measure much more than simply the topography of a surface, however. As can be seen from Fig. 25.1, the tip–sample interaction forces as the AFM tip approaches, interacts with, and retracts from the surface has a complex time dependence. This time dependence reflects the attractive and repulsive forces that act between the tip and the sample, and contains information about the chemical and physical properties of the sample.

In the remaining sections of this chapter we describe methods that enable measurements of the time-varying tip–sample force waveforms in tapping-mode AFM. We first present a simple model to calculate time-varying tip–sample force waveforms and show how these forces depend on sample properties. Then we will present two strategies to engineer the force-sensing cantilever to measure the tip–sample force waveform and its frequency components. As application examples we present: (1) time-varying force measurements that allow quantitative comparisons of material stiffness, and (2) observation of the glass transition of polymer blends with nanometer-scale lateral resolution.

After the discussion of time-varying force measurements in standard AFM tapping mode, we introduce higher-harmonic imaging in AFM with small vibration amplitudes. In small-amplitude AFM imaging, the tip is in the force field of the sample during most of its vibration cycle. Relatively low-order harmonics of the tip–sample force then contain information about the higher-order gradients of the tip–sample interaction force field. These low-order harmonics in small-amplitude dynamic AFM imaging can be measured directly, yielding excellent spatial resolution.

25.1 Modeling of Tip–Sample Interaction Forces in Tapping-Mode AFM

25.1.1 Tip–Sample Forces as a Periodic Waveform

In tapping mode the cantilever is periodically driven at its fundamental resonant frequency. Under typical operating conditions, the periodic driving force results in a periodic motion of the cantilever and a periodic tip–sample force waveform [25.3]. This allows us to use frequency-domain techniques to understand the motion of the cantilever and the tip–sample forces [25.4].

Because the tip–sample force F_{ts} is a periodic waveform we can expand it as a Fourier series as follows:

$$F_{ts}(t) = \sum_{n=0}^{\infty} a_n \cos(n\omega t) + b_n \sin(n\omega t),$$
$$n = 0, 1, 2, \ldots. \qquad (25.1)$$

Here, the frequency ω, is the driving frequency, which is chosen close to the fundamental resonance frequency of the cantilever. The coefficients a_n and b_n are given as

$$a_n = \frac{\omega}{\pi} \int_0^{2\pi/\omega} F_{ts} \cos(n\omega t)\,dt, \qquad (25.2a)$$

$$b_n = \frac{\omega}{\pi} \int_0^{2\pi/\omega} F_{ts} \sin(n\omega t)\,dt. \qquad (25.2b)$$

The k-th harmonic force can be written as

$$F_{ts_n} \cos(n\omega t + \theta_n) = a_n \cos(n\omega t) + b_n \sin(n\omega t).$$
$$(25.3)$$

Here, $F_{ts_n} = \sqrt{a_n^2 + b_n^2}$ and θ_n are the magnitude and phase of the n-th harmonic force, respectively. The

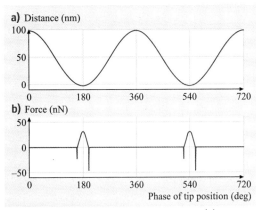

Fig. 25.1a,b Calculated tip–sample distance (**a**) and tip–sample interaction forces (**b**) over two cycles of cantilever oscillation. Negative displacements correspond to sample indentation. Attractive (negative) and repulsive (positive) forces appear during the tip–sample interaction. The magnitude and duration of these forces depend on the physical properties of the sample

phase θ_n of a higher harmonic is defined relative to a reference harmonic at the same frequency that is in phase with the fundamental displacement; i.e., if we represent the tip displacement with $A_s \cos(\omega t)$, the reference signal is $\cos(n\omega t)$.

The tip–sample forces can be seen as a superposition of harmonic forces, each at an integer multiple of the driving frequency. Later we will show how we can measure these higher harmonics by tuning a higher-order resonance of the cantilever to be an integer multiple of its fundamental resonance frequency (i.e., $n\omega = \omega_k$, where ω_k is the resonance frequency of the k-th flexural mode of the cantilever). The higher-harmonic force will then drive the higher-order resonance, and the deflection of the cantilever in the higher-order resonance is a measure of the higher-harmonic force.

25.1.2 Frequency Spectrum of the Tip–Sample Force

Figure 25.2 shows the calculated periodic tip–sample forces and their respective harmonic force components F_{ts_n} for stiff, medium, and compliant samples. Details of the tip–sample force calculations can be found in [25.4]. The harmonic force components are calculated for a cantilever with spring constant $K_1 = 10$, quality factor $Q_1 = 100$, free amplitude $A_0 = 100$ nm, and setpoint amplitude $A_s = 80$ nm. We model the tip–sample interaction forces by the equations below

$$f_{\text{ts}}(r) = \frac{HR}{6\sigma^2}\left[-\left(\frac{\sigma}{r}\right)^2 + \frac{1}{30}\left(\frac{\sigma}{r}\right)^8\right], \quad (25.4\text{a})$$

$$f_{\text{ts}}(d) = \frac{4}{3}E\sqrt{R}d^{3/2}. \quad (25.4\text{b})$$

Here, r is the tip–sample separation, and d is the sample indentation. H, R, σ, and E are the Hamaker constant, tip radius, typical atomic distance for the tip and the sample, and the reduced Young's modulus for the tip and the sample, respectively. Tip–sample forces are governed by (25.4a) when the tip is away from the sample and by (25.4b) when the tip is indenting the sample. The reduced Young's modulus E of the samples is the main factor that determines the tip–sample contact duration. In Fig. 25.2, E values for the three samples are chosen to give contact durations of 5%, 10%, and 15% of the period on the stiff, medium, and compliant samples, respectively. Attractive forces on all the samples are assumed to be equal, giving equal amounts of energy dissipation at contact. For the Hamaker constants we used $H_a = 10 \times 10^{-20}$ J for approach and $H_r = 30 \times 10^{-20}$ J for retraction. The parameters R and σ are chosen to be 10 and 0.1 nm, respectively. These values result in energy dissipation of ≈ 30 eV per contact. The parameters other than E are chosen to be the same on all three materials in order to simplify the analysis.

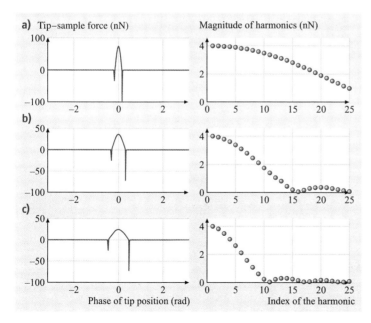

Fig. 25.2a–c Interaction forces between the tip and the sample for three different samples: (**a**) stiff, (**b**) medium, and (**c**) compliant. The amplitudes of the harmonics for the three tip–sample forces are shown *on the right*

For the specific example shown in Fig. 25.2, harmonics above the tenth are strongly dependent on the stiffness of the sample. The tip–sample force is approximately a periodic clipped sine wave, so the pulse width, or contact duration, determines the harmonic content. Shorter-duration contacts generate larger amplitudes at higher harmonics. The contact duration increases for more compliant samples, resulting in smaller magnitudes at higher harmonics. These calculations show that the first harmonics for each of the three cases have the same magnitude. This is because the magnitude of the first harmonic [$n=1$ in (25.1–25.2b)] is approximately equal to the average tapping force on the surface. The average tapping force mainly depends on the cantilever, the drive amplitude, and the setpoint amplitude, which are all equal for the three samples. This analysis shows that only the harmonics that are sufficiently high to have periods that are comparable to, or shorter than, the contact duration contain significant information on the stiffness of the samples. This information can be recovered in part by measuring the amplitude of higher-harmonic vibrations of the cantilever [25.4–7]. However, low signal-to-noise ratios of high-frequency vibrations of the cantilevers generally limit their practical applications.

25.2 Enhancing the Cantilever Response to Time-Varying Forces

In this section we will first discuss the limitations of conventional AFM cantilevers in detecting tip–sample forces that are at a frequency higher than the driving frequency. Then we will present two approaches that engineer flexural or torsional vibration modes of the cantilever to measure time-varying forces between the vibrating tip and the sample.

25.2.1 Response of the Cantilever to High-Frequency Forces

We have discussed the origin of higher-harmonic forces and how they depend on the physical properties of samples. In an experiment we can only measure the higher-harmonic forces through their effect on cantilever motion. So, it is necessary to understand how the cantilever responds to higher-harmonic force components. To obtain good signal-to-noise ratios for the measurements, the cantilever must have a good response to higher harmonics.

In determining the response of the cantilever to forces at different frequencies we need to go beyond simple harmonic oscillator models [25.8] and model the cantilever as a continuum mechanical system [25.9]. The AFM cantilever is fixed at the base and free to move at the tip end. The external forces acting on the cantilever are the drive force at the base and the tip–sample forces at the tip. The motion of the cantilever is governed by the Euler–Bernoulli equation. The solution of this equation for a rectangular cantilever can be found in [25.10]. The Euler–Bernoulli equation is linear in time, so we can describe the cantilever response in terms of the eigenmodes of the cantilever. Each of these modes has a specific resonance frequency and a mode shape.

Simulated mode shapes of a rectangular beam fixed at one end and free at the other end are given in Fig. 25.3. In a typical tapping-mode experiment, the cantilever is driven at the resonance frequency of the first flexural mode shown in Fig. 25.3. The other flexural modes are excited when the tip interacts with the surface. With tip–sample interaction as the driving force,

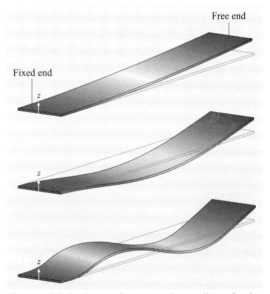

Fig. 25.3 Mode shapes of a rectangular cantilever fixed at one end and free at the other end

the motion of the cantilever can be expressed as a superposition of the responses of the eigenmodes. The response of the cantilever $y(t)$ to an external harmonic force applied to the tip at the free end can be approximated as

$$y(t) = e^{i\omega t} \frac{F}{M} \sum_{k=1}^{\infty} \frac{4}{\omega_k^2 - \omega^2 + \frac{i\omega\omega_k}{Q_k}}. \quad (25.5)$$

Here, $y(t)$ is the displacement of the tip (free end) of the cantilever at time t, and F and ω are the magnitude and frequency of the harmonic force acting on the tip, respectively. The parameters ω_k and Q_k are the resonance frequency and quality factor of the k-th eigenmode.

Figure 25.4 shows the frequency response of the optical-lever signal for a rectangular cantilever. These calculations are based on (25.5) while taking into account that the optical-lever signals in the position-sensitive detector are proportional to the slopes of the free end of the cantilever. The peaks in the response curves are the resonances of each flexural vibration mode. The frequency axis is normalized to the first resonance frequency, which equals the driving frequency in tapping-mode AFM. The higher harmonics, marked with circles on the frequency response curve, are therefore located at integer multiples of the first resonance frequency. The cantilever responds to each force harmonic with a displacement given by this frequency response. We see that the higher-order harmonics, which contain much of the information on the stiffness of the sample, yield relatively low cantilever responses, limiting our ability to measure the higher harmonics. This situation is less severe in liquid environments, where higher-harmonics-based force measurements have been demonstrated by several groups [25.11–13].

25.2.2 Improving Cantilever Response by Tuning Flexural Resonance Frequencies

Figure 25.4 shows that harmonics close to a resonance frequency will have larger deflections. A correctly designed cantilever geometry can tune a higher-order flexural resonance frequency to an integer multiple of the drive so that the corresponding harmonic is enhanced by the resonance peak. This can be done by appropriately removing mass from regions where the cantilever has high mechanical stress in that particular mode [25.14]. Recently, it was suggested that placing a concentrated mass on the cantilever could also result in a good match between higher harmonics and flexural resonances [25.15]. Engineering the cantilever geometry to improve force sensitivity of higher-order flexural modes in scanning capacitance microscopy and Kelvin probe microscopy has also been demonstrated [25.16, 17].

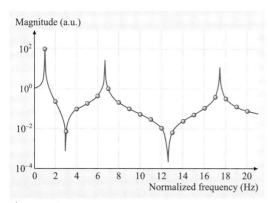

Fig. 25.4 Calculated frequency response of a rectangular cantilever. The frequency axis is normalized to the first resonance frequency. The magnitudes represent the optical signal at the position-sensitive detector. This optical signal is proportional to the slope of the cantilever at the laser spot. The *circles* are located at the integer multiples of the first resonance frequency

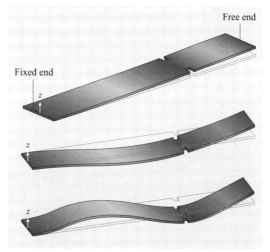

Fig. 25.5 Mode shapes of a rectangular cantilever with a notch. The notch is located at the high bending region in the third flexural mode shape and is $\approx 1/3$ of the total length away from the free end

Fig. 25.6 Scanning electron microscope (SEM) image of a harmonic cantilever. Width, length, and thickness of the cantilever are 502, 3002, and 2.2 μm respectively. The rectangular opening is 22×18 μm^2 and centered 1902 μm away from the cantilever base

Removing mass from regions of high mechanical stress reduces the elastic energy and the resonance frequency of that mode without strongly affecting the other resonant modes of the system. Figure 25.5 illustrates the first three flexural mode shapes of a rectangular cantilever with a notch. The position of the notch corresponds to a highly curved region of the third mode, but not to highly curved regions of the first two modes. Therefore, the effect of the notch in reducing the elastic energy is more prominent in the third mode. Highly curved regions of a mode are also highly displaced, so the removal of mass from these regions will also reduce the kinetic energy of that mode and increase the resonance frequency. This effect will, however, affect both the first and third mode relatively equally, because the displacements at the notch are similar for the two modes, as can be seen from Fig. 25.5. The net effect of the notch is therefore to lower the resonance frequency of the third flexural mode relative to the first flexural mode. The size of the notch can be chosen carefully to obtain an integer ratio between the resonance frequencies.

In one design intended to obtain a flexural resonance at the 16th integer multiple of the fundamental resonance frequency, we placed a hole at a bending region in the third flexural mode shape (Fig. 25.6). The effect of this hole is similar to the notch in Fig. 25.5. The hole reduces the ratio of the third resonance frequency to the fundamental to give an integer ratio. We use the name *harmonic cantilever* for cantilevers that have this property that one of their higher-order modes has a resonance frequency that is an integer multiple of the fundamental resonance frequency.

Figure 25.7 shows the measured vibration spectrum of a harmonic cantilever in tapping-mode AFM. In addition to the drive signal, two peaks (numbers 6 and 16) have relatively large signal levels compared with their neighbors. These are the harmonics that are closest to the resonance frequencies of the harmonic cantilever. Especially the 16th harmonic has a much higher signal level relative to its neighbors. This is because the frequency of this particular harmonic matches the third resonance frequency of the harmonic cantilever. Such cantilevers can be fabricated with conventional silicon-based microfabrication techniques. A more detailed discussion on the fabrication of the cantilever in Fig. 25.6 is given in [25.14].

25.2.3 Recovering the Time-Varying Tip–Sample Forces with Torsional Vibrations

AFM cantilevers have a second type of vibration modes, called torsional modes, in addition to the flexural modes discussed above. Vibrations in these modes result in angular deflections of the cantilever. These modes are excited as a result of torque acting on the cantilever. The tip of a typical cantilever is located on the longitudinal axis, preventing tip–sample forces from creating torque on the cantilever when tapping on a sample. In this section we will describe a class of cantilevers, called torsional harmonic cantilevers, that enables the

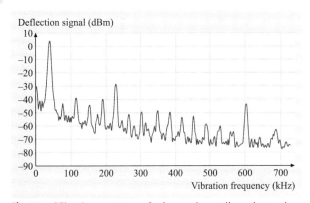

Fig. 25.7 Vibration spectrum of a harmonic cantilever in tapping-mode AFM. The cantilever is driven at its fundamental resonance frequency (37.4 kHz), and higher-harmonic generation is observed. The second (240 kHz) and third (598 kHz) harmonics coincide with higher resonances and have relatively large signal power

excitation of torsional vibration modes [25.18, 19]. Torsional vibration modes are very sensitive to tip–sample forces and allow simultaneous measurement of a large number of higher harmonics so that the tip–sample forces can be recreated with high temporal resolution. We will begin with a theoretical discussion of the torsional response of an AFM cantilever with an offset tip and then show experimental results from the vibration measurements of a torsional harmonic cantilever.

The torsional harmonic cantilever has a tip that is offset from the long axis of the cantilever. An example of such a cantilever is shown in Fig. 25.8a. When a torsional harmonic cantilever is vibrated in tapping mode, tip–sample interaction forces generate torque around the long axis of the cantilever and excite the torsional modes (Fig. 25.8b). The overall motion of the cantilever is a combination of flexural and torsional vibrations. The vibration at the fundamental flexural resonance frequency is still the dominant component. The motion of the cantilever is detected with a laser beam reflected from the backside of the cantilever falling onto a four-quadrant position-sensitive diode (Fig. 25.8c). The difference in optical powers in the upper and lower halves is proportional to longitudinal (flexural-mode) deflection and the difference in left and right halves is proportional to torsional deflection.

When the tip interacts with the surface as it approaches and retracts, the torsional vibration mode acts as a force sensor that measures the force acting on the tip. The torsional resonance frequency is much higher than the first flexural resonance frequency, so the torsional mode responds to the variations in the tip–sample force over a wide frequency range. Figure 25.9 shows the calculated frequency response of the torsional and flexural modes of the torsional har-

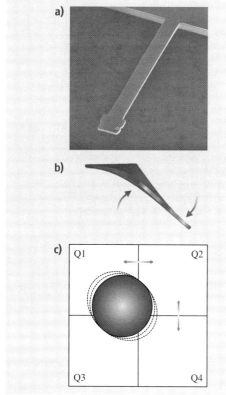

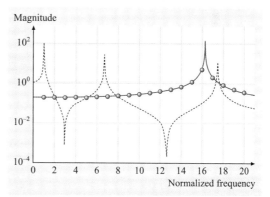

Fig. 25.8 (a) Scanning electron micrograph image of a torsional harmonic cantilever. The cantilever is nominally 300 μm long, 22 μm thick, and 302 μm wide. The tip is offset 15 μm from the centerline of the cantilever. (b) Simulated first torsional mode shape of a rectangular cantilever fixed at the base. (c) Illustration of the laser spot on the four-quadrant position-sensitive photodetector. The optical power difference $(Q1 + Q2) - (Q3 + Q4)$ is proportional to vertical cantilever deflection, and the optical power difference $(Q1 + Q3) - (Q2 + Q4)$ is proportional to torsional angle

Fig. 25.9 Calculated frequency response of a rectangular cantilever in the torsional and flexural modes. The *frequency axis* is normalized to the first flexural resonance frequency. The magnitudes represent the optical signal at the position-sensitive detector. These optical signals are proportional to the slope of the cantilever at the laser spot. The *circles* are located at the integer multiples of the first resonance frequency. Note that the torsional response (*solid curve*) is much higher than the flexural response (*dashed curve*) at higher harmonics

monic cantilever. These curves correspond to the lateral and vertical deflection signals at the position-sensitive detector.

At frequencies below the first flexural resonance frequency, the flexural response is much higher than the torsional response. This is because the effective spring constant of the first flexural mode is much smaller than that of the first torsional mode. On the other hand, at higher frequencies where the higher harmonics of the tip–sample forces are located, the torsional response is larger than the flexural response.

Figure 25.10a,b shows the flexural and torsional vibration spectra of a torsional harmonic cantilever while tapping on a polystyrene sample. The cantilever is driven near the first flexural resonance frequency (52.5 kHz) by a piezoelectric element from the base. The free vibration amplitude and setpoint amplitude are 100 and 60 nm, respectively. While tapping on the surface the cantilever simultaneously moves in flexural and torsional modes, as shown by the vibration spectra. The dominant component of the cantilever motion is the first peak in the flexural vibration spectrum. This is the motion at the drive frequency. The other harmonics are generated by the tip–sample interaction.

The higher harmonics in the flexural vibration spectrum have signal-to-noise ratios that are too low for practical measurements. In the torsional vibration spectrum in Fig. 25.10b, on the other hand, the signal-to-noise ratios are sufficient for practical measurements for the first 19 harmonics. The signal levels around the 16th harmonic increase due to the first torsional resonance of the cantilever located at 16.2 times the drive frequency. The vibration spectrum in Fig. 25.10b shows that the torsional vibrations provide good signal levels up to the 19th harmonic of the first flexural resonance frequency. This means that this particular torsional-mode force sensor can resolve tip–sample forces with a temporal resolution roughly 20 times shorter than the fundamental flexural oscillation period.

25.2.4 Time-Varying Tip–Sample Force Measurements

The mechanical bandwidth of the torsional mode determines the response to variations in the tip–sample forces as the tip vibrates. This bandwidth is determined by the first torsional resonance frequency, which is 16.2 times the drive frequency for this cantilever. In general, it is possible to measure the first few harmonics beyond the first torsional resonance frequency without significant attenuation. This high bandwidth allows the torsional mode to respond to high-frequency tip–sample forces. While the cantilever responds to harmonic forces up to 19, the magnitude and phase of the responses are different for each harmonic. This can be seen in the torsional frequency response of the cantilever shown in Fig. 25.10b. Therefore, forces at different harmonics cannot be compared directly. Instead, it is necessary to measure the frequency response and adjust the measurements by the mechanical gain introduced by the resonant response of the cantilever. (*Stark* et al. performed a similar experiment with the flexural vibrations of the cantilever and demonstrated time-varying force measurements, albeit with lower signal levels [25.20].) The first torsional resonance is typically near the 15th harmonic and the second torsional resonance is about three times higher in frequency. Therefore, the contributions of the higher-order torsional modes can be neglected, and the torsional motion can to a good approximation be described by harmonic oscillations. This approximation is even better if the laser spot is placed two-thirds of the length of the cantilever away from the base, where the second torsional mode has a node. Based on this ω assumption

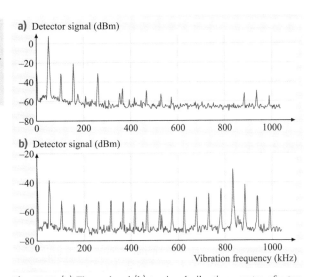

Fig. 25.10 (a) Flexural and (b) torsional vibration spectra of a torsional harmonic cantilever while tapping on a polystyrene sample. The *first peak* of the flexural spectrum in (a) is at the driving frequency. It is the largest component of the cantilever motion. The other flexural and torsional peaks are the higher harmonics generated by the tip–sample interaction forces. The torsional peaks have much higher signal levels at higher harmonics

the transfer function of the first torsional mode can then be modeled as

$$H_T(\omega) = \frac{1}{\omega_T^2 - \omega^2 + \frac{i\omega\omega_T}{Q}}. \quad (25.6)$$

Here, H_T is the mechanical gain of the torsional response as a function of frequency, ω is the angular frequency of the vibration, ω_T is the torsional resonance frequency, i is the imaginary unit, and Q_T is the quality factor of the torsional resonance. The two parameters that determine the frequency response, ω_T and Q_T, are easily measured for a given cantilever. The photodetector gain, the location of the laser spot on the cantilever, and the offset distance of the tip will all multiply H_T in a scalar fashion, but they will not affect the relative enhancement of the different harmonics.

Once the torsional frequency response is determined, it is possible to recover the time-varying tip–sample forces by measurement of torsional vibrations of the cantilever and digitally correcting for the mechanical gain introduced by the torsional frequency response.

An example of this procedure is shown in Fig. 25.11. The torsional deflection signals at the position-sensitive detector are recorded with a digital oscilloscope. The data is averaged over 128 samples to achieve an approximate noise bandwidth of 500 Hz. The measured vibration signals coming from both flexural and torsional modes over one period is shown in Fig. 25.11a. The effect of nonlinear mechanical gain due to the torsional frequency response is removed digitally with the aid of (25.6). The resulting waveform is given in Fig. 25.11b. In this waveform the tip–sample forces are not zero even when the tip is away from the surface. This error is due to the nonlinearity of the detection circuit and cross-talk from the large flexural signal that produces additional signals at the first few harmonics. Those components are removed with a signal-processing procedure that assumes that the tip–sample interaction forces are zero when the cantilever is away from the surface, and subtracts any additional signal at the first few harmonics that results in a nonzero tip–sample force. In Fig. 25.11b the computed correction signal that arises from cross-talk and nonlinearities is also given. Notice that the two curves overlap during the times when the tip is not in contact. Once this additional component is subtracted we get the corrected tip–sample force waveform in Fig. 25.11c.

It is important to note that, while the measurements of torsional vibrations are in units of volts, the

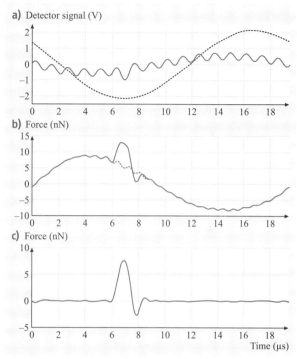

Fig. 25.11a–c Vibration signals from flexural and torsional motions, and tip–sample forces. (**a**) The signals at the four-quadrant photodetector for vertical and torsional displacements. The *solid curve* is the torsional signal. We multiplied the torsional signal by a factor of 10 to view the two curves clearly in one graph. (**b**) The torsional vibration signal after being divided by the torsional frequency response. Except for the pulse located between the 300th and 400th time steps, the tip–sample forces should have been close to zero, because the tip is far away from the surface at those times. The measured signals when not in contact come from cross-talk from the flexural deflection signal. The *dashed curve* estimates the error introduced by these sources. When it is subtracted from the *solid curve* we get the time-varying forces plotted in (**c**)

tip–sample forces are indicated in newtons. Conversions of measured voltages into force units are done by comparing the time-average forces as measured by flexural and torsional deflections. If the spring constant of the fundamental flexural mode has been calibrated with conventional methods [25.21], then the torsional mode can be calibrated against the flexural spring constant. In practice, measurements of time-average quantities are subject to drifts and misalignments of the position sensors. To overcome this difficulty, bifurcations in force–distance curves in tapping mode can be used [25.22]. Rather than comparing the absolute values, this method compares relative changes in time-average forces. Therefore, calibration is performed free from drift-related errors.

25.3 Application Examples

In this section we will present application examples of the use of harmonic forces to study various material systems. In a first experiment we show how time-varying force measurements can measure stiffness variations and reveal the origin of the hysteresis in tip–sample interaction forces that result in energy dissipation. Analysis of the measured data shows that quantitative comparisons of the stiffness of materials can be made without knowledge of cantilever spring constant, tip geometry, vibration amplitude, and the gain of the photodetector. In a second experiment we present imaging of the glass transition of a component in a binary polymer blend by mapping the magnitude of a higher harmonic across the surface at different temperatures. Then we analyze the changes in the surface with the aid of time-varying force measurements.

25.3.1 Time-Varying Force Measurements on Different Materials

The time-varying tip–sample forces obtained with a torsional harmonic cantilever tapping on high-density polyethylene ($0.92\,\mathrm{g/cm^3}$), highly oriented pyrolytic graphite, and low-density polyethylene ($0.86\,\mathrm{g/cm^3}$) are given in Fig. 25.12a. The cantilever has a nominal spring constant of $1\,\mathrm{N/m}$ and the free vibration amplitude and setpoint amplitude are 100 and 60 nm, respectively. These measurements are obtained with the signal-processing procedures illustrated in Fig. 25.11. Here, the positive forces are the repulsive tip–sample interactions that arise from the indentation of the sample. Negative forces are due to capillary forces and van der Waals forces [25.23].

These time waveforms of tip–sample forces point to many differences in these materials. First of all, the contact durations and peak forces differ from one material to another. The magnitude and duration of the attractive interactions are also different for each material. These differences are due to variations in the stiffness, van der Waals parameters, and wettability of these materials.

By plotting the tip–sample forces with respect to tip–sample separation, we get a better understanding of the differences in these samples. If the tapping amplitude, or setpoint amplitude, is known we can determine the vertical position of the tip as a function of time. We can then determine the tip–sample forces with respect to tip–sample distance as shown in Fig. 25.12b. Two important features of the curves plotted in this graph are the rate of increase of the force during indentation (negative separation) and the hysteresis in the force.

The rate of increase in the force during indentation (negative tip–sample separation) is determined

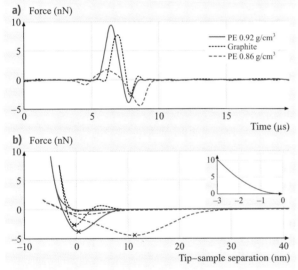

Fig. 25.12a,b Time-varying tip–sample force measurements with a torsional harmonic cantilever tapping on high-density polyethylene (PE), graphite, and low-density polyethylene (**a**). In (**b**) the same measurements as in (**a**) are plotted with respect to tip–sample separation. Negative separations mean that the sample is indented. The rates of increase in forces depend on the stiffness of the sample. Larger negative forces arise during the retraction of the tip. The *inset* shows the tip–sample forces during retraction (first 3 nm *to the left of the crosses*) with the forces on high-density polyethylene multiplied by 3.0 and low-density polyethylene multiplied by 27.0. Note that all three curves coincide. The good correspondence between the shapes of these curves indicates that contact mechanisms are the same on these samples. This allows quantitative measurements of the ratios of elastic parameters for these materials

by the stiffness of the sample. Larger loading forces are required to produce a given depth of indentation into stiffer materials. Therefore, the force increases faster on graphite than on the polyethylene samples in Fig. 25.10b. The indentation forces do not follow a linear relationship. An approximate model predicts that the repulsive forces vary with a power-law relation [25.24]

$$F_{\text{rep}} = \gamma d^n \,. \tag{25.7}$$

Here, F_{rep} is the repulsive tip–sample force, γ is a constant that depends on the reduced Young's modulus and tip diameter, d is the indentation depth, and n is a number that depends on the tip and surface geometry. For a spherical tip and flat surface, $n = 1.5$ [25.25]. If the same tip is used, the value of n is expected to be the same for experiments on different materials. The differences will arise only in the multiplicative term γ because the reduced Young's modulus values are different.

The curves in Fig. 25.12b show double values for forces mainly at positive separations. This is because tip–sample forces are different during the approach and retraction of the tip. The upper branches of the curves show the approach of the tip, whereas the lower branches show retraction. Capillary forces are larger in retraction due to attractive van der Waals forces that pull the surface and raise it above its equilibrium level during retraction, resulting in attractive forces at larger tip–sample separations. There is also the possibility of forming a liquid neck between the tip and the sample during retraction, but the samples used for the measurements are hydrophobic, so this is not likely to be the origin of the hysteresis observed in the force curves of Fig. 25.12.

25.3.2 Quantitative Comparison of Material Properties

The ratio of the Young's modulus values of different materials can be quantitatively measured by comparing the tip–sample force curves for those materials, because for a given depth of indentation, the ratio of forces is equal the ratio of the respective γ coefficients. For these quantitative measurements we do not need to assume a power-law functional dependency as in (25.7). The approximation that the ratio of the Young's modulus values equals the ratio of measured forces at a given depth of indentation is valid as long as the functional dependency is the same for the two materials.

To achieve absolute measurements, we have chosen graphite as our reference material for measurements of the indentation forces for high-density and low-density polyethylene. To compare the forces curves obtained on the three samples, we need to eliminate the contribution of attractive forces because they are not governed by the elastic properties of the materials. In the graphs of Fig. 25.12b, the points with the highest attractive forces are marked with crosses. Tip–sample mechanical contact is broken at approximately these points as the tip is retracted. There are no repulsive forces at these points, so the net negative force is due to the attractive forces. To the left of these crosses, the tip–sample forces increase as the tip indents the samples. The insert in Fig. 25.12b shows a plot of the indentation forces for the first 3 nm to the left of the crosses. We choose the points of maximum attractive forces as the origin of the graphs because this is where the indentation forces and indentation depths are all zero. By scaling the force on high-density polyethylene by a factor of 3.0 and the forces on low-density polyethylene by a factor of 27.0, they both match the reference forces on graphite. The matching of these curves supports the assumption that the functional dependency of forces with respect to indentations is similar. Graphite has an approximate elastic modulus of 5 GPa, so our measured values for the elastic moduli of high-density polyethylene and low-density polyethylene are 1.7 GPa and 180 MPa, respectively. High-density polyethylene is more ordered and is expected to be significantly stiffer than low-density polyethylene. Unfortunately, there is a relatively large spread in published values for the elastic modulus for graphite, so a well-calibrated reference material is still needed for accurate measurements.

These quantitative comparisons and absolute measurements are made without knowledge of many parameters of the AFM experiment, such as cantilever spring constant, tip geometry, drive force, setpoint amplitude, photodetector gain, position of the laser spot, and tip offset distance. This is possible because the calculations for quantitative measurements use only the voltages at the output of the photodetector, together with the flexural and torsional resonance frequencies. Because we take the ratio of force values in the force–distance curves, the ratio of measured voltages is all that is required, while knowledge of the relationship between measured voltages and actual forces is not needed. This technique eliminates all the instrument variables that are difficult to measure and control in AFM experiments.

25.3.3 Imaging the Glass Transition of a Binary Polymer Blend Film

According to the theoretical analysis carried out in the first section, the magnitudes of the higher harmonics depend on the stiffness of the samples. If carefully chosen, the amplitude of a particular harmonic will monotonically increase with increasing stiffness of the samples. We are interested in mapping mechanical property variations, so imaging the amplitude of a single higher harmonic is the simplest solution. The torsional harmonic cantilever provides the first 20 torsional harmonics with good signal levels, so we can measure any one of these harmonics with a lock-in amplifier to produce the corresponding harmonic image of the surface.

By recording the torsional harmonic amplitudes while scanning the surface in tapping mode we have generated the harmonic force images of an ultrathin (about 50 nm) binary polymer film on a silicon substrate and studied the glass transition of the two components, polystyrene (PS) and poly(methyl methacrylate) (PMMA). These components with different glass-transition temperatures (around 100 °C and 130 °C, respectively) form submicrometer domains within the film. As the temperature is elevated, polystyrene goes through the glass transition before PMMA. The regions of rubbery phase will be less stiff than the glassy regions, so that we can observe the glass transitions of individual components of the composite polymer film. For the measurements a torsional harmonic cantilever with a torsional resonance frequency about 11 times higher than the fundamental is used. The vertical spring constant and quality factor of this cantilever were 6 N/m and 91, respectively. The free amplitude and setpoint amplitude are chosen as 100 and 70 nm, respectively.

Images of the topography, phase, and tenth torsional harmonic at temperatures between 85 °C and 215 °C are presented in Fig. 25.13. The two components of

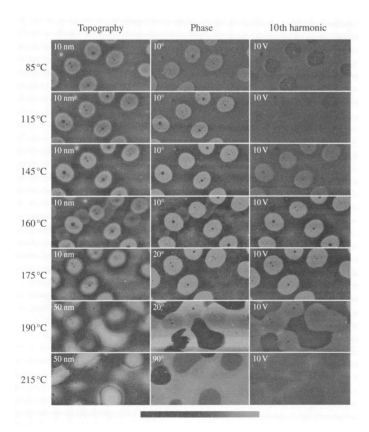

Fig. 25.13 Topography, phase and tenth-harmonic images of a thin polymer film composed of PS and PMMA. The circular features are PMMA, and the matrix is PS. *Brighter color* represents larger height, phase or harmonic amplitude. The scan area is 2.5×5 mm^2. The *color bar* below the panels corresponds to signal ranges indicated on the upper left corner of each panel (after [25.18])

the polymer blend are easily distinguishable in the topography images up to 145 °C because of the height differences. The higher regions are PMMA domains and lower regions are composed of polystyrene. Above 190 °C the boundaries of the two material components are not clear anymore. The topographical images become blurry at elevated temperatures because the glass transition is accompanied by morphological changes due to increased molecular mobility. However, the nature and magnitude of the changes cannot be extracted from the topography images.

In the same experiment, phase images distinguish the two material components but show a gradually increasing contrast up to 215 °C. The phase image is primarily determined by energy dissipation in the tip–sample contact. There are several mechanisms that are involved in tip–sample energy dissipation. In the present case, the viscous response of the sample at elevated

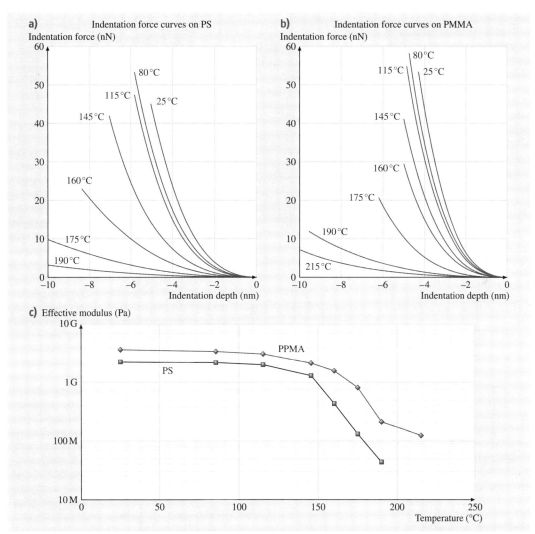

Fig. 25.14a–c Indentation force curves obtained from time-varying force measurements on polystyrene regions (**a**) and on PMMA regions (**b**). Effective Young's modulus values corresponding to each curve is plotted in (**c**) (after [25.18])

temperatures is likely to play an important role. In addition, on compliant samples, attractive forces between the tip and the sample can pull the sample and raise it above its equilibrium level. The contact is eventually broken as the tip retracts and the energy stored in the pulled sample is dissipated. These two mechanisms will result in increased energy dissipation at higher temperatures. However, once both components are in the rubbery phase, one would expect a reduction in phase contrast. The measurements show that phase contrast is even larger at elevated temperatures. It is therefore difficult to identify and quantify the changes in the material properties with temperature.

Simultaneously recorded harmonic images show contrast between the two material components first increasing and then decreasing with temperature. The contrast increases dramatically between 145 °C and 160 °C and decreases between 190 °C and 215 °C. When PS goes through its glass transition, the harmonic amplitude corresponding to PS regions reduces. PMMA goes through its glass transition at a higher temperature, which is accompanied by a reduction in the harmonic amplitude measured on PMMA. As a result the contrast in stiffness first increases and then decreases as the polymer blend is heated. This result suggests that the harmonic images can provide a qualitative explanation for the origin of the changes on the surface. A more detailed understanding and even quantification of the changes can be gained by using the full spectrum of available harmonic signals, i.e., by analyzing the force–distance curves derived from time-varying tip–sample forces.

25.3.4 Detailed Analysis with Time-Varying Nanomechanical Forces

The changes in the polymer components were studied in greater detail by measuring the time-varying forces and generating force–distance curves as previously illustrated in Fig. 25.12b. In Fig. 25.14 we plot the unloading portions of the force–distance curves as obtained from time-varying forces on polystyrene and PMMA regions at each temperature. On both samples, the rate of increase in the repulsive forces decreases with increasing temperatures. We used the Derjaguin–Muller–Toporov (DMT) contact mechanics model [25.25] to calculate the elastic modulus of these samples from the curves recorded at each temperature. The resulting values are plotted in Fig. 25.14c. We see that, at low temperatures, the effective elastic modulus values of PS and PMMA regions are 2.3 and 3.7 GPa, respectively. These values are only slightly reduced at temperatures up to 145 °C. However, the effective elastic modulus of both materials dramatically reduce around 160 °C and 190 °C. Modulus values of both materials reduce more than an order of magnitude, identifying the glass transition. From these plots we estimate the apparent glass-transition temperatures of PS and PMMA to be 160 °C and 180 °C, respectively. The relatively high glass-transition temperatures measured in these experiments are expected consequences of the frequency dependence of the glass transition. Our technique measures the mechanical properties at the drive frequency of 50 kHz. Glass transitions of bulk materials are determined either through thermal measurements or dynamic mechanical measurements with frequencies below 100 Hz [25.26].

25.4 Higher-Harmonic Force Microscopy with Small Amplitudes

25.4.1 Principle

AFM cantilevers that oscillate freely show sinusoidal motion, where the deflection of the end of the cantilever q' is described by $q'(t) = A\cos(2\pi f t)$. When the tip of the oscillating cantilever is in the force field of the sample, the potential is generally no longer harmonic, giving rise to anharmonic components where the deflection is described by a Fourier series

$$q'(t) = \sum_{n=0}^{\infty} a_n \cos(2\pi f t), \qquad (25.8)$$

with $a_1 = A$. When the oscillation amplitude of the cantilever is large compared with the range of the tip–sample potential, the lower orders of these anharmonic components are proportional to the frequency shift ([25.27, Eq. 11]), so, in principle, no new information is available over frequency-modulation (FM) AFM [25.28]. Only if higher-order components with periods comparable to, or shorter than, the contact duration are measured as outlined in the first part of this chapter, does higher-harmonic AFM with large amplitudes yield advantages over FM-AFM. For small oscillation amplitudes the tip is in the force field during all or most of the vibration cycle, so the lower-order higher harmonics are no longer proportional to the frequency shift and offer physical content in their own right. *Dürig* [25.29] has found that, in small-amplitude

AFM, the full tip–sample potential can be immediately recovered over the z-range covered by the oscillating cantilever if the amplitudes and phases of all higher harmonics are available. *Dürig* has expressed the higher harmonics as a Tchebyshev expansion of the tip–sample force [25.29]. *Sahin* et al. [25.4] and *de Lozanne* [25.30] expand the temporal dependence of the tip–sample force in a Fourier series with base frequency f and express the higher harmonics as the response of the cantilever to an excitation at frequency $n \times f$. Mathematically, these two notions are of course equivalent. We start with Dürig's formula for the amplitude of the n-th harmonic

$$a_n = \frac{2}{\pi k} \frac{1}{1-n^2} \int_{-1}^{1} F_{\text{ts}}(z+Au) \frac{T_n(u)}{\sqrt{1-u^2}} \, du , \quad (25.9)$$

where $T_n(u)$ is the n-th Tchebyshev polynomial of the first kind. We can show by applying integration by parts n-times ($\int gh' = -\int g'h + gh$) that

$$a_n = \frac{2}{\pi k} \frac{1}{1-n^2} \frac{1}{(2n+1)\cdots 3 \cdot 1} A^n$$
$$\times \int_{-1}^{1} \frac{\mathrm{d}F_{\text{ts}}^n(z+Au)}{\mathrm{d}z^n} (1-u^2)^{n-1/2} \, du , \quad (25.10)$$

because the n-th integral of $T_n(u)/(1-u^2)^{1/2}$ is $(1-u^2)^{n-1/2}/[(2n+1)\cdots 3 \cdot 1]$. Thus, the n-th harmonic

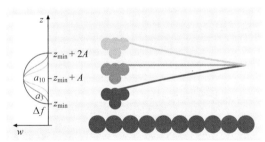

Fig. 25.15 Cantilever in three phases of its oscillation cycle: minimal distance $z_{\min}$, average distance $z_{\min} + A$, and maximal distance $z_{\min} + 2A$. The frequency shift is calculated by the convolution of the force gradient with a semispherical weight function indicated by Δf, the second-harmonic amplitude a_2 is calculated by convoluting the second-order force gradient by a weight function $(1-u^2)^{3/2}$, and the n-th harmonic a_n is computed by convoluting the n-th-order force gradient by $(1-u^2)^{n-1/2}$. The *left part* shows the weight functions for Δf (*dark brown*), a_3 (*brown*) and a_{10} (*light brown*)

can be expressed by a convolution of the n-th force gradient $\mathrm{d}F_{\text{ts}}^n/\mathrm{d}z^n$ with a bell-shaped weight function $(1-u^2)^{n-1/2}$. Figure 25.15 shows three snapshots of the oscillating cantilever in the upper turnaround point, the neutral position, and the closest sample approach. The graph on the left of the figure shows various weight functions for deriving the frequency shift and higher harmonics from the force gradient and higher-order gradients. The semicircular weight function is convoluted with the force gradient, yielding the frequency shift [25.28]. The bell-shaped weight functions are convoluted with higher-order force gradients to derive the higher harmonics (25.10). The weight functions have their maxima at $u = 0$, i.e., a distance of A further away from the surface than the minimal tip–sample distance $z_{\min}$.

If the amplitude A is small enough such that higher-order force gradients still have a reasonable magnitude at distance $z_{\min} + A$, the higher harmonics are not just proportional to the frequency shift Δf, but contain information about higher-order force gradients.

The benefit of higher-order force gradient maps is demonstrated by an elementary but instructive example illuminating the contrast achievable by tip–sample interaction potential, force, force gradient, and higher-order gradients shown in Fig. 25.16. Figure 25.16a shows a model of the charge distribution where the AFM tip is replaced by a single electron with charge $-e$. The sample ion probed by the tip is modeled by a central ion with charge $+9e$, surrounded by eight electrons with charge $-e$ that point towards the corners of a cubic lattice. This charge distribution for the sample ion is motivated by the charge density calculations by *Posternak* et al. [25.31] and *Mattheis* and *Hamann* [25.32] for the (100) surface of tungsten. While tungsten (001) is a very special surface, it is reasonable to assume that most surface atoms do not have a spherically symmetric charge distribution as free atoms with filled shells, but local charge maxima that reflect the chemical bonding symmetry. Figure 25.16b shows the potential energy as a function of lateral displacements in x and y for a constant height z. It is interesting to note that both the energy and the force shown in Fig. 25.16c are almost symmetric with respect to rotations around the z-axis. However, the symmetry of the underlying charge distribution becomes more and more apparent with increasing order of the force gradient, as shown in Fig. 25.16d–h. In theory, it is possible to record (x, y) maps of the tip–sample force for various z-values and calculate gradients and higher-order gradients from these maps later. However, experimental noise

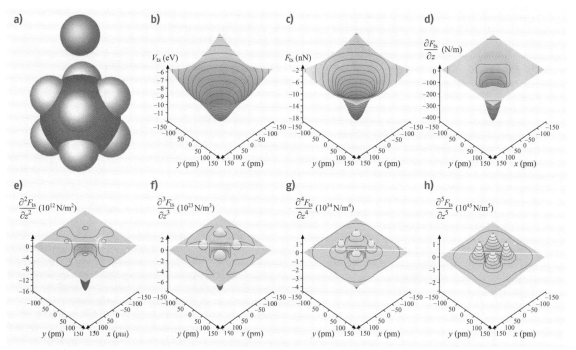

Fig. 25.16 (a) Simple electrostatic model for the tip–sample interaction in AFM: the tip is a test charge with $-e$, the sample atom that is probed consists of a central ion with charge $+9e$ and the valence charge distribution is modeled by eight surrounding charges $-e$ oriented towards the corners of a cube. The tip *atom* is located 132 pm above the center of the sample ion, and the point charges are located at a distance of 55 pm from the center of the sample ion. (b) Tip–sample interaction potential for a distance of 132 pm as a function of the lateral positions x and y. (c) Tip–sample force, (d) force gradient, (e) second derivative of force versus z, (f) third derivative, (g) fourth derivative, and (h) fifth derivative of force versus z

would render these maps totally useless for obtaining reliable data for higher-order gradients. Instead, a direct method that couples higher-order gradients to experimental observables is needed. The higher harmonics described in (25.10) are perfect for this purpose. However, the magnitude of higher harmonics is typically very small, therefore they have to be enhanced either by resonant enhancement as described in the previous sections or by using small amplitudes and enhancing the magnitude of higher harmonics by other means as outlined below.

The mechanical bending of AFM cantilevers is transformed into an electrical signal by a deflection sensor. Optical and piezoresistive deflection sensors generate an output signal that is proportional to the cantilever deflection. For example, if the mechanical deflection is composed of a sinusoidal oscillation with frequency f and amplitude A plus an oscillation at $5 \times f$ with an amplitude of 4% of the amplitude A at the base frequency, both the deflection signal and the electrical signal look like the solid curve in Fig. 25.17. However, in piezoelectric detectors such as the qPlus sensor [25.33], a charge accumulates at the electrodes of the *cantilever* that is proportional to the deflection ([25.33], Eq. 2]). When the sensor oscillates, an alternating current results that is proportional to deflection times frequency. Typically, this alternating current is transformed to a voltage by a transimpedance amplifier [25.33]. Because of the proportionality between current and deflection times frequency, higher harmonics generate greater signal strength in piezoelectric sensors. The dashed curve in Fig. 25.17 shows the current that is generated by a qPlus sensor oscillating at a base frequency and the fifth harmonic with a relative amplitude of 4%. Due to the enhancement of the sensitivity for higher harmonics, the current at frequency $5 \times f$ already amounts to 20% of the base frequency.

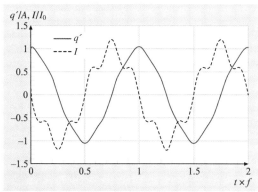

Fig. 25.17 Deflection of a cantilever versus time (*solid*) for a cantilever according to $q'(t) = A[\cos(2\pi ft) + 0.04\cos(2\pi 5 ft)]$ and current (*dashed*) $I(t) = -\text{const}A[\sin(2\pi ft) + 0.2\sin(2\pi 5 ft)]$ produced by a piezoelectric force detector such as the qPlus sensor [25.33]. Because piezoelectric detection is proportional to frequency, higher harmonics produce more current than the oscillation at the base frequency f

In principle, one can pick an individual higher harmonic by analyzing the deflection of the cantilever with a lock-in amplifier that is triggered at the base frequency f and set to the n-th harmonic of f. Of course, one could also apply a battery of lock-in amplifiers and record as many higher harmonics as practical for full potential recovery, as proposed by *Dürig* [25.29]. Because the higher force gradient images in Fig. 25.16e–h are very similar, we can just acquire the root-mean-square (RMS) sum of *all* higher harmonics by routing the deflection signal through a high-pass filter that is set to pass all frequencies above f followed by an RMS-to-direct current converter. While this high-pass technique does not allow immediate full potential recovery as proposed in [25.19], it has a good signal-to-noise ratio because it uses all the higher harmonics and it is very simple to implement and operate. This technique has been used in [25.34].

25.4.2 Application Examples

The enhanced resolution power that should be available by higher-harmonic AFM with small amplitude is best demonstrated by a direct comparison. Figure 25.18 shows simultaneous constant-height data where a graphite sample was imaged by a tungsten tip mounted on an oscillating qPlus sensor. The left

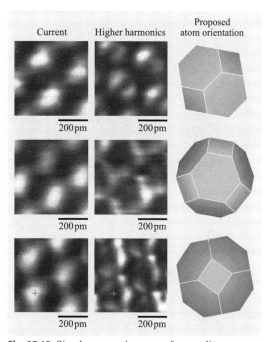

Fig. 25.18 Simultaneous images of tunneling current (*left column*) and higher harmonics (*center column*) in a constant-height measurement. A graphite sample was imaged by a tungsten tip in a 4 K STM/AFM in ultrahigh vacuum [25.34] for details. The highly resolved details in the higher-harmonic images are caused by the electronic structure of the tungsten tip atom. Because tungsten crystallizes with body-centered cubic (bcc) symmetry, the high-symmetry configurations are the twofold [110] symmetry shown in the *first row*, the threefold [111] symmetry shown in the *second row*, and the fourfold [001] symmetry shown in the *third row*. The *right column* shows the Wigner–Seitz unit cell of bcc materials such as tungsten. Typical acquisition speed for each set of images is 0.5 lines/s at 256×256 pixels, i.e., a typical time of 10 min per image

column maps the tunneling current, and the center column shows the intensity of the higher-harmonics data, clearly showing the enhanced spatial resolution. The intensity of the higher harmonics can not only be monitored in constant-height mode or when the feedback uses a tunneling current or frequency shift for distance regulation, but can also be used for z-feedback. Figure 25.19 shows an image of a Si(111)-(7×7) surface that has been acquired by higher-harmonic AFM. The deflection signal of the cantilever was fed into a lock-

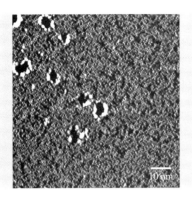

Fig. 25.19 Topographic data of higher-harmonic AFM image of damaged Si(111)-(7×7), recorded at room temperature. A qPlus sensor with stiffness of 1800 N/m, amplitude of 780 pm, and $f_0 = 16740$ Hz was used to capture this image. The z-control feedback was set such that the second-harmonic amplitude was constant at $a_2 = 4.4$ pm. The frequency shift was also recorded and had an average value of -25 Hz (Δf data not shown here). The acquisition speed was very low: 0.1 lines/s at 512×512 pixels, i.e., it took 85 min to acquire this image [25.35] ◄

in amplifier, and the second harmonic was used as the signal.

25.4.3 Conclusions

Higher-harmonic AFM with small amplitudes is an interesting AFM mode because it allows greatly increased spatial resolution. The signal levels are small, therefore the acquisition speed is very low, and low-temperature operation is helpful to reduce thermal drift to low values such that it does not harm the spatial resolution. Possibly, a combination of small-amplitude higher-harmonic AFM with the resonance-enhancement technique described in the previous section might allow to retain the high-resolution capability of small-amplitude higher-harmonic AFM while increasing the possible scanning speed.

References

25.1 Q. Zhong, D. Inniss, K. Kjoller, V.B. Elings: Fractured polymer/silica fiber surface studied by tapping mode atomic force microscopy, Surf. Sci. **280**, L688–L692 (1993)

25.2 D. Klinov, S. Magonov: True molecular resolution in tapping mode atomic force microscopy, Appl. Phys. Lett. **84**, 2697–2699 (2004)

25.3 M.V. Salapaka, D.J. Chen, J.P. Cleveland: Linearity of amplitude and phase in tapping-mode atomic force microscopy, Phys. Rev. B **61**, 1106–1115 (2000)

25.4 O. Sahin, A. Atalar, C.F. Quate, O. Solgaard: Resonant harmonic response in tapping-mode atomic force microscopy, Phys. Rev. B **69**, 5416–5424 (2004)

25.5 R. Hillenbrand, M. Stark, R. Guckenberger: Higherharmonics generation in tapping-mode atomic force microscopy: Insights into tip–sample interaction, Appl. Phys. Lett. **76**, 3478–3480 (2000)

25.6 R.W. Stark, W.M. Heckl: Higher harmonics imaging in tapping-mode atomic-force microscopy, Rev. Sci. Instrum. **74**, 5111–5114 (2003)

25.7 S. Crittenden, A. Raman, R. Reifenberger: Probing attractive forces at the nanoscale using higher-harmonic dynamic force microscopy, Phys. Rev. B **72**(13), 235422 (2005)

25.8 A.S. Paulo, R. Garcia: Unifying theory of tapping mode atomic force microscope, Phys. Rev. B **66**, 041406–041409(R) (2002)

25.9 R.W. Stark, W.M. Heckl: Fourier transformed atomic force microscopy: Tapping mode atomic force microscopy beyond the Hookian approximation, Surf. Sci. **457**, 219–228 (2000)

25.10 U. Rabe, K. Janser, W. Arnold: Vibrations of free and surface-coupled atomic force microscope cantilevers: Theory and experiment, Rev. Sci. Instrum. **67**, 3281–3293 (1996)

25.11 J. Legleiter, M. Park, B. Cusick, T. Kowalewski: Scanning probe acceleration microscopy (SPAM) in fluids: Mapping mechanical properties of surfaces at the nanoscale, Proc. Natl. Acad. Sci. USA **103**, 4813 (2006)

25.12 J. Preiner, J.L. Tang, V. Pastushenko, P. Hinterdorfer: Higher harmonic atomic force microscopy: Imaging of biological membranes in liquid, Phys. Rev. Lett. **99**, 046102 (2007)

25.13 S. Basak, A. Raman: Dynamics of tapping mode atomic force microscopy in liquids: Theory and experiments, Appl. Phys. Lett. **91**, 064107 (2007)

25.14 O. Sahin, G. Yaralioglu, R. Grow, S.F. Zappe, A. Atalar, C. Quate, O. Solgaard: High resolution imaging of elastic properties using harmonic cantilevers, Sens. Actuators A **114**, 183–190 (2004)

25.15 H. Li, Y. Chen, L. Dai: Concentrated-mass cantilever enhances multiple harmonics in tapping-mode atomic force microscopy, Appl. Phys. Lett. **92**, 151903 (2008)

25.16 S. Sadewasser, G. Villanueva, J.A. Plaza: Modified atomic force microscopy cantilever design to facili-

25.17 K. Kimura, K. Kobayashi, K. Matsushige, H. Yamada: Improving sensitivity in electrostatic force detection utilizing cantilever with tailored resonance modes, Appl. Phys. Lett. **90**, 053113 (2007)

25.18 O. Sahin, S. Magonov, C. Su, C.F. Quate, O. Solgaard: An atomic force microscope tip designed to measure time-varying nanomechanical forces, Nat. Nanotechnol. **2**, 507 (2007)

25.19 O. Sahin: Time-varying tip-sample force measurements and steady-state dynamics in tapping-mode atomic force microscopy, Phys. Rev. B **77**, 115405 (2008)

25.20 M. Stark, R.W. Stark, W.M. Heckl, R. Guckenberger: Inverting dynamic force microscopy: From signals to time resolved forces, Proc. Natl. Acad. Sci. USA **99**, 8473–8478 (2002)

25.21 J.L. Hutter, J. Bechhoefer: Calibration of atomic-force microscope tips, Rev. Sci. Instrum. **64**, 1868 (1993)

25.22 O. Sahin: Harnessing bifurcations in tapping-mode atomic force microscopy to calibrate time-varying tip-sample force measurements, Rev. Sci. Instrum. **78**, 103707 (2007)

25.23 L. Zitzler, S. Herminghaus, F. Mugele: Capillary forces in tapping-mode atomic force microscopy, Phys. Rev. B **66**, 155436–155443 (2002)

25.24 I.N. Sneddon: The relation between load and penetration in the axisymmetric Boussinesq problem for a punch of arbitrary profile, Int. J. Eng. Sci. **3**, 47–57 (1965)

25.25 J.N. Isrelachvili: *Intermolecular and Surface Forces* (Academic, London 2003)

25.26 I.M. Ward, J. Sweeney: *An Introduction to the Mechanical Properties of Solid Polymers* (Wiley, Chichester 2004)

25.27 U. Dürig: Relations between interaction force, frequency shift in large-amplitude dynamic force microscopy, Appl. Phys. Lett. **75**, 433–435 (2004)

25.28 S. Morita, F.J. Giessibl, Y. Sugawara, H. Hosoi, K. Mukasa, A. Sasahara, H. Onishi: Noncontact atomic force microscopy and related topics. In: *Springer Handbook of Nanotechnology*, 3rd edn., ed. by B. Bhushan (Springer, Berlin Heidelberg 2010) Chap. 23

25.29 U. Dürig: Interaction sensing in dynamic force microscopy, New J. Phys. **2**, 5.1–5.12 (2000)

25.30 A. de Lozanne: Music of the spheres at the atomic scale, Science **305**, 348 (2004)

25.31 M. Posternak, H. Krakauer, A.J. Freeman, D.D. Koelling: Self-consistent electronic structure of surfaces: Surface states, surface resonances on W(001), Phys. Rev. B **21**, 5601–5612 (1980)

25.32 F. Mattheiss, D.R. Hamann: Electronic structure of the tungsten (001) surface, Phys. Rev. B **20**, 5372–5381 (1984)

25.33 F.J. Giessibl: Atomic resolution on Si(111)-(7×7) by noncontact atomic force microscopy with a force sensor based on a quartz tuning fork, Appl. Phys. Lett. **76**, 1470–1472 (2000)

25.34 S. Hembacher, F.J. Giessibl, J. Mannhart: Force microscopy with light-atom probes, Science **305**, 380 (2004)

25.35 F.J. Giessibl: Higher-harmonic atomic force microscopy, Surf. Interface Anal. **38**, 1696–1701 (2006)

730

26. Dynamic Modes of Atomic Force Microscopy

André Schirmeisen, Boris Anczykowski, Hendrik Hölscher, Harald Fuchs

This chapter presents an introduction to the concept of the dynamic operational modes of the atomic force microscope (dynamic AFM). While the static (or contact-mode) AFM is a widespread technique to obtain nanometer-resolution images on a wide variety of surfaces, true atomic-resolution imaging is routinely observed only in the dynamic mode. We will explain the jump-to-contact phenomenon encountered in static AFM and present the dynamic operational mode as a solution to avoid this effect. The dynamic force microscope is modeled as a harmonic oscillator to gain a basic understanding of the underlying physics in this mode.

On closer inspection, the dynamic AFM comprises a whole family of operational modes. A systematic overview of the different modes typically found in force microscopy is presented with special attention paid to the distinct features of each mode. Two modes of operation dominate the application of dynamic AFM. First, the amplitude modulation mode (also called tapping mode) is shown to exhibit an instability, which separates the purely attractive force interaction regime from the attractive–repulsive regime. Second, the self-excitation mode is derived and its experimental realization is outlined. While the tapping mode is primarily used for imaging in air and liquid, the self-excitation mode is typically used under ultrahigh vacuum (UHV) conditions for atomic-resolution imaging. In particular, we explain the influence of different forces on spectroscopy curves obtained in dynamic force microscopy. A quantitative link between the experimental spectroscopy curves and the interaction forces is established.

Force microscopy in air suffers from small quality factors of the force sensor (i.e., the

26.1	Motivation – Measurement of a Single Atomic Bond 732
26.2	Harmonic Oscillator: a Model System for Dynamic AFM 736
26.3	Dynamic AFM Operational Modes 737
	26.3.1 Amplitude-Modulation/ Tapping-Mode AFM 738
	26.3.2 Self-Excitation Modes 745
26.4	Q-Control 750
26.5	Dissipation Processes Measured with Dynamic AFM 754
26.6	Conclusions 758
References 758

cantilever beam), which are shown to limit the resolution. Also, the above-mentioned instability in the amplitude modulation mode often hinders imaging of soft and fragile samples. A combination of the amplitude modulation with the self-excitation mode is shown to increase the quality, or Q-factor, and extend the regime of stable operation. This so-called Q-control module allows one to increase as well as decrease the Q-factor. Apart from the advantages of dynamic force microscopy as a nondestructive, high-resolution imaging method, it can also be used to obtain information about energy-dissipation phenomena at the nanometer scale. This measurement channel can provide crucial information on electric and magnetic surface properties. Even atomic-resolution imaging has been obtained in the dissipation mode. Therefore, in the last section, the quantitative relation between the experimental measurement channels and the dissipated power is derived.

26.1 Motivation – Measurement of a Single Atomic Bond

The direct measurement of the force interaction between two distinct molecules has been a challenge for scientists for many years now. The fundamental forces responsible for the solid state of matter can be directly investigated, ultimately between defined single molecules. However, it has not been until 2001 that the chemical forces could be quantitatively measured for a single atomic bond [26.1]. How can we reliably measure forces that may be as small as one billionth of 1 N? How can we identify one single pair of atoms as the source of the force interaction?

The same mechanical principle that is used to measure the gravitational force exerted by your body weight (e.g., with the scale in your bathroom) can be employed to measure the forces between single atoms. A spring with a defined elasticity is compressed by an arbitrary force (e.g., your weight). The compression Δz of the spring (with spring constant k) is a direct measure of the force F exerted, which in the regime of elastic deformation obeys Hooke's law

$$F = k \Delta z \,. \tag{26.1}$$

The only difference with regard to your bathroom scale is the sensitivity of the measurement. Typically springs with a stiffness of $0.1-10\,\text{N/m}$ are used, which will be deflected by $0.1-10\,\text{nm}$ upon application of an interatomic force of some nN. Experimentally, a laser deflection technique is used to measure the movement of the spring. The spring is a bendable cantilever microfabricated from a silicon wafer. If a sufficiently sharp tip, usually directly attached to the cantilever, is approached toward a surface within some nanometers, we can measure the interaction forces through changes in the deflected laser beam. This is a static measurement and is hence called *static AFM*. Alternatively, the cantilever can be excited to vibrate at its resonant frequency. Under the influence of tip–sample forces the resonant frequency (and consequently also the amplitude and phase) of the cantilever will change and serve as measurement parameters. This approach is called *dynamic AFM*. Due to the multitude of possible operational modes, expressions such as noncontact mode, intermittent contact mode, tapping mode, frequency modulation (FM) mode, amplitude-modulation (AM) mode, self-excitation mode, constant-excitation mode, or constant-amplitude mode are found in the literature, which will be systematically categorized in the following paragraphs.

In fact, the first AFMs were operated in dynamic mode. In 1986, *Binnig* et al. presented the concept of the atomic force microscope [26.2]. The deflection of the cantilever with the tip was measured with sub-angstrom precision by an additional scanning tunneling microscope (STM). While the cantilever was externally oscillated close to its resonant frequency, the amplitude and phase of the oscillation were measured. If the tip is approached toward the surface, the oscillation parameters, amplitude and phase, are influenced by the tip–surface interaction, and can therefore be used as feedback channels. Typically, a certain setpoint for the amplitude is defined, and the feedback loop will adjust the tip–sample distance such that the amplitude remains constant. The control parameter is recorded as a function of the lateral position of the tip with respect to the sample, and the scanned image essentially represents the surface topography.

What then is the difference between the static and dynamic modes of operation for the AFM? Static deflection AFM directly gives the interaction force between tip and sample using (26.1). In the dynamic mode, we find that the resonant frequency, amplitude, and phase of the oscillation change as a consequence of the interaction forces (and also dissipative processes, as discussed in the final section).

In order to obtain a basic understanding of the underlying physics, it is instructive to consider a highly simplified case. Assume that the vibration amplitude is small compared with the range of force interaction.

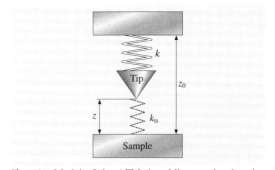

Fig. 26.1 Model of the AFM tip while experiencing tip–sample forces. The tip is attached to a cantilever with spring constant k, and the force interaction is modeled by a spring with a stiffness equal to the force gradient. Note that the force interaction spring is not constant, but depends on the tip–sample distance z

Since van der Waals forces range over typical distances of 10 nm, the vibration amplitude should be less than 1 nm. Furthermore, we require that the force gradient $\partial F_{ts}/\partial z$ does not vary significantly over one oscillation cycle. We can view the AFM setup as a coupling of two springs (Fig. 26.1). Whereas the cantilever is represented by a spring with spring constant k, the force interaction between the tip and the surface can be modeled by a second spring. The derivative of the force with respect to the tip–sample distance is the force gradient and represents the spring constant k_{ts} of the interaction spring. This spring constant k_{ts} is constant only with respect to one oscillation cycle, but varies with the average tip–sample distance as the probe is approached to the sample. The two springs are effectively coupled in parallel, since sample and tip support are rigidly connected for a given value of z_0. Therefore, we can write for the total spring constant of the AFM system

$$k_{\text{total}} = k + k_{ts} = k - \frac{\partial F_{ts}}{\partial z} . \qquad (26.2)$$

From the simple harmonic oscillator (neglecting any damping effects) we find that the resonant frequency ω of the system is shifted by $\Delta\omega$ from the free resonant frequency ω_0 due to the force interaction

$$\omega^2 = (\omega_0 + \Delta\omega)^2 = \frac{k_{\text{total}}}{m^*} = \frac{\left(k + \frac{\partial F_{ts}}{\partial z}\right)}{m^*} . \qquad (26.3)$$

Here m^* represents the effective mass of the cantilever. A detailed analysis of how m^* is related to the geometry and total mass of the cantilever can be found in the literature [26.3]. In the approximation that $\Delta\omega$ is much smaller than ω_0, we can write

$$\frac{\Delta\omega}{\omega_0} \cong -\frac{1}{2k} \frac{\partial F_{ts}}{\partial z} . \qquad (26.4)$$

Therefore, we find that the frequency shift of the cantilever resonance is proportional to the force gradient of the tip–sample interaction.

Although the above consideration is based on a highly simplified model, it shows qualitatively that in dynamic force microscopy we will find that the oscillation frequency depends on the force gradient, whereas static force microscopy measures the force itself. In principle, we can calculate the force curve from the force gradient and vice versa (neglecting a constant offset). It seems, therefore, that the two methods are equivalent, and our choice will depend on whether we can measure the beam deflection or the frequency shift with better precision at the cost of technical effort.

However, we have neglected one important issue for the operation of the AFM thus far: the mechanical stability of the measurement. In static AFM, the tip is slowly approached toward the surface. The force between the tip and the surface will always be counteracted by the restoring force of the cantilever. Figure 26.2 shows a typical force–distance curve. Upon approach of the tip toward the sample, the negative attractive forces, representing van der Waals or chemical interaction forces, increase until a maximum is reached. This turnaround point is due to the onset of repulsive forces caused by Coulomb repulsion, which will start to dominate upon further approach. The spring constant of the cantilever is represented by the slope of the straight line. The position of the z-transducer (typically a piezoelectric element), which moves the probe, is at the intersection of the line with the horizontal axis. The position of the tip, shifted from the probe's base due to the lever bending, can be found at the intersection of the cantilever line with the force curve. Hence, the to-

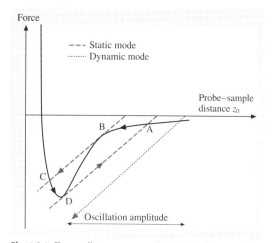

Fig. 26.2 Force–distance curve of a typical tip–sample interaction. In static-mode AFM the tip would follow the force curve until point B is reached. If the slope of the force curve becomes larger than the spring constant of the cantilever (*dashed line*) the tip will suddenly jump to position C. Upon retraction a different path will be followed along D and A again. In dynamic AFM the cantilever oscillates with amplitude. Although the equilibrium position of the oscillation is far from the surface, the tip will experience the maximum attractive force at point D during some parts of the oscillation cycle. However, the total force is always pointing away from the surface, therefore avoiding an instability

tal force is zero, i.e., the cantilever is in its equilibrium position (note that the spring constant line here shows attractive forces, although in reality the forces are repulsive, i.e., pulling the tip back from the surface). As soon as position A in Fig. 26.2 is reached, we find two possible intersection points, and upon further approach there are even three force equilibrium points. However, between points A and B the tip is at a local energy minimum and, therefore, will still follow the force curve. However, at point B, when the adhesion force upon further approach would become larger than the spring restoring force, the tip will suddenly jump to point C. We can then probe the predominantly repulsive force interaction by further reducing the tip–sample distance. When retracting the tip, we will pass point C, because the tip is still in a local energy minimum. Only at position D will the tip jump suddenly to point A again, since the restoring force now exceeds the adhesion. From Fig. 26.2 we can see that the sudden instability will happen at exactly the point where the slope of the adhesion force exceeds the slope of the spring constant. Therefore, if the negative force gradient of the tip–sample interaction will at any point exceed the spring constant, a mechanical instability occurs. Mathematically speaking, we demand that for a stable measurement

$$-\left.\frac{\partial F_{ts}}{\partial z}\right|_z < k , \quad \text{for all points } z . \tag{26.5}$$

This mechanical instability is often referred to as the *jump-to-contact* phenomenon.

Looking at Fig. 26.2, we realize that large parts of the force curve cannot be measured if the jump-to-contact phenomenon occurs. We will not be able to measure the point at which the attractive forces reach their maximum, representing the temporary chemical bonding of the tip and the surface atoms. Secondly, the sudden instability, the jump-to-contact, will often cause the tip to change the very last tip or surface atoms. A smooth, careful approach needed to measure the full force curve does not seem feasible. Our goal of measuring the chemical interaction forces of two single molecules may become impossible.

There are several solutions to the jump-to-contact problem: On the one hand, we can simply choose a sufficiently stiff spring, so that (26.5) is fulfilled at all points of the force curve. On the other hand, we can resort to a trick to enhance the counteracting force of the cantilever: We can oscillate the cantilever with large amplitude, thereby making it virtually stiffer at the point of strong force interaction.

Consider the first solution, which seems simpler at first glance. Chemical bonding forces extend over a distance range of about 0.1 nm. Typical binding energies of a couple of eV will lead to adhesion forces on the order of some nN. Force gradients will, therefore, reach values of some 10 N/m. A spring for stable force measurements will have to be as stiff as 100 N/m to ensure that no instability occurs (a safety factor of ten seems to be a minimum requirement, since usually one cannot be sure a priori that only one atom will dominate the interaction). In order to measure the nN interaction force, a static cantilever deflection of 0.01 nm has to be detected. With standard beam deflection AFM setups this becomes a challenging task.

This problem was solved by using an in situ optical interferometer measuring the beam deflection at liquid-nitrogen temperature in a UHV environment [26.4, 5]. In order to ensure that the force gradients are smaller than the lever spring constant (50 N/m), the tips were fabricated to terminate in only three atoms, thereby minimizing the total force interaction. The field ion microscope (FIM) is a tool which allows scanning probe microscopy (SPM) tips to be engineered down to atomic dimensions. This technique not only allows imaging of the tip apex with atomic precision, but also can be used to manipulate the tip atoms by field evaporation [26.6], as shown in Fig. 26.3. Atomic interaction

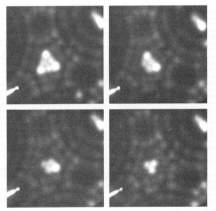

Fig. 26.3 Manipulation of the apex atoms of an AFM tip using field ion microscopy (FIM). *Images* were acquired at a tip bias of 4.5 kV. The last six atoms of the tip can be inspected in this example. Field evaporation to remove single atoms is performed by increasing the bias voltage for a short time to 5.2 kV. Each of the outer three atoms can be consecutively removed, eventually leaving a trimer tip apex

forces were measured with subnanonewton precision, revealing force curves of only a few atoms interacting without mechanical hysteresis. However, the technical effort to achieve this type of measurement is considerable, and most researchers today have resorted to the second solution.

The alternative solution can be visualized in Fig. 26.2. The straight, dashed line now represents the force values of the oscillating cantilever, with amplitude A assuming Hooke's law is valid. This is the tensile force of the cantilever spring pulling the tip away from the sample. The restoring force of the cantilever is at all points stronger than the adhesion force. For example, the total force at point D is still pointing away from the sample, although the spring has the same stiffness as before. Mathematically speaking, the measurement is stable as long as the cantilever spring force $F_{cb} = kA$ is larger than the attractive tip–sample force F_{ts} [26.7]. In the static mode we would already experience an instability at that point. However, in the dynamic mode, the spring is preloaded with a force stronger than the attractive tip–sample force. The equilibrium point of the oscillation is still far away from the point of closest contact of the tip and surface atoms. The total force curve can now be probed by varying the equilibrium point of the oscillation, i.e., by adjusting the z-piezo.

The diagram also shows that the oscillation amplitude has to be quite large if fairly soft cantilevers are to be used. With lever spring constants of $10\,\text{N/m}$, the amplitude must be at least $1\,\text{nm}$ to ensure that forces of $1\,\text{nN}$ can be reliably measured. In practical applications, amplitudes of $10–100\,\text{nm}$ are used to stay on the safe side. This means that the oscillation amplitude is much larger than the force interaction range. The above simplification, that the force gradient remains constant within one oscillation cycle, does not hold anymore. Measurement stability is gained at the cost of a simple quantitative analysis of the experiments. In fact, dynamic AFM was first used to obtain atomic resolution images of clean surfaces [26.8], and it took another 6 years [26.1] before quantitative measurements of single bond forces were obtained.

The technical realization of dynamic-mode AFMs is based on the same key components as a static AFM setup. The most common principle is the method of laser deflection sensing (Fig. 26.4). A laser beam is focused on the back side of a microfabricated cantilever. The reflected laser spot is detected with a position-sensitive diode (PSD). This photodiode is sectioned

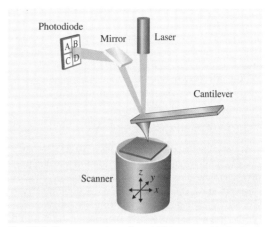

Fig. 26.4 Representation of an AFM setup with the laser beam deflection method. Cantilever and tip are microfabricated from silicon wafers. A laser beam is deflected from the back side of the cantilever and again focused on a photosensitive diode via an adjustable mirror. The diode is segmented into four quadrants, which allows measurement of vertical and torsional bending of the cantilever (artwork by D. Ebeling rendered with POV-Ray)

into two parts that are read out separately (usually even a four-quadrant diode is used to detect torsional movements of the cantilever for lateral friction measurements). With the cantilever at equilibrium, the spot is adjusted such that the two sections show the same intensity. If the cantilever bends up or down, the spot moves, and the difference signal between the upper and lower sections is a measure of the bending.

In order to enhance sensitivity, several groups have adopted an interferometer system to measure the cantilever deflection. A thorough comparison of different measurement methods with analysis of sensitivity and noise level is given in reference [26.3].

The cantilever is mounted on a device that allows the beam to be oscillated. Typically a piezo element directly underneath the cantilever beam serves this purpose. The reflected laser beam is analyzed for oscillation amplitude, frequency, and phase difference. Depending on the mode of operation, a feedback mechanism will adjust oscillation parameters and/or tip–sample distance during the scanning. The setup can be operated in air, UHV, and even fluids. This allows measurement of a wide range of surface properties from atomic-resolution imaging [26.8] up to studying biological processes in liquid [26.9, 10].

26.2 Harmonic Oscillator: a Model System for Dynamic AFM

The oscillating cantilever has three degrees of freedom: the amplitude, the frequency, and the phase difference between excitation and oscillation. Let us consider the damped driven harmonic oscillator. The cantilever is mounted on a piezoelectric element that is oscillating with amplitude A_d at frequency ω

$$z_d(t) = A_d \cos(\omega t). \quad (26.6)$$

We assume that the cantilever spring obeys Hooke's law. Secondly, we introduce a friction force that is proportional to the speed of the cantilever motion, whereas α denotes the damping coefficient (Amontons' law). With Newton's first law we find for the oscillating system the following equation of motion for the position $z(t)$ of the cantilever tip (Fig. 26.1)

$$m\ddot{z}(t) = -\alpha\dot{z}(t) - kz(t) - kz_d(t). \quad (26.7)$$

We define $\omega_0^2 = k/m^*$, which turns out to be the resonant frequency of the free (undamped, i.e., $\alpha = 0$) oscillating beam. We further define the dimensionless quality factor $Q = m^*\omega_0/\alpha$, antiproportional to the damping coefficient. The quality factor describes the number of oscillation cycles after which the damped oscillation amplitude decays to $1/e$ of the initial amplitude with no external excitation ($A_d = 0$). After some basic math, this results in

$$\ddot{z}(t) + \frac{\omega_0}{Q}\dot{z}(t) + \omega_0^2 z(t) = A_d \omega_0^2 \cos(\omega t). \quad (26.8)$$

The solution is a linear combination of two regimes [26.11]. Starting from rest and switching on the piezo excitation at $t = 0$, the amplitude will increase from zero to the final magnitude and reach a steady state, where the amplitude, phase, and frequency of the oscillation stay constant over time. The steady-state solution $z_1(t)$ is reached after $2Q$ oscillation cycles and follows the external excitation with amplitude A_0 and phase difference φ

$$z_1(t) = A_0 \cos(\omega t + \varphi). \quad (26.9)$$

The oscillation amplitude in the transient regime during the first $2Q$ cycles is

$$z_2(t) = A_t e^{\frac{-\omega_0 t}{2Q}} \sin(\omega_0 t + \varphi_t). \quad (26.10)$$

We emphasize the important fact that the exponential term causes $z_2(t)$ to decrease exponentially with time constant τ

$$\tau = \frac{2Q}{\omega_0}. \quad (26.11)$$

In vacuum conditions, only the internal dissipation due to bending of the cantilever is present, and Q reaches values of 10 000 at typical resonant frequencies of 100 000 Hz. These values result in a relatively long transient regime of $\tau \cong 30$ ms, which limits the possible operational modes for dynamic AFM (for a detailed analysis see *Albrecht* et al. [26.11]). Changes in the measured amplitude, which reflect a change of atomic forces, will have a time lag of 30 ms, which is very slow considering one wants to scan a 200×200 point image within a few minutes. In air, however, viscous damping due to air friction dominates and Q drops to less than 1000, resulting in a time constant below the millisecond level. This response time is fast enough to use the amplitude as a measurement parameter.

If we evaluate the steady-state solution $z_1(t)$ in the differential equation, we find the following well-known

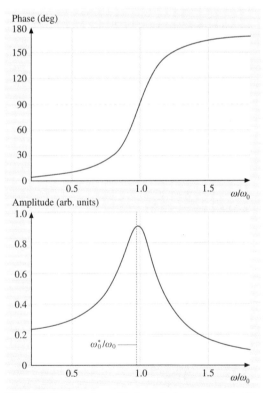

Fig. 26.5 Curves of amplitude and phase versus excitation frequency for the damped harmonic oscillator, with a quality factor of $Q = 4$

solution for amplitude and phase of the oscillation as a function of the excitation frequency ω:

$$A_0 = \frac{A_d Q \omega_0^2}{\sqrt{\omega^2 \omega_0^2 + Q^2 (\omega_0^2 - \omega^2)^2}}, \quad (26.12)$$

$$\varphi = \arctan\left(\frac{\omega \omega_0}{Q(\omega_0^2 - \omega^2)}\right). \quad (26.13)$$

Amplitude and phase diagrams are depicted in Fig. 26.5. As can be seen from (26.12), the amplitude will reach its maximum at a frequency different from ω_0 if Q has a finite value. The damping term of the harmonic oscillator causes the resonant frequency to shift from ω_0 to ω_0^*

$$\omega_0^* = \omega_0 \sqrt{1 - \frac{1}{2Q^2}}. \quad (26.14)$$

The shift is negligible for Q-factors of 100 and above, which is the case for most applications in vacuum or air. However, for measurements in liquids, Q can be smaller than 10 and ω_0 differs significantly from ω_0^*. As we will discuss later, it is also possible to enhance Q by using a special excitation method called Q-control.

In the case that the excitation frequency is equal to the resonant frequency of the undamped cantilever $\omega = \omega_0$, we find the useful relation

$$A_0 = Q A_d, \quad \text{for} \quad \omega = \omega_0. \quad (26.15)$$

Since $\omega_0^* \approx \omega_0$ for most cases, we find that (26.15) holds true for exciting the cantilever at its resonance. From a similar argument, the phase becomes approximately 90° for the resonance case. We also see that, in order to reach vibration amplitudes of some 10 nm, the excitation only has to be as small as 1 pm, for typical cantilevers operated in vacuum.

So far we have not considered an additional force term, describing the interaction between the probing tip and the sample. For typical, large vibration amplitudes of 10–100 nm the tip experiences a whole range of force interactions during one single oscillation cycle, rather than one defined tip–sample force. How this problem can be attacked will be shown in the next paragraphs.

26.3 Dynamic AFM Operational Modes

While the quantitative interpretation of force curves in contact AFM is straightforward using (26.1), we explained in the previous paragraphs that its application to assess short-range attractive interatomic forces is rather limited. The dynamic mode of operation seems to open a viable direction toward achieving this task. However interpretation of the measurements generally appears to be more difficult. Different operational modes are employed in dynamic AFM, and the following paragraphs are intended to distinguish these modes and categorize them in a systematic way.

The oscillation trajectory of a dynamically driven cantilever is determined by three parameters: the amplitude, the phase, and the frequency. Tip–sample interactions can influence all three parameters, in the following, termed the internal parameters. The oscillation is driven externally, with excitation amplitude A_d and excitation frequency ω. These variables will be referred to as the external parameters. The external parameters are set by the experimentalist, whereas the internal parameters are measured and contain the crucial information about the force interaction. In scanning probe applications, it is common to control the probe–surface distance z_0 in order to keep an internal parameter constant (i.e., the tunneling current in STM or the beam deflection in contact AFM), which represents a certain tip–sample interaction. In z-spectroscopy mode, the distance is varied in a certain range, and the change of the internal parameters is measured as a fingerprint of the tip–sample interactions.

In dynamic AFM the situation is rather complex. Any of the internal parameters can be used for feedback of the tip–sample distance z_0. However, we already realized that, in general, the tip–sample forces could only be fully assessed by measuring all three parameters. Therefore, dynamic AFM images are difficult to interpret. A solution to this problem is to establish additional feedback loops, which keep the internal parameters constant by adjusting the external variables. In the simplest setup, the excitation frequency and the excitation amplitude are set to predefined values. This is the so-called amplitude-modulation (AM) mode or tapping mode. As stated before, in principle, any of the internal parameters can be used for feedback to the tip–sample distance – in AM mode the amplitude signal is used. A certain amplitude (smaller than the free oscillation amplitude) at a frequency close to the resonance of the cantilever is chosen, the tip is approached toward the surface under investigation, and the approach is stopped as soon as the setpoint amplitude is reached. The oscillation phase is usually recorded during the scan; however, the shift of the resonant frequency of the cantilever cannot be directly accessed, since this degree of freedom is blocked by the external excitation at a fixed frequency. It turns out that this mode is simple to operate from a technical perspective. Therefore, it is one of the most commonly used modes in dynamic AFM operated in air, and even in liquid. The strength of this mode is the easy and reliable high-resolution imaging of a large variety of surfaces.

It is interesting to discuss the AM mode in the situation that the external excitation frequency is much lower than the resonant frequency [26.12, 13]. This results in a quasistatic measurement, although a dynamic oscillation force is applied, and therefore this mode can be viewed as a hybrid between static and dynamic AFM. Unfortunately, it has the drawbacks of the static mode, namely that stiff spring constants must be used and therefore the sensitivity of the deflection measurement must be very good, typically employing a high-resolution interferometer. Still, it has the advantage of the static measurement in terms of quantitative interpretation, since in the regime of small amplitudes (< 0.1 nm) direct interpretation of the experiments is possible. In particular, the force gradient at tip–sample distance z_0 is given by the change of the amplitude A and the phase angle φ

$$\left.\frac{\partial F_{ts}}{\partial z}\right|_{z_0} = k\left(1 - \frac{A_0}{A}\cos\varphi\right) . \qquad (26.16)$$

In effect, the modulated AFM technique can profit from an enhanced sensitivity due to the use of lock-in techniques, which allows the measurement of the amplitude and phase of the oscillation signal with high precision.

As stated before, the internal parameters can be fed back to the external excitation variables. One of the most useful applications in this direction is the self-excitation system. Here the resonant frequency of the cantilever is detected and selected again as the excitation frequency. In a typical setup, the cantilever is self-oscillated with a phase shift of 90° by feeding back the detector signal to the excitation piezo. In this way the cantilever is always excited at its actual resonance [26.14]. Tip–sample interaction forces then only influence the resonant frequency, but do not change the two other parameters of the oscillation (amplitude and phase). Therefore, it is sufficient to measure the frequency shift induced by the tip–sample interaction.

Since the phase remains at a fixed value, the oscillating system is much better defined than before, and the degrees of freedom for the oscillation are reduced. To even reduce the last degree of freedom an additional feedback loop can be incorporated to keep the oscillation amplitude A constant by varying the excitation amplitude A_d. Now, all internal parameters have a fixed relation to the external excitation variables, the system is well defined, and all parameters can be assessed during the measurement.

In the following section we want to discuss the two most popular operational modes, tapping mode and self-excitation mode, in more detail.

26.3.1 Amplitude-Modulation/Tapping-Mode AFM

In tapping mode, or AM-AFM, the cantilever is excited externally at a constant frequency close to its resonance. Oscillation amplitude and phase during approach of tip and sample serve as the experimental observation channels. Figure 26.6 shows a diagram of a typical tapping-mode AFM setup. The oscillation of the can-

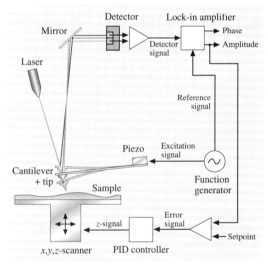

Fig. 26.6 Setup of a dynamic force microscope operated in the AM or tapping mode. A laser beam is deflected by the back side of the cantilever and the deflection is detected by a split photodiode. The excitation frequency is chosen externally with a modulation unit, which drives the excitation piezo. A lock-in amplifier analyzes the phase and amplitude of the cantilever oscillation. The amplitude is used as the feedback signal for the probe–sample distance control

tilever is detected with the photodiode, whose output signal is analyzed with a lock-in amplifier to obtain amplitude and phase information. The amplitude is then compared with the setpoint, and the resulting difference or error signal is fed into the proportional–integral–differential (PID) controller, which adjusts the z-piezo, i.e., the probe–sample distance, accordingly. The external modulation unit supplies the signal for the excitation piezo, and at the same time the oscillation signal serves as the reference for the lock-in amplifier. As shown by the following applications the tapping mode is typically used to measure surface topography and other material parameters on the nanometer scale. The tapping mode is mostly used in ambient conditions and in liquids.

High-resolution imaging has been extensively performed in the area of materials science. Due to its technical relevance the investigation of polymers has been the focus of many studies (see, e.g., a recent review about AFM imaging on polymers by *Magonov* [26.16]). In Fig. 26.7 the topography of a diblock copolymer ($BC_{0.26}$–$3A_{0.53}F_8H_{10}$) at different magnifications is shown [26.15]. On the large scan (Fig. 26.7a) the large-scale structure of the microphase-separated polystyrene (PS) cylinders (within a polyisoprene (PI) matrix) lying parallel to the substrate can be seen. In the high-resolution image (Fig. 26.7b) a surface substructure of regular domes can be seen, which were found to be related to the cooling process during the polymer preparation.

Imaging in liquids opens up the possibility of the investigation of biological samples in their natural environment. For example *Möller* et al. [26.17] have obtained high-resolution images of the topography of hexagonally packed intermediate (HPI) layer of *Deinococcus radiodurans* with tapping-mode AFM. Another interesting example is the imaging of DNA in liquid, as shown in Fig. 26.8. *Jiao* et al. [26.10] measured the time evolution of a single DNA strand interacting with a molecule as shown by a sequence of images acquired in liquid over a time period of several minutes.

For a quantitative interpretation of tip–sample forces one has to consider that during one oscillation cycle with amplitudes of 10–100 nm the tip–sample interaction will range over a wide distribution of forces, including attractive as well as repulsive forces. We will, therefore, measure a convolution of the force–distance curve with the oscillation trajectory. This complicates the interpretation of AM-AFM measurements appreciably.

At the same time, the resonant frequency of the cantilever will change due to the appearing force gradients, as could already be seen in the simplified model in (26.4). If the cantilever is excited exactly at its reso-

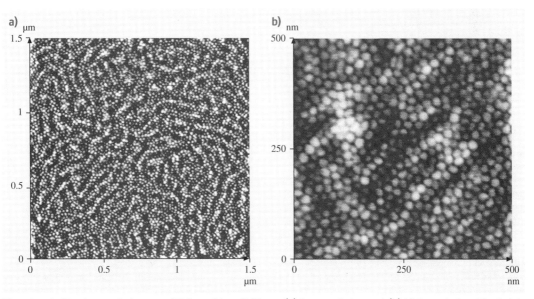

Fig. 26.7a,b Tapping-mode images of $BC_{0.26}$–$3A_{0.53}F_8H_{10}$ at (a) low resolution and (b) high resolution. The height scale is 10 nm (after [26.15], © American Chemical Society, 2001)

nant frequency before, it will be excited off resonance after interaction forces are encountered. This, in turn, changes the amplitude and phase in (26.12) and (26.13), which serve as the measurement signals. Consequently, a different amplitude will cause a change in the encountered effective force. We can see already from this simple *gedanken* experiment that the interpretation of the measured phase and amplitude curves is not straightforward.

The qualitative behavior for amplitude versus z_0-position curves is depicted in Fig. 26.9. At large distances, where the forces between tip and sample are negligible, the cantilever oscillates with its free oscillation amplitude. Upon approach of the probe toward the surface the interaction forces cause the amplitude to change, typically resulting in an amplitude that gets smaller with continuously decreased tip–sample distance. This is expected, since the force–distance curve will eventually reach the repulsive part and the tip is hindered from indenting further into the sample, resulting in smaller oscillation amplitudes.

However, in order to gain some qualitative insight into the complex relationship between forces and oscillation parameters, we resort to numerical simulations. Anczykowski et al. [26.18, 19] have calculated the oscillation trajectory of the cantilever under the influence of a given force model. van der Waals interactions were considered the only effective attractive forces, and the total interaction resembled a Lennard–Jones-type potential. Mechanical relaxations of the tip and sample surface were treated in the limits of continuum theory with the numerical Muller–Yushchenko–Derjaguin/Burgess–Hughes–White MYD/BHW [26.20, 21] approach, which allows the simulations to be compared with experiments.

The cantilever trajectory was analyzed by numerically solving the differential equation (26.7) extended by the tip–sample force. The results of the simulation for the amplitude and phase of the tip oscillation as a function of z-position of the probe are presented in Fig. 26.10. One has to keep in mind that the z-position of the probe is not equivalent to the real tip–sample distance at equilibrium position, since the cantilever might bend statically due to the interaction forces. The behavior of the cantilever can be subdivided into three different regimes. We distinguish the cases in which the beam is oscillated below its resonant frequency ω_0, exactly at ω_0, and above ω_0. In the following, we will refer

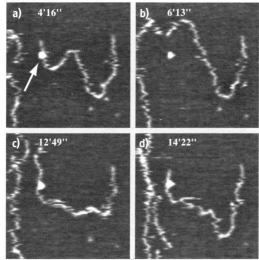

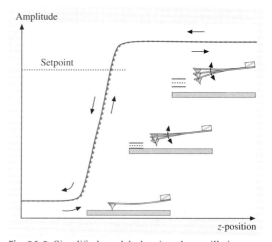

Fig. 26.8a–d Dynamic p53–DNA interactions observed by time-lapse tapping-mode AFM imaging in solution. Both p53 protein and DNA were weakly adsorbed to a mica surface by balancing the buffer conditions. (**a**) A p53 protein molecule (*arrow*) was bound to a DNA fragment. The protein (**b**) dissociated from and then (**c**) reassociated with the DNA fragment. (**d**) A downward movement of the DNA with respect to the protein occurred, constituting a *sliding* event whereby the protein changes its position on the DNA. Image size: 620 nm. Grey scale (height) range: 4 nm. Time units: min, s. (© T. Schäffer, University of Münster)

Fig. 26.9 Simplified model showing the oscillation amplitude in tapping-mode AFM for various probe–sample distances

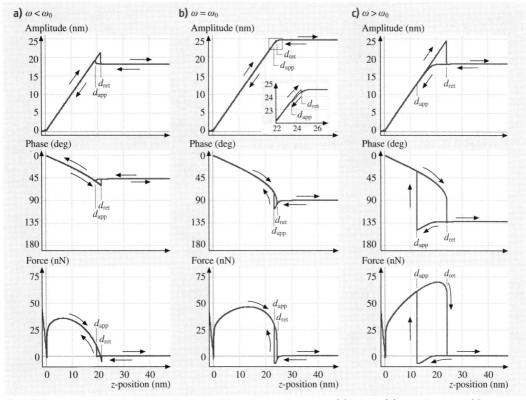

Fig. 26.10a–c Amplitude and phase diagrams with excitation frequency: (**a**) below, (**b**) exactly at, and (**c**) above the resonant frequency for tapping-mode AFM from numerical simulations. Additionally, the *bottom diagrams* show the interaction forces at the point of closest tip–sample distance, i. e., the lower turnaround point of the oscillation

to ω_0 as the resonant frequency, although the correct resonant frequency is ω_0^* if taking into account the finite Q-value.

Clearly, Fig. 26.10 exhibits more features than were anticipated from the initial, simple arguments. Amplitude and phase seem to change rather abruptly at certain points when the z_0-position is decreased. Additionally, we find hysteresis between approach and retraction.

As an example, let us start by discussing the discontinuous features in the AFM spectroscopy curves of the first case, where the excitation frequency is smaller than ω_0. Consider the oscillation amplitude as a function of excitation frequency in Fig. 26.5. Upon approach of probe and sample, attractive forces will lower the effective resonant frequency of the oscillator. Therefore, the excitation frequency will now be closer to the resonant frequency, causing the vibration amplitude to increase. This, in turn, reduces the tip–sample distance, which again gives rise to a stronger attractive force. The system becomes unstable until the point $z_0 = d_{\text{app}}$ is reached, where repulsive forces stop the self-enhancing instability. This can be clearly observed in Fig. 26.10a. Large parts of the force–distance curve cannot be measured due to this instability.

In the second case, where the excitation equals the free resonant frequency, only a small discontinuity is observed upon reduction of the z-position. Here, a shift of the resonant frequency toward smaller values, induced by the attractive force interaction, will reduce the oscillation amplitude. The distance between tip and sample is, therefore, reduced as well, and the self-amplifying effect with the sudden instability does not occur as long as repulsive forces are not encountered. However, at closer tip–sample distances, repulsive forces will cause the resonant frequency to shift again toward higher values, increasing the ampli-

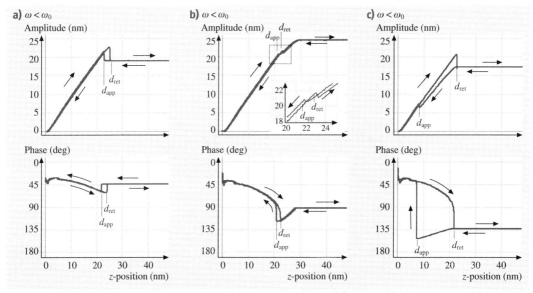

Fig. 26.11a–c Amplitude and phase diagrams with excitation frequency: (**a**) below, (**b**) exactly at, and (**c**) above the resonant frequency for tapping-mode AFM from experiments with a Si cantilever on a Si wafer in air

tude with decreasing tip–sample distance. Therefore, a self-enhancing instability will also occur in this case, but at the crossover from purely attractive forces to the regime where repulsive forces occur. Correspondingly, a small kink in the amplitude curve can be observed in Fig. 26.10b. An even clearer indication of this effect is manifested by the sudden change in the phase signal at d_{app}.

In the last case, with $\omega > \omega_0$, the effect of amplitude reduction due to the resonant frequency shift is even larger. Again, we find no instability in the amplitude signal during approach in the attractive force regime. However, as soon as the repulsive force regime is reached, the instability occurs due to the induced positive frequency shift. Consequently, a large jump in the phase curve from values smaller than 90° to values larger than 90° is observed. The small change in the amplitude curve is not resolved in the simulated curves in Fig. 26.10c; however, it can be clearly seen in the experimental curves.

Figure 26.11 depicts the corresponding experimental amplitude and phase curves. The measurements were performed in air with a Si cantilever approaching a silicon wafer, with a cantilever resonant frequency of 299.95 kHz. Qualitatively, all prominent features of the simulated curves can also be found in the experimental data sets. Hence, the above model seems to capture the important factors necessary for an appropriate description of the experimental situation.

However, what is the reason for this unexpected behavior? We have to turn to the numerical simulations again, where we have access to all physical parameters, in order to understand the underlying processes. The lower part of Fig. 26.10 also shows the interaction force between the tip and the sample at the point of closest approach, i.e., the sample-sided turnaround point of the oscillation. We see that exactly at the points of the discontinuities the total interaction force changes from the net-attractive regime to the attractive–repulsive regime, also termed the intermittent contact regime. The term net-attractive is used to emphasize that the total force is attractive, despite the fact that some minor contributions might still originate from repulsive forces. As soon as a minimum distance is reached, the tip also starts to experience repulsive forces, which completely changes the oscillation behavior. In other words, the dynamic system switches between two oscillatory states.

Directly related to this fact is the second phenomenon: the hysteresis effect. We find separate curves for the approach of the probe toward the surface and the retraction. This seems to be somewhat counterintuitive, since the tip is constantly approaching and retracting from the surface and the average values of amplitude and phase should be independent of the direction of the average tip–sample distance movement. Hysteresis between approach and retraction within one oscillation due to dissipative processes should directly influence amplitude and phase. However, no dissipation models were included in the simulation. In this case, the hysteresis in Fig. 26.11 is due to the fact that the oscillation jumps into different modes; the system exhibits bistability. This effect is often observed in oscillators under the influence of nonlinear forces (e.g., [26.22]).

For the interpretation of these effects it is helpful to look at Fig. 26.12, which shows the behavior of the simulated tip trajectory and the force during one oscillation cycle over time. The data is shown for the z-positions where hysteresis is observed, while Fig. 26.12a was taken during the approach and Fig. 26.12b during the retraction. Excitation was in resonance, where the amplitude shows small hysteresis. Also note that the amplitude is almost exactly the same in Fig. 26.12a,b. We see that the oscillation at the same z-position exhibits two different modes: Whereas in Fig. 26.12a the experienced force is net-attractive, in Fig. 26.12b the tip is exposed to attractive and repulsive interactions. Experimental and simulated data show that the change between the net-attractive and intermittent contact mode takes place at different z-positions (d_{app} and d_{ret}) for approach and retraction. Between d_{app} and d_{ret} the system is in a bistable mode. Depending on the history of the measurement, e.g., whether the position d_{app} during the approach (or d_{ret} during retraction) has been reached, the system flips to the other oscillation mode. While the amplitude might not be influenced strongly, the phase is a clear indicator of the mode switch. On the other hand, if point d_{app} is never reached during the approach, the system will stay in the net-attractive regime and no hysteresis is observed, i.e., the system remains stable.

In conclusion, we find that, although a qualitative interpretation of the interaction forces is possible from the amplitude and phase curves, they do not give direct quantitative knowledge of tip–sample force interactions. However, it is a very useful tool for imaging nanometer-sized structures in a wide variety of setups, in air or even in liquid. We find that two distinct modes exist for the externally excited oscillation – the net-attractive and the intermittent contact mode – which describe what kind of forces govern the tip–sample interaction. The phase can be used as an indicator of the current mode of the system.

In particular, it can be easily seen that, if the free resonant frequency of the cantilever is higher than the

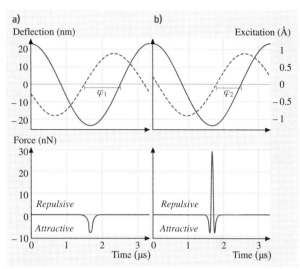

Fig. 26.12a,b Simulation of the tapping-mode cantilever oscillation in the (**a**) net-attractive and (**b**) the intermittent contact regime. The *dashed line* represents the excitation amplitude and the *solid line* is the oscillation amplitude

excitation frequency, the system cannot stay in the net-attractive regime due to a self-enhancing instability. Since in many applications involving soft and delicate biological samples strong repulsive forces should be avoided, the tapping-mode AFM should be operated at frequencies equal to or above the free resonant frequency [26.23]. Even then, statistical changes of tip–sample forces during the scan might induce a sudden jump into the intermittent contact mode, and the previously explained hysteresis will tend to keep the system in this mode. It is, therefore, of great importance to tune the oscillation parameters in such a way that the AFM stays in the net-attractive regime [26.24]. A concept that achieves this task is the Q-control system, which will be discussed in some detail in the forthcoming paragraphs.

A last word concerning the overlap of simulation and experimental data: Whereas the qualitative agreement down to the detailed shape of hysteresis and instabilities is rather striking, we still find some quantitative discrepancies between the positions of the instabilities d_{app} and d_{ret}. This is probably due to the simplified force model, which only takes into account van der Waals and repulsive forces. Especially at ambient conditions, an omnipresent water meniscus between tip and sample will give rise to much stronger attractive and also dissipative forces than considered in the model. A very interesting feature is that the simulated phase curves in the intermittent contact regime tend to have a steeper slope in the simulation than in the experiments (Fig. 26.13). We will show later that this effect is a fingerprint of an effect that had not been included in the above simulation at all: dissipative processes during the oscillation, giving rise to an additional loss of oscillation energy.

In the above paragraphs, we have outlined the influence of the tip–sample interaction on the cantilever oscillation, calculated the maximum tip–sample interaction forces based on the assumption of a specific model force, and subsequently discussed possible routes for image optimization. However, in practical imaging, the tip–sample interaction is not known a priori. In contrast, the ability to measure the continuous tip–sample interaction force as a function of both the tip–sample distance as well as the lateral location (e.g., in order to identify different bond strengths on chemically inhomogeneous surfaces) would add a tool of great value to the force-microscopist's toolbox.

Surprisingly, despite the more than 15 years during which the amplitude-modulation technique has been used, it was only recently that two solutions to this inversion problem have been suggested [26.25, 26]. As already discussed in the previous paragraphs, conventional force–distance curves suffer from a *jump-to-contact* due to attractive surface forces. As a result, the most interesting range of the tip–sample force, the last few nanometers above the surface, is left out, and conventional force–distance curves thus mainly serve to determine adhesion forces.

As shown by *Hölscher* [26.25] the tip–sample force can be calculated with the help of the integral equation

$$F_{ts}(D) = -\frac{\partial}{\partial D} \int_D^{D+2A} \frac{\kappa(z)}{\sqrt{z-D}} \, dz \,, \qquad (26.17a)$$

where

$$\kappa = \frac{kA^{3/2}}{\sqrt{2}} \left(\frac{A_d \cos\varphi}{A} - \frac{\omega_0^2 - \omega^2}{\omega^2} \right) . \qquad (26.17b)$$

It is now straightforward to recover the tip–sample force using (26.17a,b) from a *spectroscopy experiment*, i.e., an experiment where the amplitude and the phase are continuously measured as a function of the actual nearest tip–sample distance $D = z_0 - A$ at a fixed location above the sample surface. With this input, one first calculates κ as a function of D. In a second step, the tip–sample force is computed by solving the integral in (26.17a) numerically.

A verification of the algorithm is shown in Fig. 26.14, which presents computer simulations of the method by calculating numerical solutions of the equation of motion. Figure 26.14a,b shows the resulting curves of amplitude and phase versus distance during approach, respectively. The subsequent reconstruction of the tip–sample interaction based on the data provided by the curves of amplitude and phase versus distance is presented in Fig. 26.14d. The assumed tip–sample force and energy dissipation are plotted by solid lines, while the reconstructed data is indicated by symbols; the excellent agreement demonstrates the reliability of the method. Nonetheless, it is important to recognize that the often observed instability in the curves of amplitude and phase versus distance affects the reconstruction of the tip–sample force. If such an instability occurs, experimentally accessible $\kappa(D)$ values will feature a *gap* at a specific range of tip–sample distances D. This is illustrated in Fig. 26.14c, where the gap is indicated by an arrow and the question mark. Since calculation of the integral (26.17a) requires knowledge of all κ values within the oscillation range, one might be tempted to extrapolate the missing values in the gap. This could be a workable solution if, as in our example, the accessible κ values appear smooth and, in particular, the lower turning point of the $\kappa(D)$ values is clearly visible. In most realistic cases, however, the curves are unlikely to look as smooth as in our simulation and/or the lower turning point might not be reached, and we thus advise utmost caution in applying any extrapolation for missing data points.

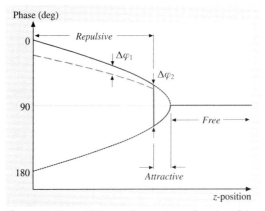

Fig. 26.13 Phase shift in tapping mode as a function of tip–sample distance

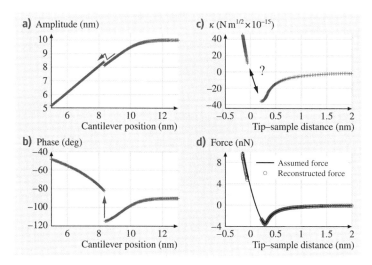

Fig. 26.14a–d Numerical verification of the proposed force spectroscopy method for the tapping mode. The numerically calculated curves of amplitude (**a**) and phase (**b**) versus distance during the approach towards the sample surface. Both curves reveal the typical instability resulting also in a gap for the κ-curve. The tip–sample interaction force (**d**) can be recalculated from this data set by the application of (26.17a). As discussed in the text, the integration over the gap has to be handled with care

26.3.2 Self-Excitation Modes

Despite the wide range of technical applications of the AM mode of dynamic AFM, it has been found unsuitable for measurements in an environment extremely useful for scientific research: vacuum or ultrahigh vacuum (UHV) with pressures reaching 10^{-10} mbar. The STM has already shown how much insight can be gained from experiments under those conditions.

Consider (26.11) from the previous section. The time constant τ for the amplitude to adjust to a different tip–sample force scales with $1/Q$. In vacuum applications, the Q-factor of the cantilever is on the order of 10 000, which means that τ is in the range of some 10 ms. This time constant is clearly too long for a scan of at least 100×100 data points. On the other hand, the resonant frequency of the system will react instantaneously to tip–sample forces. This has led *Albrecht* et al. [26.11] to use a modified excitation scheme.

The system is always oscillated at its resonant frequency. This is achieved by feeding back the oscillation signal from the cantilever into the excitation piezo element. Figure 26.15 pictures the method in a block diagram. The signal from the PSD is phase-shifted by 90° (and, therefore, always exciting in resonance) and used as the excitation signal of the cantilever. An additional feedback loop adjusts the excitation amplitude in such a way that the oscillation amplitude remains constant. This ensures that the tip–sample distance is not influenced by changes in the oscillation amplitude.

The only degree of freedom that the oscillation system still has that can react to the tip–sample forces is the change of the resonant frequency. This shift of the frequency is detected and used as the setpoint signal for surface scans. Therefore, this mode is also called the frequency-modulation (FM) mode.

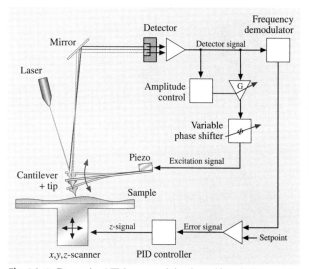

Fig. 26.15 Dynamic AFM operated in the self-excitation mode, where the oscillation signal is directly fed back to the excitation piezo. The detector signal is amplified with the variable gain G and phase-shifted by phase ϕ. The frequency demodulator detects the frequency shift due to tip–sample interactions, which serves as the control signal for the probe–sample distance

Let us take a look at the sensitivity of the dynamic AFM. If electronic noise, laser noise, and thermal drift can be neglected, the main noise contribution will come from thermal excitations of the cantilever. A detailed analysis of a dynamic system yields for the minimum detectable force gradient the relation [26.11]

$$\left.\frac{\partial F}{\partial z}\right|_{\min} = \sqrt{\frac{4kk_B TB}{\omega_0 Q \langle z_{\text{osc}}^2 \rangle}}.\qquad(26.18)$$

Here, B is the bandwidth of the measurement, T the temperature, and $\langle z_{\text{osc}}^2 \rangle$ is the mean-square amplitude of the oscillation. Please note that this sensitivity limit was deliberately calculated for the FM mode. A similar analysis of the AM mode, however, yields virtually the same result [26.27]. We find that the minimum detectable force gradient, i.e., the measurement sensitivity, is inversely proportional to the square root of the Q-factor of the cantilever. This means that it should be possible to achieve very high-resolution imaging under vacuum conditions where the Q-factor is very high.

A breakthrough in high-resolution AFM imaging was the atomic resolution imaging of the Si(111)-(7×7) surface reconstruction by *Giessibl* [26.8] under UHV conditions. Moreover, *Sugawara* et al. [26.28] observed the motion of single atomic defects on InP with true atomic resolution. However, imaging on conducting or semiconducting surfaces is also possible with the scanning tunneling microscope (STM) and these first noncontact atomic force microscopy (NC-AFM) images provided little new information on surface properties. The true potential of NC-AFM lies in the imaging of nonconducting surface with atomic precision, which was first demonstrated by *Bammerlin* et al. [26.29] on NaCl. A long-standing question about the surface reconstruction of the technological relevant material aluminum oxide could be answered by *Barth* and *Reichling* [26.30], who imaged the atomic structure of the high-temperature phase of α-Al$_2$O$_3$(0001).

The high-resolution capabilities of noncontact atomic force microscopy are nicely demonstrated by the images shown in Fig. 26.16. *Allers* et al. [26.31] imaged steps and defects on the insulator nickel oxide with atomic resolution. Recently, *Kaiser* et al. [26.32] succeeded in imaging the antiferromagnetic structure of NiO(001). Nowadays, true atomic resolution is routinely obtained by various research groups (for an overview, see [26.33–36]).

However, we are concerned with measuring atomic force potentials of a single pair of molecules. Clearly, FM-mode AFM will allow us to identify single atoms, and with sufficient care we will be able to ensure that only one atom from the tip contributes to the total force interaction. Can we, therefore, fill in the last piece of information and find a quantitative relation between the oscillation parameters and the force?

A good insight into the cantilever dynamics can be drawn from the tip potential displayed in Fig. 26.17 [26.37]. If the cantilever is far away from the sample surface, the tip moves in a symmetric parabolic potential (dotted line), and the oscillation is harmonic. In such a case, the tip motion is sinusoidal and the resonant frequency is determined by the eigenfrequency f_0 of the cantilever. If the cantilever approaches the sample surface, the potential is changed, given by an effective potential V_{eff} (solid line) which is the sum of the parabolic potential and the tip–sample interaction

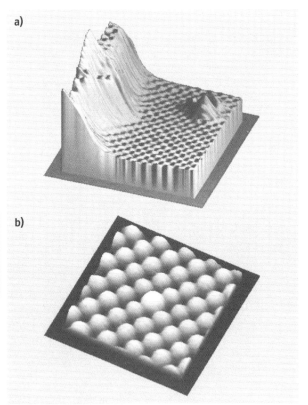

Fig. 26.16a,b Imaging of a NiO(001) sample surface with a non-contact AFM. (**a**) Surface step and an atomic defect. The lateral distance between two atoms is 4.17 Å. (**b**) A dopant atom is imaged as a light protrusion about 0.1 Å higher as the other atoms. (© of W. Allers, S. Langkat, University of Hamburg)

potential V_{ts} (dashed line). This effective potential differs from the original parabolic potential and shows an asymmetric shape. As a result the oscillation becomes inharmonic, and the resonant frequency of the cantilever depends on the oscillation amplitude.

Gotsmann and *Fuchs* [26.38] investigated this relation with a numerical simulation. During each oscillation cycle the tip experiences a whole range of forces. For each step during the approach the differential equation for the whole oscillation loop (including also the feedback system) was evaluated and finally the quantitative relation between force and frequency shift was revealed.

However, there is also an analytical relationship, if some approximations are accepted [26.39,40]. Here, we will follow the route as indicated by [26.40], although alternative ways have also been proven successful. Consider the tip oscillation trajectory reaching over a large part of force gradient curve in Fig. 26.2. We model the tip–sample interaction as a spring constant of stiffness $k_{ts}(z) = \partial F/\partial z \,|_{z_0}$ as in Fig. 26.1. For small oscillation amplitudes we already found that the frequency shift is proportional to the force gradient in (26.4). For large amplitudes, we can calculate an effective force gradient k_{eff} as a convolution of the force and the fraction of time that the tip spends between the positions x and $x + dx$

$$k_{eff}(z) = \frac{2}{\pi A^2} \int_z^{z+2A} F(x) g\left(\frac{x-z}{A} - 1\right) dx ,$$

$$\text{with } g(u) = -\frac{u}{\sqrt{1-u^2}} . \quad (26.19)$$

In the approximation that the vibration amplitude is much larger than the range of the tip–sample forces, (26.19) can be simplified to

$$k_{eff}(z) = \frac{\sqrt{2}}{\pi} A^{3/2} \int_z^{\infty} \frac{F(x)}{\sqrt{x-z}} dx . \quad (26.20)$$

This effective force gradient can now be used in (26.4), the relation between frequency shift and force gradient. We find

$$\Delta f = \frac{f_0}{\sqrt{2\pi} k A^{3/2}} \int_z^{\infty} \frac{F(x)}{\sqrt{x-z}} dx . \quad (26.21)$$

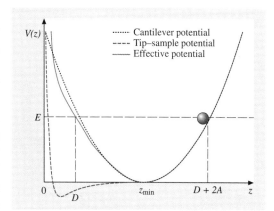

Fig. 26.17 The frequency shift in dynamic force microscopy is caused by the tip–sample interaction potential (*dashed line*), which alters the harmonic cantilever potential (*dotted line*). Therefore, the tip moves in an anharmonic and asymmetric effective potential (*solid line*). Here z_{min} is the minimum position of the effective potential (after [26.37])

If we separate the integral from other parameters, we can define

$$\Delta f = \frac{f_0}{k A^{3/2}} \gamma(z) ,$$

$$\text{with } \gamma(z) = \frac{1}{\sqrt{2\pi}} \int_z^{\infty} \frac{F(x)}{\sqrt{x-z}} dx . \quad (26.22)$$

This means we can define $\gamma(z)$, which is only dependent on the shape of the force curve $F(z)$ but independent of the external parameters of the oscillation. The function $\gamma(z)$ is also referred to as the *normalized frequency shift* [26.7], a very useful parameter, which allows us to compare measurements independent of resonant frequency, amplitude, and spring constant of the cantilever.

The dependence of the frequency shift on the vibration amplitude is an especially useful relation, since this parameter can be easily varied during one experiment. A nice example is depicted in Fig. 26.18, where frequency shift curves for different amplitudes were found to collapse into one curve in the $\gamma(z)$-diagram [26.41].

This relationship has been nicely exploited for the calibration of the vibration amplitude by *Guggisberg* [26.42], which is a problem often encountered in dynamic AFM operation and worthy of discussion. One approaches tip and sample and records curves of frequency shift versus distance, which show a repro-

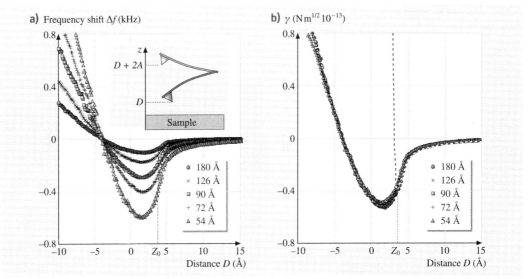

Fig. 26.18 (a) Frequency-shift curves for different oscillation amplitudes for a silicon tip on a graphite surface in UHV, (b) γ curves calculated from the Δf curves in (a) (after [26.41], © The American Physical Society)

ducible shape. Then, the z-feedback is disabled, and several curves with different amplitudes are acquired. The amplitudes are typically chosen by adjusting the amplitude setpoint in volts. One has to take care that drift in the z-direction is negligible. An analysis of the corresponding $\gamma(z)$-curves will show the same curves (as in Fig. 26.18), but the curves will be shifted in the horizontal axis. These shifts correspond to the change in amplitude, allowing one to correlate the voltage values with the z-distances.

For the often encountered force contributions from electrostatic, van der Waals, and chemical binding forces the frequency shift has been calculated from the force laws. In the approximation that the tip radius R is larger than the tip–sample distance z, an electrostatic potential V will yield a normalized frequency shift of [26.43]

$$\gamma(z) = \frac{\pi\varepsilon_0 R V^2}{\sqrt{2}} z^{-1/2} . \quad (26.23)$$

For van der Waals forces with Hamaker constant H and also with R larger than z we find accordingly

$$\gamma(z) = \frac{HR}{12\sqrt{2}} z^{-3/2} . \quad (26.24)$$

Finally, short-range chemical forces represented by the Morse potential (with the parameters binding energy U_0, decay length λ, and equilibrium distance z_{equ}) yield

$$\gamma(z) = \frac{U_0 \sqrt{2}}{\sqrt{\pi\lambda}} \exp\left(-\frac{(z - z_{\text{equ}})}{\lambda}\right) . \quad (26.25)$$

These equations allow the experimentalist to directly interpret the spectroscopic measurements. For example, the contributions of the electrostatic and van der Waals forces can be easily distinguished by their slope in a double-logarithmic plot [26.43].

Alternatively, if the force law is not known beforehand, the experimentalist wants to analyze the experimental frequency-shift data curves and extract the force or energy potential curves. We therefore have to invert the integral in (26.21) to find the tip–sample interaction potential V_{ts} from the $\gamma(z)$-curves [26.40]

$$V_{\text{ts}}(z) = \sqrt{2} \int_z^\infty \frac{\gamma(x)}{\sqrt{x - z}} \, dx . \quad (26.26)$$

Using this method, quantitative force curves were extracted from Δf spectroscopy measurements on different, atomically resolved sites of the Si(111)-(7×7) reconstruction [26.1]. Comparison with theoretical molecular dynamics (MD) simulations showed good quantitative agreement with theory and confirmed the assumption that force interactions were governed by a single atom at the tip apex. Our initially formulated goal seems to be achieved: With FM-AFM we have found a powerful method that allows us to measure the chemical bond formation of single molecules. The last uncertainty, the exact shape and identity of the tip apex atom, can possibly be resolved by employing the FIM technique to characterize the tip surface in combination with FM-AFM.

All the above equations are only valid in the approximation that the oscillation amplitudes are much larger than the distance range of the encountered forces. However, for amplitudes of, e.g., 10 nm and long-range forces such as electrostatic interactions this approximation is no longer valid. Several approaches have been proposed by different authors to solve this issue [26.44–46]. The matrix method [26.45, 47] uses the fact that in a real experiment the frequency shift curve is not continuous, but rather a set of discrete values acquired at equidistant points. Therefore the integral in (26.18) can be substituted by a sum and the equation can be rewritten as a linear equation system, which in return can be easily inverted by appropriate matrix operations. This *matrix method* is a very simple and general method for the AFM user to extract force curves from experimental frequency-shift curves without the restrictions of the large-amplitude approximation.

The concept of dynamic force spectroscopy can be also extended to three-dimensional (3-D) force spectroscopy by mapping the complete force field above the sample surface [26.48]. Figure 26.19a shows a schematic of the measurement principle. Curves of frequency shift versus distance are recorded on a matrix of points perpendicular to the sample surface. From this frequency shift data the complete three-dimensional force field between tip and sample can be recovered with atomic resolution. Figure 26.19b shows a cut through the force field as a two-dimensional map. The 3-D force technique has been applied also to a NaCl(100) surface, where not only conservative but also the dissipative tip–sample interaction could be measured in full space [26.49]. On the one hand, the forces were measured in the attractive as well as repulsive regime, allowing for the determination of the local minima in the corresponding potential energy curves in Fig. 26.20. This information is directly related to the atomic energy barriers responsible for a multitude of dynamic phenomena in surface science, such as diffusion, faceting, and crystalline growth. The direct comparison of conservative with the simultaneously acquired dissipative processes furthermore allowed the determination of atomic-scale mechanical relaxation processes.

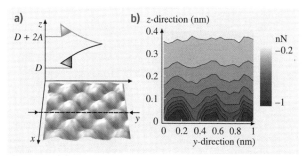

Fig. 26.19 (a) Principle of 3-D force spectroscopy. The cantilever oscillates near the sample surface and measure the frequency shift in an xyz-box. The three-dimensional surface shows the topography of the sample (image size $1 \times 1\,\text{nm}^2$) obtained immediately before the recording of the spectroscopy field. (b) The reconstructed force field of NiO(001) shows atomic resolution. The data are taken along the *line* shown in (a)

In this context it is worth pointing out a slightly different dynamic AFM method. While in the typical FM-AFM setup the oscillation amplitude is controlled to stay constant by a dedicated feedback circuit, one could simply keep the excitation amplitude constant; this has been termed the *constant-excitation* (CE) mode, as opposed to the *constant-amplitude* (CA) mode. It is expected that this mode will be gentler to the surface, because any dissipative interaction will reduce the amplitude and therefore prevent a further reduction of the effective tip–sample distance. This mode has been employed to image soft biological molecules such as DNA or thiols in UHV [26.50].

At first glance, quantitative interpretation of the obtained frequency spectra seems more complicated, since

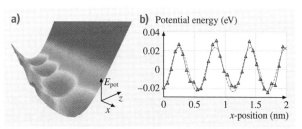

Fig. 26.20 (a) Three-dimensional representation of the interaction energy map determine from 3-D force spectroscopy experiments on a NaCl(100) crystal surface. The *circular depressions* represent the local energy minima. (b) Potential energy profile obtained from (a) by collecting the minimum-energy values along the x-axis. This curve thus directly reveals the potential energy barrier of $\Delta E = 48\,\text{meV}$ separating the local energy minima

the amplitude as well as the tip–sample distance is altered during the measurement. However, it was found by *Hölscher* et al. [26.51] that for the CE mode in the large-amplitude approximation the distance and the amplitude channel can be decoupled by calculating the effective tip–sample distance from the piezo-controlled tip–sample distance z_0 and the change in the amplitude with distance $A(z) : D(z_0) = z_0 - A(z_0)$. As a result, (26.22) can then be directly used to calculate the normalized frequency shift $\gamma(D)$ and consequently the force curve can be obtained from (26.26). This concept has been verified in experiments by *Schirmeisen* et al. [26.52] through a direct comparison of spectroscopy curves acquired in the CE mode and CA mode.

Until now, we have always associated the self-excitation scheme with vacuum applications. Although it is difficult to operate the FM-AFM in constant-amplitude mode in air, since large dissipative effects make it difficult to ensure a constant amplitude, it is indeed possible to use the constant-excitation FM-AFM in air or even in liquid [26.51, 53, 54]. Interestingly, a low-budget construction set (employing a tuning-fork force sensor) for a CE-mode dynamic AFM setup has been published on the internet (http://sxm4.uni-muenster.de).

If it is possible to measure atomic-scale forces with the NC-AFM, it should vice versa also be possible to exert forces with similar precision. In fact, the new and exciting field of nanomanipulation would be driven to a whole new dimension if defined forces could be reliably applied to single atoms or molecules. In this respect, *Loppacher* et al. [26.55] managed to push different parts of an isolated Cu-tetra-3,5 di-tertiary-butyl-phenyl porphyrin (Cu-TBBP) molecule, which is known to possess four rotatable legs. They measured the force–distance curves while pushing one of the legs with the AFM tip. From the force curves they were able to determine the energy which was dissipated during the *switching* process of the molecule. The manipulation of single silicon atoms with NC-AFM was demonstrated by *Oyabu* et al. [26.56], who removed single atoms from a Si(111)-7×7 surface with the AFM tip and could subsequently deposit atoms from the tip on the surface again. This technique was further improved by *Sugimoto* et al. [26.57], who wrote artificial atomic structures with single Sn atoms. The possibility to exert and measure forces simultaneously during single atom or molecule manipulation is an exciting new application of high-resolution NC-AFM experiments.

26.4 Q-Control

We have already discussed the virtues of a high Q value for high-sensitivity measurements: The minimum detectable force gradient was inversely proportional to the square root of Q. In vacuum, Q mainly represents the internal dissipation of the cantilever during oscillation, an internal damping factor. Low damping is obtained by using high-quality cantilevers, which are cut (or etched) from defect-free, single-crystal silicon wafers. Under ambient or liquid conditions, the quality factor is dominated by dissipative interactions between the cantilever and the surrounding medium, and Q values can be as low as 100 for air or even 5 in liquid. Still, we ask if it is somehow possible to compensate for the damping effect by exciting the cantilever in a sophisticated way.

It turns out that the shape of the resonance curves in Fig. 26.5 can be influenced toward higher (or lower) Q values by an amplitude feedback loop. In principle, there are several mechanisms to couple the amplitude signal back to the cantilever, e.g., by the photothermal effect [26.58] or capacitive forces [26.59]. Figure 26.21 shows a method in which the amplitude feedback is mediated directly by the excitation piezo [26.60]. This has the advantage that no additional mechanical setups are necessary.

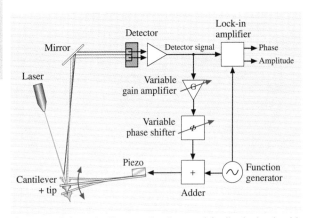

Fig. 26.21 Schematic diagram of a Q-control feedback circuit with an externally driven dynamic AFM. The tapping-mode setup is in effect extended by an additional feedback loop

The working principle of the feedback loop can be understood by analyzing the equation of motion of the modified dynamic system

$$m^*\ddot{z}(t) + \alpha\dot{z}(t) + kz(t) - F_{ts}[z_0 + z(t)]$$
$$= F_{ext}\cos(\omega t) + G e^{i\phi} z(t) \,. \tag{26.27}$$

This ansatz takes into account the feedback of the detector signal through a phase shifter, amplifier, and adder as an additional force, which is linked to the cantilever deflection $z(t)$ through the gain G and the phase shift $e^{i\phi}$. We assume that the oscillation can be described by a harmonic oscillation trajectory. With a phase shift of $\phi = \pm\pi/2$ we find

$$e^{\pm i\pi/2} z(t) = \pm\frac{1}{\omega}\dot{z}(t) \,. \tag{26.28}$$

This means, that the additional feedback force signal $G e^{i\phi} z(t)$ is proportional to the velocity of the cantilever, just like the damping term in the equation of motion. We can define an effective damping constant α_{eff}, which combines the two terms

$$m^*\ddot{z}(t) + \alpha_{eff}\dot{z}(t) + kz(t) - F_{ts}[z_0 + z(t)]$$
$$= F_{ext}\cos(\omega t) \,,$$
$$\text{with } \alpha_{eff} = \alpha \mp \frac{1}{\omega} G \,, \quad \text{for } \phi = \pm\frac{\pi}{2} \,. \tag{26.29}$$

Equation (26.28) shows that the damping of the oscillator can be enhanced or weakened by choosing $\phi = +\frac{\pi}{2}$ or $\phi = -\frac{\pi}{2}$, respectively. The feedback loop therefore allows us to vary the effective quality factor $Q_{eff} = m\omega_0/\alpha_{eff}$ of the complete dynamic system. Hence, this system was termed Q-control. Figure 26.22 shows experimental data regarding the effect of Q-control on the amplitude and phase as a function of the external excitation frequency [26.60]. In this example, Q-control was able to increase the Q-value by a factor of > 40.

The effect of improved image contrast is demonstrated in Fig. 26.23. Here, a computer hard disk was analyzed with a magnetic tip in tapping mode, where the magnetic contrast is observed in the phase image. The upper part shows the magnetic data structures recorded in standard mode, whereas in the lower part of the image Q-control feedback was activated, giving rise to an improved signal, i.e., magnetic contrast. A more detailed analysis of measurements on a magnetic tape shows that the signal amplitude (upper diagrams in Fig. 26.24) was increased by a factor of 12.4 by the Q-control feedback. The lower image shows a noise analysis of the signal, indicating an improvement of the signal-to-noise ratio by a factor of 2.3.

It might be interesting to note that Q-control can also be applied in FM mode, which might be counterintuitive at first sight. However, it has been shown by *Ebeling* et al. [26.61, 62] that the increase of the Q-factor in liquids helps to increase the imaging features of the FM mode in liquids.

The diagrams represent measurements in air with an AFM operated in AM mode. Only then can we make a distinction between excitation and vibration frequency, since in the FM mode these two frequencies are equal by definition. Although the relation between sensitivity and Q-factor in (26.17a) is the same for AM and FM mode, it must be critically investigated to see whether the enhanced quality factor by Q-control can be inserted into the equation for FM-mode AFM. In vacuum applications, Q is already very high, which

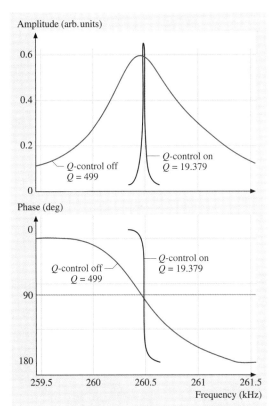

Fig. 26.22 Amplitude and phase diagrams measured in air with a Si cantilever far away from the sample. The quality factor can be increased from 450 to 20 000 by using the Q-control feedback method

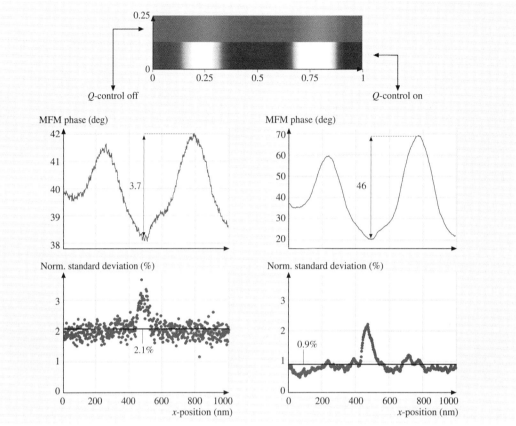

Fig. 26.23 Enhancement of the contrast in the phase channel due to Q-control on a magnetic hard disk measured with a magnetic tip in tapping-mode AFM in air. Scan size $5 \times 5\,\mu$m, phase range 10 (www.nanoanalytics.com) (MFM – magnetic force microscopy)

makes it unnecessary to operate an additional Q-control module.

As stated before, we can also use Q-control to enhance the damping in the oscillating system. This would decrease the sensitivity of the system. However, on the other hand, the response time of the amplitude change is decreased as well. For tapping-mode applications, where high-speed scanning is the goal, Q-control was able to reduce the scan speed limiting relaxation time [26.63].

A large quality factor Q does not only have the virtue of increasing the force sensitivity of the in-

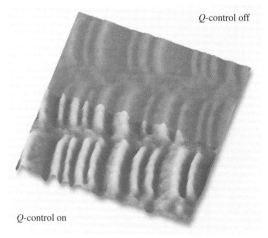

Fig. 26.24 Signal-to-noise analysis with a magnetic tip in tapping-mode AFM on a magnetic tape sample with Q-control ▶

Fig. 26.25 Imaging of a delicate organic surface with Q-control. Sample was a Langmuir–Blodgett film (ethyl-2,3-dihydroxyoctadecanoate) on a mica substrate. The topographical image clearly shows that the highly sensitive sample surface can only be imaged nondestructively with active Q-control, whereas the periodic repulsive contact with the probe in standard operation without Q-control leads to significant modification or destruction of the surface structure (© L. Chi and coworkers, University of Münster) ▶

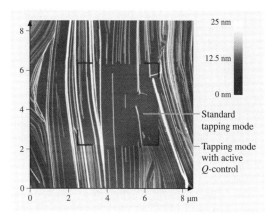

strument. It also has the advantage of increasing the parameter space of stable AFM operation in AM-mode AFM. Consider the resonance curve of Fig. 26.5. When approaching the tip toward the surface there are two competing mechanisms: On the one hand, we bring the tip closer to the sample, which results in an increase in attractive forces (Fig. 26.2). On the other hand, for the case $\omega > \omega_0$, the resonant frequency of the cantilever is shifted toward smaller values due to the attractive forces, which causes the amplitude to become smaller. This is the desirable regime, where stable

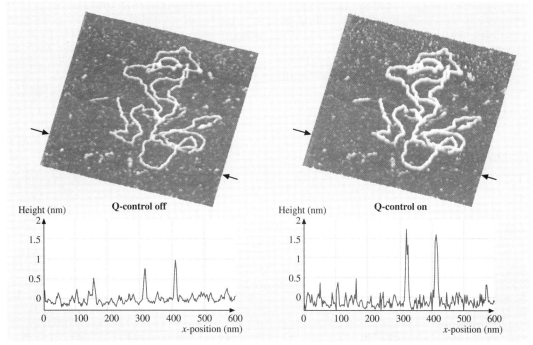

Fig. 26.26 AFM images of DNA on mica scanned in buffer solution ($600 \times 600\,\text{nm}^2$). Each scan line was scanned twice – in standard tapping mode during the first scan of the line (*left data*) and with Q-control being activated by a trigger signal during the subsequent scan of the same line (*right data*). This interleave technique allows direct comparison of the results of the two modes obtained on the same surface area while minimizing drift effects. Cross-sections of the topographic data reveal that the observed DNA height is significantly higher in the case of imaging under Q-control (© D. Ebeling, University of Münster)

operation of the AFM is possible in the net-attractive regime. However, as explained before, below a certain tip–sample separation d_{app}, the system switches suddenly into intermittent contact mode, where surface modifications are more likely due to the onset of strong repulsive forces. The steeper the amplitude curve, the larger the regime of stable, net-attractive AFM operation. Looking at Fig. 26.22 we find that the slope of the amplitude curve is governed by the quality factor Q. A high Q, therefore, facilitates stable operation of the AM-AFM in the net-attractive regime. (A more detailed discussion about this topic can be found in [26.64].)

An example can be seen in Fig. 26.25, which shows a surface scan of an ultrathin organic film acquired in tapping mode under ambient conditions. First, the inner square was scanned without the Q enhancement, and then a wider surface area was scanned with applied Q-control. The high quality factor provides a larger parameter space for operating the AFM in the net-attractive regime, allowing good resolution of the delicate organic surface structure. Without the Q-control the surface structures are deformed and even destroyed due to the strong repulsive tip–sample interactions [26.65–67]. This also allowed imaging of DNA structures without predominantly depressing the soft material during imaging. It was then possible to observe a DNA diameter close to the theoretical value with the Q-control feedback [26.68].

The same technique has been successfully employed to minimize the interaction forces during scanning in liquids. This is of special relevance for imaging delicate biological samples in environments such as water or buffer solution. When the AFM probe is submerged in a liquid medium, the oscillation of the AFM cantilever is strongly affected by hydrodynamic damping. This typically leads to quality factors < 10 and accordingly to a loss in force sensitivity. However, the Q-control technique allows the effective quality factor to be increased by about three orders of magnitude in liquids. Figure 26.26 shows results of scanning DNA structures on a mica substrate under buffer solution [26.69]. Comparison of the topographic data obtained in standard tapping mode and under Q-control, in particular the difference in the observed DNA height, indicates that the imaging forces were successfully reduced by employing Q-control.

In conclusion, we have shown that, by applying an additional feedback circuit to the dynamic AFM system, it is possible to influence the quality factor Q of the oscillator system. High-resolution, high-speed, or low-force scanning is then possible.

26.5 Dissipation Processes Measured with Dynamic AFM

Dynamic AFM methods have proven their great potential for imaging surface structures at the nanoscale, and we have also discussed methods that allow the assessment of forces between distinct single molecules. However, there is another physical mechanism that can be analyzed with the dynamic mode and has been mentioned in some previous paragraphs: energy dissipation.

In Fig. 26.12 we have already shown an example where the phase signal in tapping mode cannot be explained by conservative forces alone; dissipative processes must also play a role. In constant-amplitude FM mode, where the quantitative interpretation of experiments has proven to be less difficult, an intuitive distinction between conservative and dissipative tip–sample interaction is possible. We have shown the correlation between forces and frequency shifts of the oscillating system, but we have neglected one experimental input channel. The excitation amplitude, which is necessary to keep the oscillation amplitude constant, is a direct indication of the energy dissipated during one oscillation cycle. *Dürig* [26.70] and *Hölscher* et al. [26.71] have shown that, in self-excitation mode (with an excitation–oscillation phase difference of 90°), conservative and dissipative interactions can be strictly separated. Part of this energy is dissipated in the cantilever itself; another part is due to external viscous forces in the surrounding medium. However, more interestingly, some energy is dissipated at the tip–sample junction. This mechanism is the focus of the following paragraphs.

In contrast to conservative forces acting at the tip–sample junction, which at least in vacuum can be understood in terms of van der Waals, electrostatic, and chemical interactions, the dissipative processes are poorly understood. *Stowe* et al. [26.72] have shown that, if a voltage potential is applied between tip and sample, charges are induced in the sample surface, which will follow the tip motion (in their setup the oscillation was parallel to the surface). Due to the finite resistance of the sample material, energy will be dissipated during the charge movement. This effect has been exploited to image the doping level of semiconductors. Energy dissipation has also been observed in imaging magnetic materials. *Liu* and *Grütter* [26.73] found that energy dissipation due to magnetic interactions was enhanced at the boundaries of magnetic domains, which was attributed to domain wall oscillations. Even a simple system such as two clean metal surfaces which are moved in close proximity can give rise to frictional forces. *Stipe* et al. [26.74] have measured the energy dissipation due to fluctuating electromagnetic fields between two closely spaced gold surfaces, which was later interpreted by *Volokitin* and *Persson* [26.75] in terms of *van der Waals friction*.

However, also in the absence of external electromagnetic fields, energy dissipation was observed in close proximity of tip and sample, within 1 nm. Clearly, mechanical surface relaxations must give rise to energy losses. One could model the AFM tip as a small hammer, hitting the surface at high frequency, possibly resulting in phonon excitations. From a continuum-mechanics point of view, we assume that the mechanical relaxation of the surface is not only governed by elastic responses. Viscoelastic effects of soft surfaces will also render a significant contribution to energy dissipation. The whole area of phase imaging in tapping mode is concerned with those effects [26.76–79].

In the atomistic view, the last tip atom can be envisaged to change position while experiencing the tip–sample force field. A strictly reversible change of position would not result in a loss of energy. Still, it has been pointed out by *Sasaki* and *Tsukada* [26.80] that a change in atom position would result in a change in the force interaction itself. Therefore, it is possible that the tip atom changes position at different tip–surface distances during approach and retraction, effectively causing atomic-scale hysteresis to develop. *Hoffmann* et al. [26.13] and *Hembacher* et al. [26.81] have measured the short-range energy dissipation for different combinations of tip and surface materials in UHV. For atomic-resolution experiments at low temperatures on graphite [26.81] it was found that the energy dissipation is a step-like function. A similar shape of dissipation curves was found in a theoretical analysis by *Kantorovich* and *Trevethan* [26.82], where the energy dissipation was directly associated with atomic instabilities at the sample surface.

The dissipation channel has also been used to image surfaces with atomic resolution [26.83]. Instead of feeding back the distance on the frequency shift, the excitation amplitude in FM mode has been used as the control signal. The Si(111)-(7×7) reconstruction was successfully imaged in this mode. The step edges of monatomic NaCl islands on single-crystalline copper have also rendered atomic-resolution contrast in the dissipation channel [26.84]. The dissipation processes discussed so far are mostly in the configuration in which the tip is oscillated perpendicular to the surface. Friction is usually referred to as the energy loss due to lateral movement of solid bodies in contact. It is interesting to note in this context that *Israelachivili* [26.85] has pointed out a quantitative relationship between lateral and vertical (with respect to the surface) dissipation. He states that the hysteresis in vertical force–distance curves should equal the energy loss in lateral friction. An experimental confirmation of this conjecture at the molecular level is still lacking.

Physical interpretation of energy-dissipation processes at the atomic scale seems to be a daunting task at this point. Notwithstanding, we can find a quantitative relation between the energy loss per oscillation cycle and the experimental parameters in dynamic AFM, as will be shown in the following section.

In static AFM it was found that permanent changes of the sample surface by indentations can cause hysteresis between approach and retraction. The area between the approach and retraction curves in a force–distance diagram represents the lost or dissipated energy caused by the irreversible change of the surface structure. In dynamic-mode AFM, the oscillation parameters such as amplitude, frequency, and phase must contain the information about the dissipated energy per cycle. So far, we have resorted to a treatment of the equation of motion of the cantilever vibration in order to find a quantitative correlation between forces and the experimental parameters. For the dissipation it is useful to treat the system from the energy-conservation point of view.

Assuming that a dynamic system is in equilibrium, the average energy input must equal the average energy output or dissipation. Applying this rule to an AFM running in dynamic mode means that the average power fed into the cantilever oscillation by an external driver, denoted by $\bar{P}_{\mathrm{in}}$, must equal the average power dissipated by the motion of the cantilever beam $\bar{P}_0$ and by tip–sample interaction $\bar{P}_{\mathrm{tip}}$

$$\bar{P}_{\mathrm{in}} = \bar{P}_0 + \bar{P}_{\mathrm{tip}} \,. \tag{26.30}$$

The term $\bar{P}_{\mathrm{tip}}$ is what we are interested in, since it gives us a direct physical quantity to characterize the tip–sample interaction. Therefore, we have first to calculate and then measure the two other terms in (26.30) in order to determine the power dissipated when the tip periodically probes the sample surface. This requires an

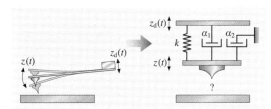

Fig. 26.27 Rheological models applied to describe the dynamic AFM system, comprising the oscillating cantilever and tip interacting with the sample surface. The movement of the cantilever base and the tip is denoted by $z_d(t)$ and $z(t)$, respectively. The cantilever is characterized by the spring constant k and the damping constant α. In a first approach, damping is broken into two pieces α_1 and α_2: first, intrinsic damping caused by the movement of the cantilever's tip relative to its base, and second, damping related to the movement of the cantilever body in a surrounding medium, e.g., air damping

appropriate rheological model to describe the dynamic system. Although there are investigations in which the complete flexural motion of the cantilever beam has been considered [26.86], a simplified model, comprising a spring and two dashpots (Fig. 26.27), represents a good approximation in this case [26.87].

The spring, characterized by the constant k according to Hooke's law, represents the only channel through which power P_{in} can be delivered to the oscillating tip $z(t)$ by the external driver $z_d(t)$. Therefore, the instantaneous power fed into the dynamic system is equal to the force exerted by the driver times the velocity of the driver (the force which is necessary to move the base side of the dashpot can be neglected, since this power is directly dissipated and therefore does not contribute to the power delivered to the oscillating tip)

$$P_{\text{in}}(t) = F_d(t)\dot{z}_d(t) = k[z(t) - z_d(t)]\dot{z}_d(t). \quad (26.31)$$

Assuming a sinusoidal steady-state response and that the base of the cantilever is driven sinusoidally (26.6) with amplitude A_d and frequency ω, the deflection from equilibrium of the end of the cantilever follows (26.9), where A and $0 \leq \varphi \leq \pi$ are the oscillation amplitude and phase shift, respectively. This allows us to calculate the average power input per oscillation cycle by integrating (26.30) over one period $T = 2\pi/\omega$

$$\bar{P}_{\text{in}} = \frac{1}{T} \int_0^T P_{\text{in}}(t)\,dt = \frac{1}{2} k\omega A_d A \sin\varphi . \quad (26.32)$$

This contains the familiar result that the maximum power is delivered to an oscillator when the response is $90°$ out of phase with the drive.

The simplified rheological model as depicted in Fig. 26.27 exhibits two major contributions to the damping term $\bar{P}_0$. Both are related to the motion of the cantilever body and assumed to be well modeled by viscous damping with coefficients α_1 and α_2. The dominant damping mechanism in UHV conditions is intrinsic damping, caused by the deflection of the cantilever beam, i.e., the motion of the tip relative to the cantilever base. Therefore the instantaneous power dissipated by such a mechanism is given by

$$P_{01}(t) = |F_{01}(t)\dot{z}(t)| = |\alpha_1 [\dot{z}(t) - \dot{z}_d(t)] \dot{z}(t)| . \quad (26.33)$$

Note that the absolute value has to be calculated, since all dissipated power is *lost* and therefore cannot be returned to the dynamic system.

However, when running an AFM in ambient conditions an additional damping mechanism has to be considered. Damping due to the motion of the cantilever body in the surrounding medium, e.g., air damping, is in most cases the dominant effect. The corresponding instantaneous power dissipation is given by

$$P_{02}(t) = |F_{02}(t)\dot{z}(t)| = \alpha_2 \dot{z}^2(t) . \quad (26.34)$$

In order to calculate the average power dissipation, (26.33) and (26.34) have to be integrated over one complete oscillation cycle. This procedure yields

$$\bar{P}_{01} = \frac{1}{T} \int_0^T P_{01}(t)\,dt$$

$$= \frac{1}{\pi}\alpha_1 \omega^2 A \left[(A - A_d \cos\varphi) \arcsin \right.$$

$$\left. \times \left(\frac{A - A_d \cos\varphi}{\sqrt{A^2 + A_d^2 - 2AA_d \cos\varphi}} \right) + A_d \sin\varphi \right]$$

$$(26.35)$$

and

$$\bar{P}_{02} = \frac{1}{T} \int_0^T P_{02}(t)\,dt = \frac{1}{2}\alpha_2 \omega^2 A^2 . \quad (26.36)$$

Considering the fact that commonly used cantilevers exhibit a quality factor of at least several hundreds (in

UHV even several tens of thousands), we can assume that the oscillation amplitude is significantly larger than the drive amplitude when the dynamic system is driven at or near its resonance frequency: $A \gg A_d$. Therefore (26.34) can be simplified in first-order approximation to an expression similar to (26.35). Combining the two equations yields the total average power dissipated by the oscillating cantilever

$$\bar{P}_0 = \frac{1}{2}\alpha\omega^2 A^2, \quad \text{with } \alpha = \alpha_1 + \alpha_2, \quad (26.37)$$

where α denotes the overall effective damping constant.

We can now solve (26.30) for the power dissipation localized to the small interaction volume of the probing tip with the sample surface, represented by the question mark in Fig. 26.27. Furthermore by expressing the damping constant α in terms of experimentally accessible quantities such as the spring constant k, the quality factor Q, and the natural resonant frequency ω_0 of the free oscillating cantilever, $\alpha = k/Q\omega_0$, we obtain

$$\bar{P}_{tip} = \bar{P}_{in} - \bar{P}_0$$
$$= \frac{1}{2}\frac{k\omega}{Q}\left(Q_{cant} A_d A \sin\varphi - A^2 \frac{\omega}{\omega_0}\right). \quad (26.38)$$

Note that so far no assumptions have been made on how the AFM is operated, except that the motion of the oscillating cantilever has to remain sinusoidal to a good approximation. Therefore (26.38) is applicable to a variety of different dynamic AFM modes.

For example, in FM-mode AFM the oscillation frequency ω changes due to tip–sample interaction while at the same time the oscillation amplitude A is kept constant by adjusting the drive amplitude A_d. By measuring these quantities, one can apply (26.38) to determine the average power dissipation related to tip–sample interaction. In spectroscopy applications usually $A_d(z)$ is not measured directly, but a signal $G(z)$ proportional to $A_d(z)$ is acquired, representing the gain factor applied to the excitation piezo. With the help of (26.15) we can write

$$A_d(z) = \frac{A_0 G(z)}{QG_0}, \quad (26.39)$$

where A_0 and G_0 are the amplitude and gain at large tip–sample distances where the tip–sample interactions are negligible.

Now let us consider the tapping-mode AFM. In this case the cantilever is driven at a fixed frequency and with constant drive amplitude, while the oscillation amplitude and phase shift may change when the probing

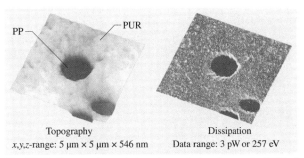

Fig. 26.28 Topography and phase image in tapping-mode AFM of a polymer blend composed of polypropylene (PP) particles embedded in a polyurethane (PUR) matrix. The dissipation image shows a strong contrast between the harder PP (little dissipation, *dark*) and the softer PUR (large dissipation, *bright*) surface

tip interacts with the sample surface. Assuming that the oscillation frequency is chosen to be ω_0, (26.37) can be further simplified again by employing (26.15) for the free oscillation amplitude A_0. This calculation yields

$$\bar{P}_{tip} = \frac{1}{2}\frac{k\omega_0}{Q_{cant}}\left(A_0 A \sin\varphi - A^2\right). \quad (26.40)$$

Equation (26.40) implies that, if the oscillation amplitude A is kept constant by a feedback loop, as is commonly done in tapping mode, simultaneously acquired phase data can be interpreted in terms of energy dissipation [26.77, 79, 88, 89]. When analyzing such phase images [26.90–92] one has also to consider the fact that the phase may also change due to the transition from net-attractive ($\varphi > 90°$) to intermittent contact ($\varphi < 90°$) interaction between the tip and the sample [26.19, 60, 93, 94]. For example, consider the phase shift in tapping mode as a function of z-position (Fig. 26.12). If phase measurements are performed close to the point where the oscillation switches from the net-attractive to the intermittent contact regime, a large contrast in the phase channel is observed. However, this contrast is not due to dissipative processes. Only a variation of the phase signal within the intermittent contact regime will give information about the tip–sample dissipative processes.

An example of dissipation measurement is depicted in Fig. 26.28. The surface of a polymer blend was imaged in air, simultaneously acquiring the topography and dissipation. The dissipation on the softer polyurethane matrix is significantly larger than on the embedded, mechanically stiffer polypropylene particles.

26.6 Conclusions

Dynamic force microscopy is a powerful tool, which is capable of imaging surfaces with atomic precision. It also allows us to look at surface dynamics and can operate in vacuum, air or even liquid. However, the oscillating cantilever system introduces a level of complexity which disallows straightforward interpretation of acquired images. An exception is the self-excitation mode, where tip–sample forces can be successfully extracted from spectroscopic experiments. However, not only conservative forces can be investigated with dynamic AFM; energy dissipation also influences the cantilever oscillation and can therefore serve as a new information channel.

Open questions are still concerned with the exact geometric and chemical identity of the probing tip, which significantly influences the imaging and spectroscopic results. Using predefined tips such as single-walled nanotubes or using atomic-resolution techniques such as field ion microscopy to image the tip itself are possible approaches addressing this issue.

References

26.1   M.A. Lantz, H.J. Hug, R. Hoffmann, P.J.A. van Schendel, P. Kappenberger, S. Martin, A. Baratoff, H.-J. Güntherodt: Quantitative measurement of short-range chemical bonding forces, Science **291**, 2580–2583 (2001)

26.2   G. Binnig, C.F. Quate, C. Gerber: Atomic force microscope, Phys. Rev. Lett. **56**, 930–933 (1986)

26.3   O. Marti: AFM Instrumentation and Tips. In: *Handbook of Micro/Nanotribology*, 2nd edn., ed. by B. Bushan (CRC, Boca Raton 1999) pp. 81–144

26.4   G. Cross, A. Schirmeisen, A. Stalder, P. Grütter, M. Tschudy, U. Dürig: Adhesion interaction between atomically defined tip and sample, Phys. Rev. Lett. **80**, 4685–4688 (1998)

26.5   A. Schirmeisen, G. Cross, A. Stalder, P. Grütter, U. Dürig: Metallic adhesion and tunneling at the atomic scale, New J. Phys. **2**, 1–29 (2000)

26.6   A. Schirmeisen: Metallic adhesion and tunneling at the atomic scale. Ph.D. Thesis (McGill University, Montréal 1999) pp. 29–38

26.7   F.J. Giessibl: Forces and frequency shifts in atomicresolution dynamic-force microscopy, Phys. Rev. B **56**, 16010–16015 (1997)

26.8   F.J. Giessibl: Atomic resolution of the silicon (111)-(7×7) surface by atomic force microscopy, Science **267**, 68–71 (1995)

26.9   M. Bezanilla, B. Drake, E. Nudler, M. Kashlev, P.K. Hansma, H.G. Hansma: Motion and enzymatic degradation of DNA in the atomic forcemicroscope, Biophys. J. **67**, 2454–2459 (1994)

26.10 Y. Jiao, D.I. Cherny, G. Heim, T.M. Jovin, T.E. Schäffer: Dynamic interactions of p53 with DNA in solution by time-lapse atomic force microscopy, J. Mol. Biol. **314**, 233–243 (2001)

26.11 T.R. Albrecht, P. Grütter, D. Horne, D. Rugar: Frequency modulation detection using high-Q cantilevers for enhanced force microscopy sensitivity, J. Appl. Phys. **69**, 668–673 (1991)

26.12 S.P. Jarvis, M.A. Lantz, U. Dürig, H. Tokumoto: Off resonance ac mode force spectroscopy and imaging with an atomic force microscope, Appl. Surf. Sci. **140**, 309–313 (1999)

26.13 P.M. Hoffmann, S. Jeffery, J.B. Pethica, H.Ö. Özer, A. Oral: Energy dissipation in atomic force microscopy and atomic loss processes, Phys. Rev. Lett. **87**, 265502–265505 (2001)

26.14 H. Hölscher, B. Gotsmann, W. Allers, U.D. Schwarz, H. Fuchs, R. Wiesendanger: Comment on damping mechanism in dynamic force microscopy, Phys. Rev. Lett. **88**, 019601 (2002)

26.15 E. Sivaniah, J. Genzer, G.H. Fredrickson, E.J. Kramer, M. Xiang, X. Li, C. Ober, S. Magonov: Periodic surface topology of three-arm semifluorinated alkane monodendron diblock copolymers, Langmuir **17**, 4342–4346 (2001)

26.16 S.N. Magonov: Visualization of polymer structures with atomic force microscopy. In: *Applied Scanning Probe Methods*, ed. by H. Fuchs, M. Hosaka, B. Bhushan (Springer, Berlin 2004) pp. 207–250

26.17 C. Möller, M. Allen, V. Elings, A. Engel, D.J. Müller: Tapping-mode atomic force microscopy produces faithful high-resolution images of protein surfaces, Biophys. J. **77**, 1150–1158 (1999)

26.18 B. Anczykowski, D. Krüger, H. Fuchs: Cantilever dynamics in quasinoncontact force microscopy: Spectroscopic aspects, Phys. Rev. B **53**, 15485–15488 (1996)

26.19 B. Anczykowski, D. Krüger, K.L. Babcock, H. Fuchs: Basic properties of dynamic force spectroscopy with the scanning force microscope in experiment and simulation, Ultramicroscopy **66**, 251–259 (1996)

26.20 V.M. Muller, V.S. Yushchenko, B.V. Derjaguin: On the influence of molecular forces on the deformation of an elastic sphere and its sticking to a rigid plane, J. Colloid Interface Sci. **77**, 91–101 (1980)

26.21 B.D. Hughes, L.R. White: 'Soft' contact problems in linear elasticity, Q. J. Mech. Appl. Math. **32**, 445–471 (1979)

26.22 P. Gleyzes, P.K. Kuo, A.C. Boccara: Bistable behavior of a vibrating tip near a solid surface, Appl. Phys. Lett. **58**, 2989–2991 (1991)

26.23 A. San Paulo, R. Garcia: High-resolution imaging of antibodies by tapping-mode atomic force microscopy: Attractive and repulsive tip-sample interaction regimes, Biophys. J. **78**, 1599–1605 (2000)

26.24 D. Krüger, B. Anczykowski, H. Fuchs: Physical properties of dynamic force microscopies in contact and noncontact operation, Ann. Phys. **6**, 341–363 (1997)

26.25 H. Hölscher: Quantitative measurement of tip-sample interactions in amplitude modulation atomic force microscopy, Appl. Phys. Lett. **89**, 123109 (2006)

26.26 M. Lee, W. Jhe: General theory of amplitude-modulation atomic force microscopy, Phys. Rev. Lett. **97**, 036104 (2006)

26.27 Y. Martin, C.C. Williams, H.K. Wickramasinghe: Atomic force microscope – force mapping and profiling on a sub 100-Å scale, J. Appl. Phys. **61**, 4723–4729 (1987)

26.28 Y. Sugawara, M. Otha, H. Ueyama, S. Morita: Defect motion on an InP(110) surface observed with noncontact atomic force microscopy, Science **270**, 1646–1648 (1995)

26.29 M. Bammerlin, R. Lüthi, E. Meyer, A. Baratoff, J. Lue, M. Guggisberg, C. Gerber, L. Howald, H.-J. Güntherodt: True atomic resolution on the surface of an insulator via ultrahigh vacuum dynamic dynamic force microscopy, Probe Microsc. **1**, 3–9 (1996)

26.30 C. Barth, M. Reichling: Imaging the atomic arrangement on the high-temperature reconstructed α-Al$_2$O$_3$(0001) Surface, Nature **414**, 54–57 (2001)

26.31 W. Allers, S. Langkat, R. Wiesendanger: Dynamic low-temperature scanning force microscopy on nickel oxide(001), Appl. Phys. A [Suppl.] **72**, S27–S30 (2001)

26.32 U. Kaiser, A. Schwarz, R. Wiesendanger: Magnetic exchange force microscopy with atomic resolution, Nature **446**, 522–525 (2007)

26.33 S. Morita, R. Wiesendanger, E. Meyer: *Noncontact Atomic Force Microscopy* (Springer, Berlin Heidelberg 2002)

26.34 R. García, R. Pérez: Dynamic atomic force microscopy methods, Surf. Sci. Rep. **47**, 197–301 (2002)

26.35 F.J. Giessibl: Advances in atomic force microscopy, Rev. Mod. Phys. **75**, 949–983 (2003)

26.36 H. Hölscher, A. Schirmeisen: Dynamic force microscopy and spectroscopy, Adv. Imaging Electron Phys. **135**, 41–101 (2005)

26.37 H. Hölscher, U.D. Schwarz, R. Wiesendanger: Calculation of the frequency shift in dynamic force microscopy, Appl. Surf. Sci. **140**, 344–351 (1999)

26.38 B. Gotsmann, H. Fuchs: Dynamic force spectroscopy of conservative and dissipative forces in an Al-Au(111) tip-sample system, Phys. Rev. Lett. **86**, 2597–2600 (2001)

26.39 H. Hölscher, W. Allers, U.D. Schwarz, A. Schwarz, R. Wiesendanger: Determination of tip-sample interaction potentials by dynamic force spectroscopy, Phys. Rev. Lett. **83**, 4780–4783 (1999)

26.40 U. Dürig: Relations between interaction force and frequency shift in large-amplitude dynamic force microscopy, Appl. Phys. Lett. **75**, 433–435 (1999)

26.41 H. Hölscher, A. Schwarz, W. Allers, U.D. Schwarz, R. Wiesendanger: Quantitative analysis of dynamicforce-spectroscopy data on graphite (0001) in the contact and noncontact regime, Phys. Rev. B **61**, 12678–12681 (2000)

26.42 M. Guggisberg: Lokale Messung von atomaren Kräften. Ph.D. Thesis (University of Basel, Basel 2000) pp. 9–11, in German

26.43 M. Guggisberg, M. Bammerlin, E. Meyer, H.-J. Güntherodt: Separation of interactions by noncontact force microscopy, Phys. Rev. B **61**, 11151–11155 (2000)

26.44 U. Dürig: Extracting interaction forces and complementary observables in dynamic probemicroscopy, Appl. Phys. Lett. **76**, 1203–1205 (2000)

26.45 F.J. Giessibl: A direct method to calculate tip-Sample forces from frequency shifts in frequencymodulation atomic force microscopy, Appl. Phys. Lett. **78**, 123–125 (2001)

26.46 J.E. Sader, S.P. Jarvis: Accurate formulas for interaction force and energy in frequency modulation force spectroscopy, Appl. Phys. Lett. **84**, 1801–1803 (2004)

26.47 O. Pfeiffer: Quantitative dynamische Kraft- und Dissipationsmikroskopie auf molekularer Skala. Ph.D. Thesis (Universität Basel, Basel 2004), , in German

26.48 H. Hölscher, S.M. Langkat, A. Schwarz, R. Wiesendanger: Measurement of three-dimensional force fields with atomic resolution using dynamic force spectroscopy, Appl. Phys. Lett. **81**, 4428 (2002)

26.49 A. Schirmeisen, D. Weiner, H. Fuchs: Single-atom contact mechanics: From atomic scale energy barrier to mechanical relaxation hysteresis, Phys. Rev. Lett. **97**, 136101 (2006)

26.50 T. Uchihashi, T. Ishida, M. Komiyama, M. Ashino, Y. Sugawara, W. Mizutani, K. Yokoyama, S. Morita, H. Tokumoto, M. Ishikawa: High-resolution imaging of organic monolayers using noncontact AFM, Appl. Surf. Sci. **157**, 244–250 (2000)

26.51 H. Hölscher, B. Gotsmann, A. Schirmeisen: Dynamic force spectroscopy using the frequency modulation technique with constant excitation, Phys. Rev. B **68**, 153401-1–153401-4 (2003)

26.52 A. Schirmeisen, H. Hölscher, B. Anczykowski, D. Weiner, M.M. Schäfer, H. Fuchs: Dynamic force spectroscopy using the constant-excitation and

26.53 T. Uchihashi, M.J. Higgins, S. Yasuda, S.P. Jarvis, S. Akita, Y. Nakayama, J.E. Sader: Quantitative force measurements in liquid using frequency modulation atomic force microscopy, Appl. Phys. Lett. **85**, 3575 (2004)

26.54 J.-E. Schmutz, H. Hölscher, D. Ebeling, M.M. Schäfer, B. Ancyzkowski: Mapping the tip-sample interactions on DPPC and DNA by dynamic force spectroscopy under ambient conditions, Ultramicroscopy **107**, 875–881 (2007)

26.55 C. Loppacher, M. Guggisberg, O. Pfeiffer, E. Meyer, M. Bammerlin, R. Lüthi, R. Schlittler, J.K. Gimzewski, H. Tang, C. Joachim: Direct determination of the energy required to operate a single molecule switch, Phys. Rev. Lett. **90**, 066107-1–066107-4 (2003)

26.56 N. Oyabu, O. Custance, I. Yi, Y. Sugawara, S. Morita: Mechanical vertical manipulation of selected single atoms by soft nanoindentation using near contact atomic force microscopy, Phys. Rev. Lett. **90**, 176102 (2003)

26.57 Y. Sugimoto, M. Abe, S. Hirayama, N. Oyabu, O. Custance, S. Morita: Atom inlays performed at room temperature using atomic force microscopy, Nat. Mater. **4**, 156–159 (2005)

26.58 J. Mertz, O. Marti, J. Mlynek: Regulation of a microcantilever response by force feedback, Appl. Phys. Lett. **62**, 2344–2346 (1993)

26.59 D. Rugar, P. Grütter: Mechanical parametric amplification and thermomechanical noise squeezing, Phys. Rev. Lett. **67**, 699–702 (1991)

26.60 B. Anczykowski, J.P. Cleveland, D. Krüger, V.B. Elings, H. Fuchs: Analysis of the interaction mechanisms in dynamic mode SFM by means of experimental data and computer simulation, Appl. Phys. A **66**, 885 (1998)

26.61 D. Ebeling, H. Hölscher, B. Anczykowski: Increasing the Q-factor in the constant-excitation mode of frequency-modulation atomic force microscopy in liquid, Appl. Phys. Lett. **89**, 203511 (2006)

26.62 D. Ebeling, H. Hölscher: Analysis of the constant-excitation mode in frequency-modulation atomic force microscopy with active Q-control applied in ambient conditions and liquids, J. Appl. Phys. **102**, 114310 (2007)

26.63 T. Sulchek, G.G. Yaralioglu, C.F. Quate, S.C. Minne: Characterization and optimisation of scan speed for tapping-mode atomic force microscopy, Rev. Sci. Instrum. **73**, 2928–2936 (2002)

26.64 H. Hölscher, U.D. Schwarz: Theory of amplitude modulation atomic force microscopy with and without Q-control, Int. J. Nonlinear Mech. **42**, 608–625 (2007)

26.65 L.F. Chi, S. Jacobi, B. Anczykowski, M. Overs, H.-J. Schäfer, H. Fuchs: Supermolecular periodic structures in monolayers, Adv. Mater. **12**, 25–30 (2000)

26.66 S. Gao, L.F. Chi, S. Lenhert, B. Anczykowski, C. Niemeyer, M. Adler, H. Fuchs: High-quality mapping of DNA–protein complexes by dynamic scanning forcemicroscopy, ChemPhysChem **6**, 384–388 (2001)

26.67 B. Zou, M. Wang, D. Qiu, X. Zhang, L.F. Chi, H. Fuchs: Confined supramolecular nanostructures of mesogen-bearing amphiphiles, Chem. Commun. **9**, 1008–1009 (2002)

26.68 B. Pignataro, L.F. Chi, S. Gao, B. Anczykowski, C. Niemeyer, M. Adler, H. Fuchs: Dynamic scanning force microscopy study of self-assembled DNA–protein nanostructures, Appl. Phys. A **74**, 447–452 (2002)

26.69 D. Ebeling, H. Hölscher, H. Fuchs, B. Anczykowski, U.D. Schwarz: Imaging of biomaterials in liquids: A comparison between conventional and Q-controlled amplitude modulation ('tapping mode') atomic force microscopy, Nanotechnology **17**, S221–S226 (2005)

26.70 U. Dürig: Interaction sensing in dynamic force microscopy, New J. Phys. **2**, 1–5 (2000)

26.71 H. Hölscher, B. Gotsmann, W. Allers, U.D. Schwarz, H. Fuchs, R. Wiesendanger: Measurement of conservative and dissipative tip-sample interaction forces with a dynamic force microscope using the frequency modulation technique, Phys. Rev. B **64**, 075402 (2001)

26.72 T.D. Stowe, T.W. Kenny, D.J. Thomson, D. Rugar: Silicon dopant imaging by dissipation force microscopy, Appl. Phys. Lett. **75**, 2785–2787 (1999)

26.73 Y. Liu, P. Grütter: Magnetic dissipation force microscopy studies of magnetic materials, J. Appl. Phys. **83**, 7333–7338 (1998)

26.74 B.C. Stipe, H.J. Mamin, T.D. Stowe, T.W. Kenny, D. Rugar: Noncontact friction and force fluctuations between closely spaced bodies, Phys. Rev. Lett. **87**, 96801-1–96801-4 (2001)

26.75 A.I. Volokitin, B.N.J. Persson: Resonant photon tunneling enhancement of the van der Waals friction, Phys. Rev. Lett. **91**, 106101-1–106101-4 (2003)

26.76 J. Tamayo, R. Garcia: Effects of elastic and inelastic interactions on phase contrast images in tapping-mode scanning force microscopy, Appl. Phys. Lett. **71**, 2394–2396 (1997)

26.77 J.P. Cleveland, B. Anczykowski, A.E. Schmid, V.B. Elings: Energy dissipation in tapping-mode atomic force microscopy, Appl. Phys. Lett. **72**, 2613–2615 (1998)

26.78 R. García, J. Tamayo, M. Calleja, F. García: Phase contrast in tapping-mode scanning force microscopy, Appl. Phys. A **66**, S309–S312 (1998)

26.79 B. Anczykowski, B. Gotsmann, H. Fuchs, J.P. Cleveland, V.B. Elings: How to measure energy dissipation in dynamic mode atomic force microscopy, Appl. Surf. Sci. **140**, 376–382 (1999)

26.80 N. Sasaki, M. Tsukada: Effect of microscopic nonconservative process on noncontact atomic force microscopy, Jpn. J. Appl. Phys. **39**, 1334 (2000)

26.81 S. Hembacher, F.J. Giessibl, J. Mannhart, C.F. Quate: Local spectroscopy and atomic imaging of tunneling current, forces, and dissipation on graphite, Phys. Rev. Lett. **94**, 056101-1–056101-4 (2005)

26.82 L.N. Kantorovich, T. Trevethan: General theory of microscopic dynamical response in surface probe microscopy: From imaging to dissipation, Phys. Rev. Lett. **93**, 236102-1–236102-4 (2004)

26.83 R. Lüthi, E. Meyer, M. Bammerlin, A. Baratoff, L. Howald, C. Gerber, H.-J. Güntherodt: Ultrahigh vacuum atomic force microscopy: True atomic resolution, Surf. Rev. Lett. **4**, 1025–1029 (1997)

26.84 R. Bennewitz, A.S. Foster, L.N. Kantorovich, M. Bammerlin, C. Loppacher, S. Schär, M. Guggisberg, E. Meyer, A.L. Shluger: Atomically resolved edges and kinks of NaCl islands on Cu(111): Experiment and theory, Phys. Rev. B **62**, 2074–2084 (2000)

26.85 J. Israelachvili: *Intermolecular and Surface Forces* (Academic, London 1992)

26.86 U. Rabe, J. Turner, W. Arnold: Analysis of the high-frequency response of atomic force microscope cantilevers, Appl. Phys. A **66**, 277 (1998)

26.87 T.R. Rodriguez, R. García: Tip motion in amplitude modulation (tapping-mode) atomic-force microscopy: Comparison between continuous and point-mass models, Appl. Phys. Lett. **80**, 1646–1648 (2002)

26.88 J. Tamayo, R. García: Relationship between phase shift and energy dissipation in tapping-mode scanning force microscopy, Appl. Phys. Lett. **73**, 2926–2928 (1998)

26.89 R. García, J. Tamayo, A. San Paulo: Phase contrast and surface energy hysteresis in tapping mode scanning force micropcopy, Surf. Interface Anal. **27**(5/6), 312–316 (1999)

26.90 S.N. Magonov, V.B. Elings, M.H. Whangbo: Phase imaging and stiffness in tapping-mode atomic force microscopy, Surf. Sci. **375**, 385–391 (1997)

26.91 J.P. Pickering, G.J. Vancso: Apparent contrast reversal in tapping mode atomic force microscope images on films of polystyrene-b-polyisoprene-b-polystyrene, Polym. Bull. **40**, 549–554 (1998)

26.92 X. Chen, S.L. McGurk, M.C. Davies, C.J. Roberts, K.M. Shakesheff, S.J.B. Tendler, P.M. Williams, J. Davies, A.C. Dwakes, A. Domb: Chemical and morphological analysis of surface enrichment in a biodegradable polymer blend by phase-detection imaging atomic force microscopy, Macromolecules **31**, 2278–2283 (1998)

26.93 A. Kühle, A.H. Sørensen, J. Bohr: Role of attractive forces in tapping tip force microscopy, J. Appl. Phys. **81**, 6562–6569 (1997)

26.94 A. Kühle, A.H. Sørensen, J.B. Zandbergen, J. Bohr: Contrast artifacts in tapping tip atomic force microscopy, Appl. Phys. A **66**, 329–332 (1998)

27. Molecular Recognition Force Microscopy: From Molecular Bonds to Complex Energy Landscapes

Peter Hinterdorfer, Andreas Ebner, Hermann Gruber, Ruti Kapon, Ziv Reich

Atomic force microscopy (AFM), developed in the late 1980s to explore atomic details on hard material surfaces, has evolved into a method capable of imaging fine structural details of biological samples. Its particular advantage in biology is that measurements can be carried out in aqueous and physiological environments, which opens the possibility to study the dynamics of biological processes in vivo. The additional potential of the AFM to measure ultralow forces at high lateral resolution has paved the way for measuring inter- and intramolecular forces of biomolecules on the single-molecule level. Molecular recognition studies using AFM open the possibility to detect specific ligand–receptor interaction forces and to observe molecular recognition of a single ligand–receptor pair. Applications include biotin–avidin, antibody–antigen, nitrilotriacetate (NTA)–hexahistidine 6, and cellular proteins, either isolated or in cell membranes.

The general strategy is to bind ligands to AFM tips and receptors to probe surfaces (or vice versa). In a force–distance cycle, the tip is first approached towards the surface, whereupon a single receptor–ligand complex is formed due to the specific ligand receptor recognition. During subsequent tip–surface retraction a temporarily increasing force is exerted on the ligand–receptor connection, thus reducing its lifetime until the interaction bond breaks at a critical (unbinding) force. Such experiments allow for estimation of affinity, rate constants, and structural data of the binding pocket. Comparing them with values obtained from ensemble-average techniques and

27.1	Ligand Tip Chemistry	764
27.2	Immobilization of Receptors onto Probe Surfaces	766
27.3	Single-Molecule Recognition Force Detection	767
27.4	Principles of Molecular Recognition Force Spectroscopy	769
27.5	Recognition Force Spectroscopy: From Isolated Molecules to Biological Membranes	771
	27.5.1 Forces, Energies, and Kinetic Rates	771
	27.5.2 Complex Bonds and Energy Landscapes	774
	27.5.3 Live Cells and Membranes	778
27.6	Recognition Imaging	779
27.7	Concluding Remarks	781
References		781

binding energies is of particular interest. The dependences of unbinding force on the rate of load increase exerted on the receptor–ligand bond reveal details of the molecular dynamics of the recognition process and energy landscapes. Similar experimental strategies have also been used for studying intramolecular force properties of polymers and unfolding–refolding kinetics of filamentous proteins. Recognition imaging, developed by combing dynamic force microscopy with force spectroscopy, allows for localization of receptor sites on surfaces with nanometer positional accuracy.

Molecular recognition plays a pivotal role in nature. Signaling cascades, enzymatic activity, genome replication and transcription, cohesion of cellular structures, interaction of antigens and antibodies, and metabolic pathways all rely critically on specific recognition. In fact, every process which requires molecules to interact with each other in a specific manner requires that they be able to recognize each other.

Molecular recognition studies emphasize specific interactions between receptors and their cognitive ligands. Despite a growing body of literature on the structure and function of receptor–ligand complexes, it is still not possible to predict reaction kinetics or energetics for any given complex formation, even when the structures are known. Additional insights, in particular into the molecular dynamics within the complex during the association and dissociation process, are needed. The high-end strategy is to probe the energy landscape that underlies the interactions between molecules whose structures are known with atomic resolution.

Receptor–ligand complexes are usually formed by a few, noncovalent weak interactions between contacting chemical groups in complementary determining regions, supported by framework residues providing structurally conserved scaffolding. Both the complementary determining regions and the framework have a considerable amount of plasticity and flexibility, allowing for conformational movements during association and dissociation. In addition to knowledge about structure, energies, and kinetic constants, information about these movements is required to understand the recognition process. Deeper insight into the nature of these movements as well as the spatiotemporal action of the many weak interactions, in particular the cooperativity of bond formation, is the key to understanding receptor–ligand recognition.

For this, experiments at the single-molecule level, and on time scales typical for receptor–ligand complex formation and dissociation, are required. The methodology described in this chapter for investigating molecular dynamics of receptor–ligand interactions, molecular recognition force microscopy (MRFM) [27.1–3], is based on atomic force microscope (AFM) technology [27.4]. The ability of the AFM [27.4] to measure ultralow forces at high lateral resolution together with its unique capability to operate in an aqueous and physiological environment opens the possibility of studying biological recognition processes in vivo. The interaction between a receptor and a ligand complex is studied by exerting a force on the complex and following the dissociation process over time. Dynamic aspects of recognition are addressed in force spectroscopy (FS) experiments, where distinct force–loading rate profiles are used to provide insight into the energy landscape underlying the reaction. It is also possible to investigate the force–time behavior to unravel changes of conformation which occur during the dissociation process. It will be shown that MRFM is a versatile tool to explore kinetic and structural details of receptor–ligand recognition.

27.1 Ligand Tip Chemistry

In MRFM experiments, the binding of ligands immobilized on AFM tips to surface-bound receptors (or vice versa) is studied by applying a force to the receptor–ligand complex. The force reduces the lifetime of the bond, ultimately leading to its disassociation. The distribution of forces at which rupture occurs, and its dependence on parameters such as loading rate and temperature, can be used to provide insight into the interaction. This type of setup requires careful AFM tip sensor design, including tight attachment of the ligands to the tip surface. In the first pioneering demonstrations of single-molecule recognition force measurements [27.1, 2], strong physical adsorption of bovine serum albumin (BSA) was used to directly coat the tip [27.2] or a glass bead glued to it [27.1]. This physisorbed protein layer then served as a matrix for biochemical modifications with chemically active ligands (Fig. 27.1). In spite of the large number of probe molecules on the tip (10^3–10^4 /nm^2) the low fraction of properly oriented molecules, or internal blocks of

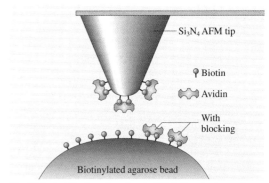

Fig. 27.1 Avidin-functionalized AFM tip. A dense layer of biotinylated BSA was adsorbed onto the tip and subsequently saturated with avidin. The biotinylated agarose bead opposing the tip also contained a high surface density of reactive sites. These were partly blocked with avidin to achieve single-molecule binding events (after [27.2])

most reactive sites (Fig. 27.1), allowed measurement of single receptor–ligand unbinding forces. Nevertheless, parallel breakage of multiple bonds was predominately observed with this configuration.

To measure interactions between isolated receptor–ligand pairs, strictly defined conditions need to be fulfilled. Covalently coupling ligands to gold-coated tip surfaces via freely accessible SH groups guarantees sufficiently stable attachment because these bonds are about ten times stronger than typical ligand–receptor interactions [27.5]. This chemistry has been used to detect the forces between complementary deoxyribonucleic acid (DNA) strands [27.1] as well as between isolated nucleotides [27.6]. Self-assembled monolayers of dithio-bis(succinimidylundecanoate) were formed to enable covalent coupling of biomolecules via amines [27.7] and were used to study the binding strength between cell adhesion proteoglycans [27.8] and between biotin-directed immunoglobulin G (IgG) antibodies and biotin [27.9]. Vectorial orientation of Fab molecules on gold tips was achieved by site-directed chemical binding via their SH groups [27.10], without the need for additional linkers. To this end, antibodies were digested with papain and subsequently purified to generate Fab fragments with freely accessible SH groups in the hinge region.

Gold surfaces exhibit a unique and selective affinity for thiols, although the adhesion strength of the resulting bonds is comparatively weak [27.5]. Since all commercially available AFM tips are etched from silicon nitride or silicon oxide material, deposition of a gold layer onto the tip surface is required prior to using this chemistry. Therefore, designing a sensor with covalent attachments of biomolecules to the silicon surface may be more straightforward. Amine functionalization procedures, a strategy widely used in surface biochemistry, were applied using ethanolamine [27.3, 11] and various silanization methods [27.12–15], as a first step in thoroughly developed surface anchoring protocols suitable for single-molecule experiments. Since the amine surface density determines, to a large extent, the number of ligands on the tip which can specifically bind to the receptors on the surface, it has to be sufficiently low to guarantee single-molecular recognition events [27.3, 11]. Typically, these densities are kept between 200 and 500 molecules/μm^2, which for AFM tips with radii of $\approx 5-20$ nm, amounts to about one molecule per effective tip area. A striking example of a minimally ligated tip was given by *Wong* et al. [27.16], who derivatized a few carboxyl groups present at the open end of carbon nanotubes attached to the tips of gold-coated Si cantilevers.

In a number of laboratories, a distensible and flexible linker was used to distance the ligand molecule from the tip surface (e.g., [27.3, 13]) (Fig. 27.2). At a given low number of spacer molecules per tip, the ligand can freely orient and diffuse within a certain volume, provided by the length of the tether, to achieve unconstrained binding to its receptor. The unbinding process occurs with little torque and the ligand molecule escapes the danger of being squeezed between the tip and the surface. This approach also opens the possibility of site-directed coupling for a defined orientation of the ligand relative to the receptor at receptor–ligand unbinding. As a cross-linking element, polyethylene glycol (PEG), a water-soluble nontoxic polymer with a wide range of applications in surface technology and clinical research, was often used [27.17]. PEG is known to prevent surface adsorption of proteins and lipid structures and therefore appears ideally suited for this purpose. Glutaraldehyde [27.12] and DNA [27.1] were also successfully applied as molecular spacers in recognition force studies. Cross-linker lengths, ideally arriving at a good compromise between high tip molecule mobility and narrow lateral resolution of the target recognition site, varied from 2 to 100 nm.

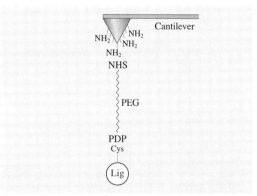

Fig. 27.2 Linkage of ligands to AFM tips. Ligands were covalently coupled to AFM tips via a heterobifunctional polyethylene glycol (PEG) derivative of 8 nm length. Silicon tips were first functionalized with ethanolamine ($NH_2-C_2H_4OH \cdot HCl$). Then, the N-hydroxy-succinimide (NHS)-end of the PEG linker was covalently bound to amines on the tip surface before ligands were attached to the pyridyldithiopropionate (PDP) end via a free thiol or cysteine

For coupling to the tip surface and to the ligand, the cross-linker typically carries two different functional ends, e.g., an amine reactive *N*-hydroxysuccinimidyl (NHS) group on one end, and a thiol reactive 2-pyridyldithiopropionyl group (PDP) [27.18, 19] on the other (Fig. 27.2). This sulfur chemistry is highly advantageous, since it is very reactive and readily enables site-directed coupling. However, free thiols are hardly available on native ligands and must often be added by chemical derivatization.

Different strategies have been used to achieve this goal. Lysine residues were derivatized with the short heterobifunctional linker *N*-succinimidyl-3-(*S*-acetylthio)propionate (SATP) [27.18]. Subsequent deprotection with NH_2OH led to reactive SH groups. Alternatively, lysins can be directly coupled via aldehyde groups [27.15]. The direct coupling of proteins via an NHS–PEG–aldehyde linker allows binding via lysine groups without prederivatization. Nevertheless, since both ends are reactive against amino groups, loop formation can occur between adjacent NH_2 groups on the tip. The probability for this side-effect is significantly lowered (1) by the much higher amino reactivity of the NHS ester in comparison with the aldehyde function and (2) by high linker concentration. Another disadvantage of the latter two methods is that it does not allow for site-specific coupling of the cross-linker, since lysine residues are quite abundant. Several protocols are commercially available (Pierce, Rockford, IL) to generate active antibody fragments with free cysteines. Half-antibodies are produced by cleaving the two disulfide bonds in the central region of the heavy chain using 2-mercaptoethylamine HCl [27.20], and Fab fragments are generated from papain digestion [27.10]. The most elegant methods are to introduce a cysteine into the primary sequence of proteins or to append a thiol group to the end of a DNA strand [27.21], allowing for well-defined sequence-specific coupling of the ligand to the cross-linker.

An attractive alternative for covalent coupling is provided by the widely used nitrilotriacetate (NTA)-His_6 system. The strength of binding in this system, which is routinely used in chromatographic and biosensor matrices, is significantly larger than that between most ligand–receptor pairs [27.22–24]. For receptor–ligand interactions with very high unbinding force, NTA can be substituted with a recently developed Tris-NTA linker [27.25, 26]. Since a His_6 tag can be readily introduced in recombinant proteins, a cross-linker containing an (Tris-)NTA residue is ideally suited for coupling proteins to the AFM tip. This generic, site-specific coupling strategy also allows rigorous and ready control of binding specificity by using Ni^{2+} as a molecular switch of the NTA–His_6 bond. A detailed description of actual coupling strategies can by found in [27.26].

27.2 Immobilization of Receptors onto Probe Surfaces

To enable force detection, the receptors recognized by the ligand-functionalized tip need to be firmly attached to the probed surface. Loose association will unavoidably lead to pull-off of the receptors from the surface by the tip-immobilized ligands, precluding detection of the interaction force.

Freshly cleaved muscovite mica is a perfectly pure and atomically flat surface and, therefore, ideally suited for MRFM studies. The strong negative charge of mica also accomplishes very tight electrostatic binding of various biomolecules; for example, lysozyme [27.20] and avidin [27.27] strongly adhere to mica at pH < 8. In such cases, simple adsorption of the receptors from solution is sufficient, since attachment is strong enough to withstand pulling. Nucleic acids can also be firmly bound to mica through mediatory divalent cations such as Zn^{2+}, Ni^{2+} or Mg^{2+} [27.28]. The strongly acidic sarcoplasmic domain of the skeletal muscle calcium release channel (RYR1) was likewise absorbed to mica via Ca^{2+} bridges [27.29]. By carefully optimizing buffer conditions, similar strategies were used to deposit protein crystals and bacterial layers onto mica in defined orientations [27.30, 31].

The use of nonspecific electrostatic-mediated binding is however quite limited and generally offers no means to orient the molecules over the surface in a desirable direction. Immobilization by covalent attachment must therefore be frequently explored. When glass, silicon or mica are used as probe surfaces, immobilization is essentially the same as described above for tip functionalization. The number of reactive SiOH groups of the chemically relatively inert mica can be optionally increased by water plasma treatment [27.32]. As with tips, cross-linkers are also often used to provide receptors with motional freedom and to prevent surface-induced protein denaturation [27.3].

Immobilization can be controlled, to some extent, by using photoactivatable cross-linkers such as N-5-azido-2-nitrobenzoyloxysuccinimide [27.33].

A major limitation of silicon chemistry is that it does not allow for high surface densities, i.e., $> 1000/\mu m^2$. By comparison, the surface density of a monolayer of streptavidin is about $60\,000$ molecules/μm^2 and that of a phospholipid monolayer may exceed 10^6 molecules/μm^2. The latter high density is also achievable by chemisorption of alkanethiols to gold. Tightly bound functionalized alkanethiol monolayers formed on ultraflat gold surfaces provide excellent probes for AFM [27.9] and readily allow for covalent and noncovalent attachment of biomolecules [27.9, 34] (Fig. 27.3).

Kada et al. [27.35] reported on a new strategy to immobilize proteins on gold surfaces using phosphatidyl choline or phosphatidyl ethanolamine analogues containing dithiophospholipids at their hydrophobic tail. Phosphatidyl ethanolamine, which is chemically reactive, was derivatized with a long-chain biotin for molecular recognition of streptavidin molecules in an initial study [27.35]. These self-assembled phospholipid monolayers closely mimic the cell surface and minimize nonspecific adsorption. Additionally, they can be spread as insoluble monolayers at an air–water interface. Thereby, the ratio of functionalized thiolipids to host lipids accurately defines the surface density of bioreactive sites in the monolayer. Subsequent transfer onto gold substrates leads to covalent, and hence tight, attachment of the monolayer.

MRFM has also been used to study the interactions between ligands and cell surface receptors in situ, on fixed or unfixed cells. In these studies, it was found that the immobilization of cells strongly depends on cell type. Adherent cells are readily usable for MRFM whereas cells that grow in suspension need to be adsorbed onto the probe surface. Various protocols for tight immobilization of cells over a surface are available. For adherent cells, the easiest approach is to grow the cells directly on glass or other surfaces suitable for MRFM [27.36]. Firm immobilization of non- and weakly adhering cells can be achieved by various adhesive coatings such as Cell-Tak [27.37], gelatin, or polylysine. Hydrophobic surfaces such as gold or carbon are also very useful to immobilize nonadherent cells or membranes [27.38]. Covalent attachment of cells to surfaces can be accomplished by cross-linkers that carry reactive groups, such as those used for immobilization of molecules [27.37]. Alternatively, one can use cross-linkers carrying a fatty-acid moiety that can penetrate into the lipid bilayer of the cell membrane. Such linkers provide sufficiently strong fixation without interference with membrane proteins [27.37].

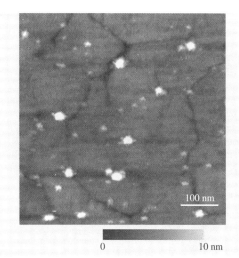

Fig. 27.3 AFM image of hisRNAP molecules specifically bound to nickel-NTA domains on a functionalized gold surface. Alkanethiols terminated with ethylene glycol groups to resist unspecific protein adsorption served as a host matrix and were doped with 10% nickel-NTA alkanthiols. The sample was prepared to achieve full monolayer coverage. Ten individual hisRNAP molecules can be clearly visualized bound to the surface. The more abundant, smaller, lower features are NTA islands with no bound molecules. The underlying morphology of the gold can also be distinguished (after [27.34])

27.3 Single-Molecule Recognition Force Detection

Measurements of interaction forces traditionally rely on ensemble techniques such as shear flow detachment (SFD) [27.39] and the surface force apparatus (SFA) [27.40]. In SFD, receptors are fixed to a surface to which ligands carried by beads or presented on the cell surface bind specifically. The surface-bound particles are then subjected to a fluid shear stress that disrupts the ligand–receptor bonds. However, the force acting between single molecular pairs can only be estimated because the net force applied to the particles can only be approximated and the number of bonds per particle is unknown.

SFA measures the forces between two surfaces to which different interacting molecules are attached, using a cantilever spring as force probe and interferometry for detection. The technique, which has a distance resolution of ≈ 1 Å, allows the measurement of adhesive and compressive forces and rapid transient effects to be followed in real time. However, the force sensitivity of the technique (≈ 10 nN) does not allow for single-molecule measurements of noncovalent interaction forces.

The biomembrane force probe (BFP) technique uses pressurized membrane capsules rather than mechanical springs as a force transducer (Fig. 27.4; see, for example, [27.41]). To form the transducer, a red blood cell or a lipid bilayer vesicle is pressurized into the tip of a glass micropipette. The spring constant of the capsule can then be varied over several orders of magnitude by suction. This simple but highly effective configuration enables the measurement of forces ranging from $0.1-1000$ pN with a force resolution of about 1 pN, allowing probing of single-molecular bonds.

In optical tweezers (OT), small dielectric particles (beads) are manipulated by electromagnetic traps [27.42, 43]. Three-dimensional light intensity gradients of a focused laser beam are used to pull or push

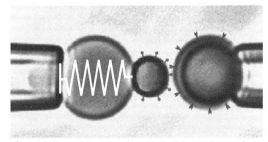

Fig. 27.4 Experimental setup of the biomembrane force probe (BFP). The spring in the BFP is a pressurized membrane capsule. Its spring constant is set by membrane tension, which is controlled by micropipette suction. The BFP tip is formed by a glass microbead with diameter of $1-2$ μm chemically glued to the membrane. The BFP (*on the left*) was kept stationary and the test surface, formed by another microbead (*on the right*), was translated to or from contact with the BFP tip by precision piezoelectric control (after [27.41])

particles with nanometer positional accuracy. Using this technique, forces in the range of $10^{-13}-10^{-10}$ N can be measured accurately. Optical tweezers have been used extensively to measure the force-generating properties of various molecular motors at the single-molecule level [27.44–46] and to obtain force–extension profiles of single DNA [27.47] or protein [27.48] molecules. Defined, force-controlled twisting of DNA using rotating magnetically manipulated particles gave further insights into DNA's viscoelastic properties [27.49, 50].

AFM has successfully been used to measure the interaction forces between various single-molecular pairs [27.1–3]. In these measurements, one of the binding partners is immobilized onto a tip mounted at the end of a flexible cantilever that functions as a force transducer and the other is immobilized over a hard surface such as mica or glass. The tip is initially brought to, and subsequently retracted from the surface, and the interaction (unbinding) force is measured by following the cantilever deflection, which is monitored by measuring the reflection of a laser beam focused on the back of the cantilever using a split photodiode. Approach and retract traces obtained from the unbinding of a single-molecular pair is shown in Fig. 27.5 [27.3]. In this experiment, the binding partners were immobilized onto their respective surfaces through a distensible PEG tether.

Cantilever deflection, Δx, relates directly to the force F acting on it through Hook's law $F = k\Delta x$, where k is the spring constant of the cantilever. During most of the approach phase (trace, and points 1–5), when the tip and the surface are sufficiently far away from each other (1–4), cantilever deflection remains zero because the molecules are still unbound from each other. Upon contact (4) the cantilever bends upwards (4–5) due to a repulsive force that increases linearly as the tip is pushed further into the surface. If the cycle was futile, and no binding had occurred, retraction of the tip from the surface (retrace, 5–7) will lead to a gradual relaxation of the cantilever to its rest position (5–4). In such cases, the retract curve will look very much like the approach curve. On the other hand, if binding had occurred, the cantilever will bend downwards as the cantilever is retracted from the surface (retrace, 4–7). Since the receptor and ligand were tethered to the surfaces through flexible cross-linkers, the shape of the attractive force–distance profile is nonlinear, in contrast to the profile obtained during contact (4–7). The exact shape of the retract curve depends on the elastic properties of the cross-linker used for immobilization [27.17, 51] and exhibits parabolic-like characteristics, reflecting an increase of the spring constant of the cross-linker during extension. The downward bending of the retracting cantilever continues until the ramping force reaches a critical value that dissociates the ligand–receptor complex (unbinding force, 7). Unbinding of the complex is indicated by a sharp spike in the retract curve that reflects an abrupt recoil of the cantilever to its rest position. Specificity of binding is usually demonstrated by block experiments in which free ligands are added to mask receptor sites over the surface.

The force resolution of the AFM, $\Delta F = (k_B T k)^{1/2}$, is limited by the thermal noise of the cantilever which, in turn, is determined by its spring constant. A way to reduce thermal fluctuations of cantilevers without changing their stiffness or lowering the temperature is to increase the apparent damping constant. Applying an actively controlled external dissipative force to cantilevers to achieve such an increase, *Liang* et al. [27.52] reported a 3.4-fold decrease in thermal noise amplitude. The smallest forces that can be detected with commercially available cantilevers are in the range of a few piconewtons. Decreasing cantilever dimensions enables the range of detectable forces to be pushed to smaller forces since small cantilevers have lower coefficients of viscous damping [27.53]. Such miniaturized cantilevers also have much higher resonance frequencies than conventional cantilevers and, therefore, allow for faster measurements.

The atomic force microscope (AFM) [27.4] is the force-measuring method with the smallest sensor and therefore provides the highest lateral resolution. Radii of commercially available AFM tips vary between 2 and 50 nm. In contrast, the particles used for force sensing in SFD, BFP, and OT are in the 1–10 μm range, and the surfaces used in SFA exceed millimeter extensions. The small apex of the AFM tip allows visualization

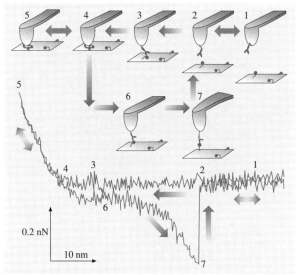

Fig. 27.5 Single-molecule recognition event detected with AFM: a force–distance cycle, measured with an amplitude of 100 nm at a sweep frequency of 1 Hz, for an antibody–antigen pair in PBS. Binding of the antibody immobilized on the tip to the antigen on the surface, which occurs during the approach (*trace points 1–5*), results in a parabolic retract force curve (*points 6–7*) reflecting the extension of the distensible cross-linker antibody–antigen connection. The force increases until unbinding occurs at a force of 268 pN (*points 7 to 2*) (after [27.3])

of single biomolecules with molecular to submolecular resolution [27.28, 30, 31].

Besides the detection of intermolecular forces, the AFM also shows great potential for measuring forces acting within molecules. In these experiments, the molecule is clamped between the tip and the surface and its viscoelastic properties are studied by force–distance cycles.

27.4 Principles of Molecular Recognition Force Spectroscopy

Molecular recognition is mediated by a multitude of noncovalent interactions whose energy is only slightly higher than that of thermal energy. Due to the power-law dependence of these interactions on distance, the attractive forces between noncovalently interacting molecules are extremely short-ranged. A close geometrical and chemical fit within the binding interface is therefore a prerequisite for productive association.

The weak bonds that govern molecular cohesion are believed to be formed in a spatially and temporarily correlated fashion. Protein binding often involves structural rearrangements that can be either localized or global. These rearrangements often bear functional significance by modulating the activity of the interactants. Signaling pathways, enzyme activity, and the activation and inactivation of genes all depend on conformational changes induced in proteins by ligand binding.

The strength of binding is usually given by the binding energy E_b, which amounts to the free energy difference between the bound and the free state, and which can readily be determined by ensemble measurements. E_b determines the ratio of bound complexes [RL] to the product of free reactants [R][L] at equilibrium and is related to the equilibrium dissociation constant K_D through $E_b = -RT \ln(K_D)$, where R is the gas constant. K_D itself is related to the empirical association (k_{on}) and dissociation (k_{off}) rate constants through $K_D = k_{off}/k_{on}$. In order to obtain an estimate for the interaction force f, from the binding energy E_b, the depth of the binding pocket may be used as a characteristic length scale l. Using typical values of $E_b = 20k_B T$ and $l = 0.5$ nm, an order-of-magnitude estimate of $f(= E_b/l) \approx 170$ pN is obtained for the binding strength of a single-molecular pair. Classical mechanics describes bond strength as the gradient in energy along the direction of separation. Unbinding therefore occurs when the applied force exceeds the steepest gradient in energy. This purely mechanical description of molecular bonds, however, does not provide insights into the microscopic determinants of bond formation and rupture.

Noncovalent bonds have limited lifetimes and will therefore break even in the absence of external force on characteristic time scales needed for spontaneous dissociation $(\tau(0) = k_{off}^{-1})$. When pulled faster than $\tau(0)$, however, bonds will resist detachment. Notably, the unbinding force may approach and even exceed the adiabatic limit given by the steepest energy gradient of the interaction potential, if rupture occurs in less time than needed for diffusive relaxation (10^{-10}–10^{-9} s for biomolecules in viscous aqueous medium) and friction effects become dominant [27.55]. Therefore, unbinding forces do not resemble unitary values and the dynamics of the experiment critically affects the measured bond strengths. On the time scale of AFM experiments (milliseconds to seconds), thermal impulses govern the unbinding process. In the thermal activation model, the lifetime of a molecular complex in solution is described by a Boltzmann ansatz $\tau(0) = \tau_{osc} \exp[E_b/(k_B T)]$ [27.56], where τ_{osc} is the inverse of the natural oscillation frequency and E_b is the height of the energy barrier for dissociation. This gives a simple Arrhenius dependency of dissociation rate on barrier height.

A force acting on a complex deforms the interaction free energy landscape and lowers barriers for dissocia-

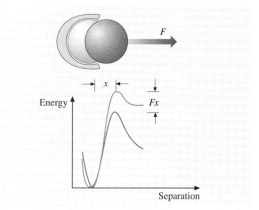

Fig. 27.6 Dissociation over a single sharp energy barrier. Under a constant force, the barrier is lowered by the applied force F. This gives rise to a characteristic length scale x_β that is interpreted as the distance of the energy barrier from the energy minimum along the projection of the force (after [27.54])

tion (Fig. 27.6). As a result of the latter, bond lifetime is shortened. The lifetime $\tau(f)$ of a bond loaded with a constant force f is given by $\tau(f) = \tau_{osc} \exp[(E_b - x_\beta f)/(k_B T)]$ [27.56], where x_β marks the thermally averaged projection of the energy barrier along the direction of the force. A detailed analysis of the relation between bond strength and lifetime was performed by *Evans* and *Ritchie* [27.57], using Kramers' theory for overdamped kinetics. For a sharp barrier, the lifetime $\tau(f)$ of a bond subjected to a constant force f relates to its characteristic lifetime $\tau(0)$ according to $\tau(f) = \tau(0) \exp[-x_\beta f/(k_B T)]$ [27.3]. However, in most pulling experiments the applied force is not constant. Rather, it increases in a complex, nonlinear manner, which depends on the pulling velocity, the spring constant of the cantilever, and the force–distance profile of the molecular complex. Nevertheless, contributions arising from thermal activation manifest themselves mostly near the point of detachment. Therefore, the change of force with time or the loading rate $r (= df/dt)$ can be derived from the product of the pulling velocity and the effective spring constant at the end of the force curve, just before unbinding occurs.

The dependence of the rupture force on the loading rate (force spectrum) in the thermally activated regime was first derived by *Evans* and *Ritchie* [27.57] and described further by *Strunz* et al. [27.54]. Forced dissociation of receptor–ligand complexes using AFM or BFP can often be regarded as an irreversible process because the molecules are kept away from each other after unbinding occurs (rebinding can be safely neglected when measurements are made with soft springs). Rupture itself is a stochastic process, and the likelihood of bond survival is expressed in the master equation as a time-dependent probability $N(t)$ to be in the bound state under a steady ramp of force, namely $\mathrm{d}N(t)/\mathrm{d}t = -k_{\mathrm{off}}(rt)N(t)$ [27.54]. This results in a distribution of unbinding forces $P(F)$ parameterized by the loading rate [27.54, 57, 58]. The most probable force for unbinding f^*, given by the maximum of the distribution, relates to the loading rate through $f^* = f_\beta \ln\left(rk_{\mathrm{off}}^{-1}/f_\beta\right)$, where the force scale f_β is set by the ratio of thermal energy to x_β [27.54, 57]. Thus, the unbinding force scales linearly with the logarithm of the loading rate. For a single barrier, this would give rise to a simple, linear force spectrum f^* versus $\log(r)$. In cases where the escape path is traversed by several barriers, the curve will follow a sequence of linear regimes, each marking a particular barrier [27.41, 57, 58]. Transition from one regime to the other is associated with an abrupt change of slope determined by the characteristic barrier length scale and signifying that a crossover between barriers has occurred.

Dynamic force spectroscopy (DFS) exploits the dependence of bond strength on the loading rate to obtain detailed insights into intra- and intermolecular interactions. By measuring bond strength over a broad range of loading rates, length scales and relative heights of energy barriers traversing the free energy surface can be readily obtained. The lifetime of a bond at any given force is likewise contained in the complete force distribution [27.3]. Finally, one may attempt to extract dissociation rate constants by extrapolation to zero force [27.59]. However, the application of force acts to select the dissociation path. Since the kinetics of reactions is pathway dependent, such a selection implies that kinetic parameters extracted from force–probe experiments may differ from those obtained from assays conducted in the absence of external force. For extremely fast complexation/decomplexation kinetics the forces can be independent of the loading rate, indicating that the experiments were carried out under thermodynamic equilibrium [27.60].

27.5 Recognition Force Spectroscopy: From Isolated Molecules to Biological Membranes

27.5.1 Forces, Energies, and Kinetic Rates

Conducted at fixed loading rates, pioneering measurements of interaction forces provided single points in continuous spectra of bond strengths [27.41]. Not unexpectedly, the first interaction studied was that between biotin and its extremely high-affinity receptors avidin [27.2] and streptavidin [27.1]. The unbinding forces measured for these interactions were 250–300 pN and 160 pN, for streptavidin and avidin, respectively. During this initial phase it was also revealed that different unbinding forces can be obtained for the same pulling velocity if the spring constant of the cantilever is varied [27.1], consistent with the aforementioned dependency of bond strength on the loading rate. The interaction force between several biotin analogues and avidin or streptavidin [27.61] and between biotin and a set of streptavidin mutants [27.62] was investigated and found to generally correlate with the equilibrium binding enthalpy and the enthalpic activation barrier. No correlation with the equilibrium free energy of binding or the activation free energy barrier to dissociation was observed, suggesting that internal energies rather than entropic contributions were probed by the force measurements [27.62].

In another pioneering study, *Lee* et al. [27.21] measured the forces between complementary 20-base DNA strands covalently attached to a spherical probe and surface. The interaction forces fell into three different distributions amounting to the rupture of duplexes consisting of 12, 16, and 20 base pairs. The average rupture force per base pair was $\approx 70\,\mathrm{pN}$. When a long, single-stranded DNA was analyzed, both intra- and interchain forces were observed, the former probing the elastic properties of the molecule. Hydrogen bonds between nucleotides have been probed for all 16 combinations of the four DNA bases [27.6]. Directional hydrogen-bonding interactions were measured only when complementary bases were present on tip and probe surface, indicating that AFM can be used to follow specific pairing of DNA strands.

Strunz et al. [27.14] measured the forces required to separate individual double-stranded DNA molecules of 10, 20, and 30 base pairs (Fig. 27.7). The parameters describing the energy landscape, i.e., the distance from the energy barrier to the minimum energy along the

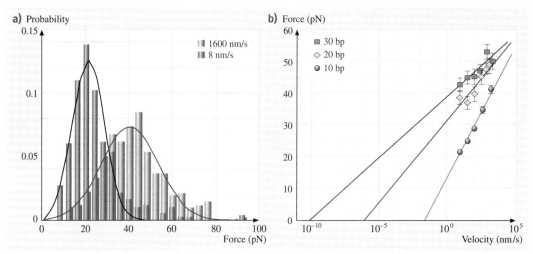

Fig. 27.7a,b Dependence of the unbinding force between DNA single-strand duplexes on the retract velocity. In addition to the expected logarithmic behavior on the loading rate, the unbinding force scales with the length of the strands, increasing from the 10- to 20- to 30-base-pair duplexes (after [27.14])

separation path and the logarithm of the thermal dissociation rate, were found to be proportional to the number of base pairs of the DNA duplex. Such scaling suggests that unbinding proceeds in a highly cooperative manner characterized by one length scale and one time scale. Studying the dependence of rupture forces on temperature, it was proposed by *Schumakovitch* et al. [27.63] that entropic contributions play an important role in the unbinding of complementary DNA strands [27.63].

Prevalent as it is, molecular recognition has mostly been discussed in the context of interactions between antibodies and antigens. To maximize motional freedom and to overcome problems associated with misorientation and steric hindrance, antibodies and antigens were immobilized onto the AFM tip and probe surface via flexible molecular spacers [27.3, 9, 12, 13]. By optimizing the antibody density over the AFM tip [27.3, 11], the interaction between individual antibody–antigen pairs could be examined. Binding of antigen to the two Fab fragments of the antibody was shown to occur independently and with equal probability. Single antibody–antigen recognition events were also recorded with tip-bound antigens interacting with intact antibodies [27.9, 12] or with single-chain Fv fragments [27.13]. The latter study also showed that an Fv mutant whose affinity to the antigen was attenuated by about 10-fold dissociated from the antigen under applied forces that were 20% lower than those required to unbind the wild-type (Fv) antibody.

Besides measurements of interaction forces, single-molecule force spectroscopy also allows estimation of association and dissociation rate constants, notwithstanding the concern stated above [27.3, 11, 23, 59, 64, 65], and measurement of structural parameters of the binding pocket [27.3, 11, 14, 64, 65]. Quantification of the association rate constant k_{on} requires determination of the interaction time needed for half-maximal probability of binding ($t_{1/2}$). This can be obtained from experiments where the encounter time between receptor and ligand is varied over a broad range [27.64]. Given that the concentration of ligand molecules on the tip available for interaction with the surface-bound receptors c_{eff} is known, the association rate constant can be derived from $k_{on} = t_{0.5}^{-1} c_{eff}^{-1}$. Determination of the effective ligand concentration requires knowledge of the effective volume V_{eff} explored by the tip-tethered ligand which, in turn, depends on the tether length. Therefore, only order-of-magnitude estimates of k_{on} can be obtained from such measurements [27.64].

Additional information about the unbinding process is contained in the distributions of the unbinding forces. Concomitant with the shift of maxima to higher unbinding forces, increasing the loading rate also leads to an increase in the width σ of the distributions [27.23, 41], indicating that at lower loading rates the system adjusts closer to equilibrium. The lifetime $\tau(f)$ of a bond under an applied force was estimated by the time the cantilever spends in the force window spanned by the standard deviation of the most probable force for unbinding [27.3]. In the case of Ni^{2+}-His_6, the lifetime of the complex decreased from 17 to 2.5 ms when the force was increased from 150 to 194 pN [27.23]. The data fit well to Bell's model, confirming the predicted exponential dependence of bond lifetime on the applied force, and yielded an estimated lifetime at zero force of about 15 s. A more direct measurement of τ is afforded by force-clamp experiments in which the applied force is kept constant by a feedback loop. This configuration was first adapted for use with AFM by *Oberhauser* et al. [27.66], who employed it to study the force dependence of the unfolding probability of the I27 and I28 modules of cardiac titin as well as of the full-length protein [27.66].

However, as discussed above, in most experiments the applied force is not constant but varies with time, and the measured bond strength depends on the loading rate [27.55, 57, 67]. In accordance with this, experimentally measured unbinding forces do not assume unitary values but rather vary with both pulling velocity [27.59, 64] and cantilever spring constant [27.1]. The slopes of the force versus loading rate curves contain information about the length scale x_β of prominent energy barriers along the force-driven dissociation pathway, which may be related to the depth of the binding pocket of the interaction [27.64]. The predicted logarithmic dependence of the unbinding force on the loading rate holds well when the barriers are stationary with force, as confirmed by a large number of unbinding and unfolding experiments [27.14, 23, 41, 59, 64, 65, 68]. However, if the position of the transition state is expected to vary along the reaction coordinate with the force, as for example when the curvature at the top of the barrier is small, the strict logarithmic dependence gives way to more complex forms.

Schleiff and *Rief* [27.69] used a Kramers diffusion model to calculate the probability force distributions when the barrier cannot be assumed to be stationary. Notably, although the position of the transition state predicted by the Bell model was smaller than that predicted by the Kramer analysis by 6 Å, the most probable unfolding forces showed an almost perfect logarithmic dependence on the pulling velocity, indicating that great care should be taken before the linear theory of DFS is applied. Failure to fit force distributions at high loading rates using a Bell model was also reported by *Neuert* et al. [27.70] for the interaction of digoxigenin and antidigoxigenin. Poor matches were observed in the crossover region between the two linear regimes of the force spectrum as well. *Klafter*

et al. also suggested a method to analyze force spectra which also does not assume stationarity of the energy barrier [27.71]. In their treatment, they find a $(\ln r)^{2/3}$ dependence of the mean force of dissociation, where r is the loading rate. They also find the distribution of unbinding forces to be asymmetric, as indeed observed many times. *Evstigneev* and *Reimann* [27.72] suggest that the practice of fitting this asymmetric distribution with a Gaussian one in order to extract the mean rupture leads to the latter's overestimation and consequently to an overestimate of the force-free dissociation rate. They suggest an optimized statistical data analysis which overcomes this limitation by combining data at many pulling rates into a single distribution of the probability of rupture versus force.

The force spectra may also be used to derive the dissociation rate constant k_{off} by extrapolation to zero force [27.59, 64, 65]. As mentioned above, values derived in this manner may differ from those obtained from bulk measurements because only a subset of dissociation pathways defined by the force is sampled. Nevertheless, a simple correlation between unbinding forces and thermal dissociation rates was obtained for a set consisting of nine different Fv fragments constructed from point mutations of three unrelated antifluorescein antibodies [27.65, 70]. This correlation, which implies a close similarity between the force- and thermally driven pathways explored during dissociation, was probably due to the highly rigid nature of the interaction, which proceeds in a lock-and-key fashion. The force spectra obtained for the different constructs exhibited a single linear regime, indicating that in all cases unbinding was governed by a single prominent energy barrier (Fig. 27.8). Interestingly, the position of the energy barrier along the forced-dissociation pathway was found to be proportional to the height of the barrier and, thus, most likely includes contributions arising from elastic stretching of the antibodies during the unbinding process.

A good correspondence between dissociation rates derived from mechanical unbinding experiments and from bulk assays was also reported by *Neuert* et al. [27.70]. In this case, the experimental system consisted of digoxigenin and its specific antibody. This pair is used as a noncovalent coupler in various applications, including forced-unbinding experiments. The force spectra obtained for the complex suggested that the unbinding path is traversed by two activation energy barriers located at $x_\beta = 0.35$ nm and $x_\beta = 1.15$ nm. Linear fit of the low-force regime revealed a dissociation

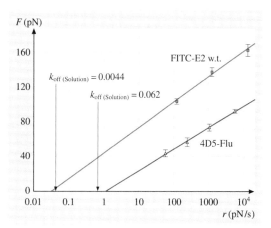

Fig. 27.8 The dependence of the unbinding force on the loading rate for two antifluorescein antibodies. For both FITC-E2 w.t. and 4D5-Flu a strictly single-exponential dependence was found in the range accessed, indicating that only a single energy barrier was probed. The same energy barrier dominates dissociation without forces applied because extrapolation to zero force matches kinetic off-rates determined in solution (indicated by the *arrow*) (after [27.65])

rate at zero force of $0.015\,\text{s}^{-1}$, in close agreement with the $0.023\,\text{s}^{-1}$ value obtained from bulk measurements made on antidigoxigenin Fv fragments.

27.5.2 Complex Bonds and Energy Landscapes

The energy landscapes that describe proteins are generally not smooth. Rather, they are traversed by multiple energy barriers of various heights that render them highly corrugated or rugged. All these barriers affect the kinetics and conformational dynamics of proteins and any one of them may govern interaction lifetime and strength on certain time scales. Dynamic force spectroscopy provides an excellent tool to detect energy barriers which are difficult or impossible to detect by conventional, near-equilibrium assays and to probe the free energy surface of proteins and protein complexes. It also provides a natural means to study interactions which are normally subjected to varying mechanical loads [27.59, 64, 73–75].

A beautiful demonstration of the ability of dynamic force spectroscopy to reveal hidden barriers was provided by *Merkel* et al. [27.41], who used BFP to probe bond formation between biotin and streptavidin or avidin over a broad range of loading rates. In contrast to early studies which reported fixed values of bond strength [27.61, 62], a continuous spectrum of unbinding forces ranging from 5 to 170 pN was obtained (Fig. 27.9). Concomitantly, interaction lifetime decreased from about 1 min to 0.001 s, revealing the reciprocal relation between bond strength and lifetime expected for thermally activated kinetics under a ris-

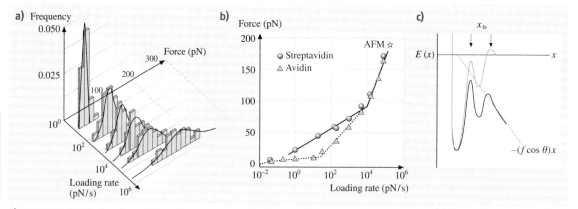

Fig. 27.9a–c Unbinding force distributions and energy landscape of a complex molecular bond. (**a**) Force histograms of single biotin–streptavidin bonds recorded at different loading rates. The shift in peak location and the increase in width with increasing loading rate is clearly demonstrated. (**b**) Dynamic force spectra for biotin–streptavidin (*circles*) and biotin–avidin (*triangles*). The slopes of the linear regimes mark distinct activation barriers along the direction of force. (**c**) Conceptual energy landscape traversed along a reaction coordinate under force. The external force f adds a mechanical potential that tilts the energy landscape and lowers the barriers. The inner barrier starts to dominate when the outer has fallen below it due to the applied force (after [27.41])

ing force. Most notably, depending on the loading rate, unbinding kinetics was dominated by different activation energy barriers positioned along the force-driven unbinding pathway. Barriers emerged sequentially, with the outermost barrier appearing first, each giving rise to a distinct linear regime in the force spectrum. Going from one linear regime to the next was associated with an abrupt change in slope, indicating that a crossover between an outer to (more) inner barrier had occurred. The position of two of the three barriers identified in the force spectra was consistent with the location of prominent transition states revealed by molecular dynamics simulations [27.55, 67]. However, as was mentioned earlier, unbinding is not necessarily confined to a single, well-defined path, and may take different routes even when directed by an external force. Molecular dynamics simulations of force-driven unbinding of an antibody–antigen complex characterized by a highly flexible binding pocket revealed a large heterogeneity of enforced dissociation pathways [27.76].

The rolling of leukocytes on activated endothelium is a first step in the emergence of leukocytes out of the blood stream into sites of inflammation. This rolling, which occurs under hydrodynamic shear forces, is mediated by selectins, a family of extended, calcium-dependent lectin receptors present on the surface of endothelial cells. To fulfill their function, selectins and their ligands exhibit a unique combination of mechanical properties: they associate rapidly and avidly and can tether cells over very long distances by their long, extensible structure. In addition, complexes formed between selectins and their ligands can withstand high tensile forces and dissociate in a controllable manner, which allows them to maintain rolling without being pulled out of the cell membrane.

Fritz et al. [27.59] used dynamic force spectroscopy to study the interaction between P-selectin and its leukocyte-expressed surface ligand P-selectin glycoprotein ligand-1 (PSGL-1). Modeling both intermolecular and intramolecular forces, as well as adhesion probability, they were able to obtain detailed information on rupture forces, elasticity, and the kinetics of the interaction. Complexes were able to withstand forces up to 165 pN and exhibited a chain-like elasticity with a molecular spring constant of 5.3 pN/nm and a persistence length of 0.35 nm. Rupture forces and the lifetime of the complexes exhibited the predicted logarithmic dependence on the loading rate.

An important characteristics of the interaction between P-selectin and PSGL-1, which is highly relevant to the biological function of the complex, was found by investigating the dependence of the adhesion probability between the two molecules on the velocity of the AFM probe. Counterintuitively and in contrast to experiments with avidin–biotin [27.61], antibody–antigen [27.3], or cell adhesion proteoglycans [27.8], the adhesion probability between P-selectin and PSGL-1 was found to *increase* with increasing velocities [27.59]. This unexpected dependency explains the increase in leukocyte tethering probability with increased shear flow observed in rolling experiments. Since the adhesion probability approached 1, it was concluded that binding occurs instantaneously as the tip reaches the surface and, thus, proceeds with a very fast on-rate. The complex also exhibited a fast forced off-rate. Such a fast-on/fast-off kinetics is probably important for the ability of leukocytes to bind and detach rapidly from the endothelial cell surface. Likewise, the long contour length of the complex together with its high elasticity reduces the mechanical loading on the complex upon binding and allows leukocyte rolling even at high shear rates.

Evans et al. [27.73] used BPF to study the interaction between PSGL-1 and another member of the selectin family, L-selectin. The force spectra, obtained over a range of loading rates extending from 10 to 100 000 pN/s, revealed two prominent energy barriers along the unbinding pathway: an outer barrier, probably constituted by an array of hydrogen bonds, that impeded dissociation under slow detachment, and an inner, Ca^{2+}-dependent barrier that dominated dissociation under rapid detachment. The observed hierarchy of inner and outer activation barriers was proposed to be important for multibond recruitment during selectin-mediated function.

Using force-clamp AFM [27.66], bond lifetimes were directly measured in dependence on a constantly applied force. For this, lifetime–force relations of P-selectin complexed to two forms of P-selectin glycoprotein ligand 1 (PSGL-1) and to G1, a blocking monoclonal antibody against P-selectin, respectively, were determined [27.75]. Both monomeric (sPSGL-1) and dimeric PSGL-1 exhibited a biphasic relationship between lifetime and force in their interaction to P-selectin (Fig. 27.10a,b). The bond lifetimes initially increased, indicating the presence of catch bonds. After reaching a maximum, the lifetimes decreased with force, indicating a catch bond. In contrast, the P-selectin/G1 bond lifetimes decreased exponentially with force (Fig. 27.10c), displaying typical slip bond characteristics that are well described by the single-energy-barrier Bell model. The curves of lifetime

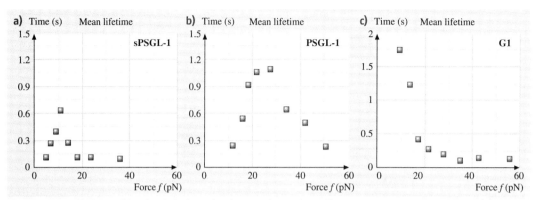

Fig. 27.10a–c Lifetimes of bonds of single-molecular complexes, depending on a constantly applied force. (**a**) sPSGL-1/P-selectin: catch bond and slip bond. (**b**) PSGL-1/P-selectin: catch bond and slip bond. (**c**) G1/P-selectin: slip bond only (after [27.75])

against force for the two forms of PSGL1-1 had similar biphasic shapes (Fig. 27.10a,b), but the PSGL-1 curve (Fig. 27.10b) was shifted relative to the sPSGL-1 curve (Fig. 27.10a), approximately doubling the force and the lifetime. These data suggest that sPSGL-1 forms monomeric bonds with P-selectin, whereas PSGL-1 forms dimeric bonds with P-selectin. In agreement with the studies describes above, it was concluded that the use of force-induced switching from catch to slip bonds might be physiologically relevant for the tethering and rolling process of leukocytes on selectins [27.75].

Baumgartner et al. [27.64] used AFM to probe specific trans-interaction forces and conformational changes of recombinant vascular endothelial (VE)-cadherin strand dimers. VE-cadherins are cell-surface proteins that mediate the adhesion of cells in the vascular endothelium through Ca^{2+}-dependent homophilic interactions of their N-terminal extracellular domains. Acting as such they play an important role in the regulation of intercellular adhesion and communication in the inner surface of blood vessels. Unlike selectin-mediated adhesion, association between trans-interacting VE dimers was slow and independent of probe velocity, and complexes were ruptured at relatively low forces. These differences were attributed to the fact that, as opposed to selectins, cadherins mediate adhesion between resting cells. Mechanical stress on the junctions is thus less intense and high-affinity binding is not required to establish and maintain intercellular adhesion. Determination of Ca^{2+} dependency of recognition events between tip- and surface-bound VE-cadherins revealed a surprisingly high K_D (1.15 mM), which is very close to the free extracellular Ca^{2+} concentration in the body. Binding also revealed a strong dependence on calcium concentration, giving rise to an unusually high Hill coefficient of ≈ 5. This steep dependency suggests that local changes of free extracellular Ca^{2+} in the narrow intercellular space may facilitate rapid remodeling of intercellular adhesion and permeability.

Odorico et al. [27.77] used DFS to explore the energy landscape underlying the interaction between a chelated uranyl compound and a monoclonal antibody raised against the uranyl-dicarboxy-phenanthroline complex. To isolate contributions of the uranyl moiety to the binding interaction, measurements were performed with and without the ion in the chelating ligand. In the presence of uranyl, the force spectra contained two linear regimes, suggesting the presence of at least two major energy barriers along the unbinding pathway. To relate the experimental data to molecular events, the authors constructed a model with a variable fragment of the antibody and used computational graphics to dock the chelated uranyl ion into the binding pocket. The analysis suggested that the inner barrier ($x_\beta = 0.5$ Å) reflects the rupture of coordination bonds between the uranium atom and an Asp residue, whereas the outer barrier ($x_\beta = 3.9$ Å) amounts to the detachment of the entire ligand from the Ab binding site.

Nevo et al. [27.78, 79] used single-molecule force spectroscopy to discriminate between alternative mechanisms of protein activation (Fig. 27.11). The activation of proteins by other proteins, protein domains or small ligands is a central process in biology, e.g., in signalling pathways and enzyme activity. Moreover, activation and deactivation of genes both depend on the switching of proteins between alternative functional states. Two general mechanisms have been proposed. The induced-fit model assigns changes in protein activity to conformational changes triggered by effector binding. The population-shift model, on the other hand, ascribes these changes to a redistribution of *preexisting* conformational isomers. According to this model, also known as the preequilibrium or conformational selection model, protein structure is regarded as an ensemble of conformations existing in equilibrium. The ligand binds to one of these conformations, i.e., the one to which it is most complementary, thus shifting the equilibrium in favor of this conformation. Discrimination between the two models of activation requires that the distribution of conformational isomers in the ensemble is known. Such information, however, is very hard to obtain from conventional bulk methods because of ensemble averaging.

Using AFM, Nevo and coworkers measured the unbinding forces of two related protein complexes in the absence or presence of a common effector. The complexes consisted of the nuclear transport receptor importin β(impβ) and the small GTPase Ran. The difference between them was the nucleotide-bound state of Ran, which was either guanosine diphosphate (GDP) or guanosine-5'-triphosphate (GTP). The effector molecule was the Ran-binding protein RanBP1. Loaded with GDP, Ran associated weakly with impβ to form a single bound state characterized by unimodal distributions of small unbinding forces (Fig. 27.11a, dotted line). Addition of Ran BP1 resulted in a marked shift of the distribution to higher unbinding forces (Fig. 27.11b, dashed to solid line). These results were interpreted to be consistent with an induced-fit mechanism where binding of RanBP1 induces a conformational change in the complex, which, in turn, strengthens the interaction between impβ and Ran(GDP). In contrast, association of RanGTP with impβ was found to lead to alternative bound states of relatively low and high adhesion strength represented by partially overlapping force distributions (Fig. 27.11a, solid line). When RanBP1 was added to the solution, the higher-strength population, which predominated the ensemble in the absence of the effector (Fig. 27.11c, dashed lines), was diminished, and the lower-strength conformation became correspondingly more populated (Fig. 27.11c, solid line). The means of the distributions, however, remain unchanged, indicating that the strength of the interaction in the two states of the complex had not been altered by the effector. These data fit a dynamic population-shift mechanism in which RanBP1 binds selectively to the lower-strength conformation of RanGTP–impβ, changing the properties and function of the complex by shifting the equilibrium between its two states.

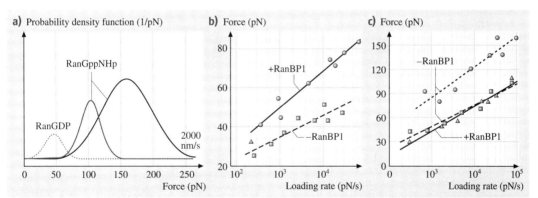

Fig. 27.11a–c Protein activation revealed by force spectroscopy. Ran and importin β (impβ) were immobilized onto the AFM cantilevered tip and mica, respectively, and the interaction force was measured at different loading rates in the absence or presence of RanBP1, which was added as a mobile substrate to the solution in the AFM liquid cell. Unbinding force distributions obtained for impβ–Ran complexes at pulling velocity of 2000 nm/s. Association of impβ with Ran loaded with GDP (**a**) or with nonhydrolyzable GTP analogue (GppNHp) (**b**) gives rise to uni- or bimodal force distributions, respectively, reflecting the presence of one or two bound states. (**b–c**) Force spectra obtained for complexes of impβ with RanGDP or with RanGppNHp, in the absence (*dashed lines*) or presence (*solid lines*) of RanBP1. The results indicate that activation of impβ–RanGDP and imp–RanGTP complexes by RanBP1 proceeds through induced-fit and dynamic population-shift mechanisms, respectively (see text for details) (after [27.78, 79])

The complex between impβ and RanGTP was also used in studies aimed to measure the energy landscape roughness of proteins. The roughness of the energy landscapes that describe proteins has numerous effects on their folding and binding as well as on their behavior at equilibrium, since undulations in the free energy surface can attenuate diffusion dramatically. Thus, to understand how proteins fold, bind, and function, one needs to know not only the energy of their initial and final states, but also the roughness of the energy surface that connects them. However, for a long time, knowledge of protein energy landscape roughness came solely from theory and simulations of small model proteins

Adopting Zwanzig's theory of diffusion in rough potentials [27.80], *Hyeon* and *Thirumalai* [27.81] proposed that the energy landscape roughness of proteins can be measured from single-molecule mechanical unfolding experiments conducted at different temperatures. In particular, their simulations showed that at a constant loading rate the most probable force for unfolding increases because of roughness that acts to attenuate diffusion. Because this effect is temperature dependent, an overall energy scale of roughness, ε, can be derived from plots of force versus loading rate acquired at two arbitrary temperatures. Extending this theory to the case of unbinding, and performing single-molecule force spectroscopy measurements, *Nevo* et al. [27.82] extracted the overall energy scale of roughness ε for RanGTP–impβ. The results yielded $\varepsilon > 5k_BT$, indicating a bumpy energy surface, which is consistent with the unusually high structural flexibility of impβ and its ability to interact with different, structurally distinct ligands in a highly specific manner. This mechanistic principle may also be applicable to other proteins whose function demands highly specific and regulated interactions with multiple ligands.

More recently, the same type of analysis using three temperatures and pulling speeds in the range of 100 to 38 000 nm/s, was applied to derive ε for the well-studied streptavidin–biotin interaction [27.83]. Analysis of the Bell parameters revealed considerable widening of the inner barrier for the transition with temperature, reflecting perhaps a softening of the dominant hydrogen-bond network that stabilizes the ground state of the complex. In contrast, the position of the outer barrier did not change significantly upon increase of the temperature. Estimations of ε were made at four different forces, 75, 90, 135, and 156 pN, with the first two forces belonging to the first linear loading regime of the force spectrum (outer barrier) and the last two to the second (inner barrier). The values obtained were consistent *within each of the two regimes*, averaging at 7.5 and $\approx 5.5k_BT$ along the outer and inner barriers of the transition, respectively. The difference was attributed to contributions from the intermediate state of the reaction, which is suppressed (along with the outer barrier) at high loading rates. The origin of roughness was attributed to competition of solvent water molecules with some of the hydrogen bonds that stabilize the complex and to the aforementioned 3–4 loop of streptavidin, which is highly flexible and, therefore, may induce the formation of multiple conformational substates in the complex. It was also proposed by the authors that the large roughness detected in the energy landscape of streptavidin–biotin is a significant contributor to the unusually slow dissociation kinetics of the complex and may account for the discrepancies in the unbinding forces measured for this pair.

27.5.3 Live Cells and Membranes

Thus far, there have been only a few attempts to apply recognition force spectroscopy to cells. In one of the early studies, *Lehenkari* and *Horton* [27.84] measured the unbinding forces between integrin receptors present on the surface of intact cells and several RGD-containing (Arg–Gly–Asp) ligands. The unbinding forces measured were found to be cell and amino acid sequence specific, and sensitive to pH and the divalent cation composition of the cellular culture medium. In contrast to short linear RGD hexapeptides, larger peptides and proteins containing the RGD sequence showed different binding affinities, demonstrating that the context of the RGD motif within a protein has a considerable influence upon its interaction with the receptor. In another study, *Chen* and *Moy* [27.85] used AFM to measure the adhesive strength between concanavalin A (Con A) coupled to an AFM tip and Con A receptors on the surface of NIH3T3 fibroblasts. Cross-linking of receptors with either glutaraldehyde or 3, 3′-dithio-bis(sulfosuccinimidylproprionate) (DTSSP) led to an increase in adhesion that was attributed to enhanced cooperativity among adhesion complexes. The results support the notion that receptor cross-linking can increase adhesion strength by creating a shift towards cooperative binding of receptors. *Pfister* et al. [27.86] investigated the surface localization of HSP60 on stressed and unstressed human umbilical venous endothelial cells (HUVECs). By detecting specific single-molecule binding events between the monoclonal antibody AbII-13 tethered to AFM tips and HSP60 molecules on cells, clear evidence was found for the occurrence of HSP60 on the surface of stressed HUVECs, but not on unstressed HUVECs.

The sidedness and accessibility of protein epitopes of the Na^{2+}/D-glucose cotransporter 1 (SGLT1) was probed in intact brush border membranes by a tip-bound antibody directed against an amino acid sequence close to the glucose binding site [27.38]. Binding of glucose and transmembrane transport altered both the binding probability and the most probable unbinding force, suggesting changes in the orientation and conformation of the transporter. These studies were extended to live SGLT1-transfected CHO cells [27.87]. Using AFM tips carrying the substrate 1-β-thio-D-glucose, direct evidence could be obtained that, in the presence of sodium, a sugar binding site appears on the SGLT1 surface. It was shown that this binding site accepts the sugar residue of the glucoside phlorizin, free D-glucose and D-galactose, but not free L-glucose. The data indicate the importance of stereoselectivity for sugar binding and transport.

Studies on the interaction between leukocyte function-associated antigen-1 (LFA-1) and its cognate ligand, intercellular adhesion molecules 1 and 2 (ICAM-1 and ICAM-2), which play a crucial role in leukocyte adhesion, revealed two prominent barriers [27.74, 88]. The experimental system consisted of LFA-1-expressing Jurkat T-cells attached to the end of the AFM cantilever and surface-immobilized ICAM-1 or -2. For both ICAM-1 and ICAM-2, the force spectra exhibited fast and slow loading regimes, amounting to a sharp, inner energy barrier ($x_\beta \approx 0.56$ Å and 1.5 Å, for complexes formed with ICAM-1 and ICAM-2) and a shallow, outer barrier ($x_\beta \approx 3.6$ Å and 4.9 Å), respectively. Addition of Mg^{2+} led to an increase of the rupture forces measured in the slow loading regime, indicating an increment of the outer barrier in the presence of the divalent cation. Comparison between the force spectra obtained for the complexes formed between LFA-1 and ICAM-1 or ICAM-2 indicated that, in the fast loading regime, the rupture of LFA-1–ICAM-1 depends more steeply on the loading rate than that of LFA-1–ICAM-2. The difference in dynamic strength between the two interactions was attributed to the presence of wider barriers in the LFA-1–ICAM-2 complex, which render the interaction more receptive to the applied load. The enhanced sensitivity of complexes with ICAM-2 to pulling forces was proposed to be important for the ability of ICAM-2 to carry out routine immune surveillance, which might otherwise be impeded due to frequent adhesion events.

27.6 Recognition Imaging

Besides measuring interaction strengths, locating binding sites over biological surfaces such as cells or membranes is of great interest. To achieve this goal, force detection must be combined with high-resolution imaging.

Ludwig et al. [27.89] used chemical force microscopy to image a streptavidin pattern with a biotinylated tip. An approach–retract cycle was performed at each point of a raster, and topography, adhesion, and sample elasticity were extracted from the local force ramps. This strategy was also used to map binding sites on cells [27.90, 91] and to differentiate between red blood cells of different blood groups (A and 0) using AFM tips functionalized with a group A-specific lectin [27.92].

Identification and localization of single antigenic sites was achieved by recording force signals during the scanning of an AFM tip coated with antibodies along a single line across a surface immobilized with a low density of antigens [27.3, 11]. Using this method, antigens could be localized over the surface with positional accuracy of 1.5 nm. A similar configuration used by *Willemsen* et al. [27.93] enabled the simultaneous acquisition of height and adhesion-force images with near molecular resolution.

The aforementioned strategies of force mapping either lack high lateral resolution [27.89] and/or are much slower [27.3, 11, 93] than conventional topographic imaging since the frequency of the force-sensing retract–approach cycles is limited by hydrodynamic damping. In addition, the ligand needs to be detached from the receptor in each retract–approach cycle, necessitating large working amplitudes (50 nm). Therefore, the surface-bound receptor is inaccessible to the tip-immobilized ligand on the tip during most of the time of the experiment. This problem, however, should be overcome with the use of small cantilevers [27.53], which should increase the speed for force mapping because the hydrodynamic forces are significantly reduced and the resonance frequency is higher than that of commercially available cantilevers. Short cantilevers were recently applied to follow the association and dissociation of individual chaperonin proteins, GroES to GroEL, in real time using dynamic force microscopy topography imaging [27.94].

An imaging method for mapping antigenic sites on surfaces was developed [27.20] by combining molecular recognition force spectroscopy [27.3] with dynamic force microscopy (DFM) [27.28,96]. In DFM, the AFM tip is oscillated across a surface and the amplitude reduction arising from tip–surface interactions is held constant by a feedback loop that lifts or lowers the tip according to the detected amplitude signal. Since the tip contacts the surface only intermittently, this technique provides very gentle tip–surface interactions and the specific interaction of the antibody on the tip with the antigen on the surface can be used to localize antigenic sites for recording recognition images. The AFM tip is magnetically coated and oscillated by an alternating magnetic field at very small amplitudes while being scanned along the surface. Since the oscillation frequency is more than a hundred times faster than typical frequencies in conventional force mapping, the data acquisition rate is much higher. This method was recently extended to yield fast, simultaneous acquisition of two independent maps, i.e., a topography image and a lateral map of recognition sites, recorded with nm resolution at experimental times equivalent to normal AFM imaging [27.95, 97, 98].

Topography and recognition images were simultaneously obtained (TREC imaging) using a special electronic circuit (PicoTrec, Agilent, Chandler, AZ) (Fig. 27.12a). Maxima (U_{up}) and minima (U_{down}) of each sinusoidal cantilever deflection period were depicted in a peak detector, filtered, and amplified. Direct-current (DC) offset signals were used to compensate for the thermal drifts of the cantilever. U_{up} and U_{down} were fed into the AFM controller, with U_{down} driving the feedback loop to record the height (i.e., topography) image and U_{up} providing the data for constructing the recognition image (Fig. 27.12a). Since we used cantilevers with low Q-factor (≈ 1 in liquid) driven at frequencies below resonance, the two types of information were independent. In this way, topography and recognition image were recorded simultaneously and independently.

The circuit was applied to mica containing singly distributed avidin molecules using a biotinylated AFM tip [27.95]. The sample was imaged with an antibody-containing tip, yielding the topography (Fig. 27.12b, left image) and the recognition image (Fig. 27.12b, right image) at the same time. The tip oscillation amplitude (5 nm) was chosen to be slightly smaller than the extended cross-linker length (8 nm), so that both the antibody remained bound while passing a binding site and the reduction of the upwards deflection was of sufficient significance compared with the thermal noise. Since the spring constant of the polymeric cross-linker increases nonlinearly with the tip–surface distance (Fig. 27.5), the binding force is only sensed close to full extension of the cross-linker (given at the maxima of the oscillation period). Therefore, the recognition signals were well separated from the topographic signals arising from the surface, in both

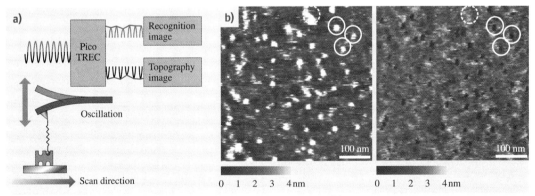

Fig. 27.12a,b Simultaneous topography and recognition (TREC) imaging. (**a**) Principle: the cantilever oscillation is split into lower and upper parts, resulting in simultaneously acquired topography and recognition images. (**b**) Avidin was electrostatically adsorbed to mica and imaged with a biotin-tethered tip. Good correlation between topography (*left image, bright spots*) and recognition (*right image, dark spots*) was found (*solid circles*). Topographical spots without recognition denote structures lacking specific interaction (*dashed circle*). Scan size was 500 nm (after [27.95])

space ($\Delta z \approx 5$ nm) and time (half-oscillation period ≈ 0.1 ms).

The bright dots with 2–3 nm height and 15–20 nm diameter visible in the topography image (Fig. 27.12b, left image) represent single avidin molecules stably adsorbed onto the flat mica surface. The recognition image shows black dots at positions of avidin molecules (Fig. 27.12b, right image) because the oscillation maxima are lowered due to the physical avid–biotin connection established during recognition. That the lateral positions of the avidin molecules obtained in the topography image are spatially correlated with the recognition signals of the recognition image is indicated by solid circles in the images (Fig. 27.12). Recognition between the antibody on the tip and the avidin on the surface took place for almost all avidin molecules, most likely because avidin contains four biotin binding sites, two on either side. Thus, one would assume to have always binding epitopes oriented away from the mica surface and accessible to the biotinylated tip, resulting in a high binding efficiency. Structures observed in the topography image and not detected in the recognition image were very rare (dotted circle in Fig. 27.12b).

It is important to note that topography and recognition images were recorded at speeds typical for standard AFM imaging and were therefore considerably faster than conventional force mapping. With this methodology, topography and recognition images can be obtained at the same time and distinct receptor sites in the recognition image can be assigned to structures from the topography image. This method is applicable to any ligand, and therefore it should prove possible to recognize many types of proteins or protein layers and carry out epitope mapping on the nm scale on membranes, cells, and complex biological structures. In a striking recent example, histone proteins H3 were identified and localized in a complex chromatin preparation [27.98].

Recently, TREC imaging was applied to gently fixed microvascular endothelial cells from mouse myocardium (MyEnd) in order to visualize binding sites of VE-cadherin, known to play a crucial role in homophilic cell adhesion [27.99]. TREC images were acquired with AFM tips coated with a recombinant VE-cadherin. The recognition images revealed prominent, irregularly shaped *dark* spots (domains) with size from 30 to 250 nm. The domains enriched in VE-cadherins molecules were found to be collocated with the cytoskeleton filaments supporting the anchorage of VE-cadherins to F-actin. Compared with conventional techniques such as immunochemistry or single-molecule optical microscopy, TREC represents an alternative method to quickly obtain the local distribution of receptors on cell surface with unprecedented lateral resolution of several nm.

27.7 Concluding Remarks

Atomic force microscopy has evolved to become an imaging method that can yield the finest structural details on live, biological samples in their native, aqueous environment under ambient conditions. Due to its high lateral resolution and sensitive force detection capability, it is now possible to measure molecular forces of biomolecules on the single-molecule level. Well beyond the proof-of-principle stage of the pioneering experiments, AFM has now developed into a high-end analysis method for exploring kinetic and structural details of interactions underlying protein folding and molecular recognition. The information obtained from force spectroscopy, being on a single-molecule level, includes physical parameters not accessible by other methods. In particular, it opens up new perspectives to explore the dynamics of biological processes and interactions.

References

27.1 G.U. Lee, D.A. Kidwell, R.J. Colton: Sensing discrete streptavidin–biotin interactions with atomic force microscopy, Langmuir **10**, 354–357 (1994)

27.2 E.L. Florin, V.T. Moy, H.E. Gaub: Adhesion forces between individual ligand receptor pairs, Science **264**, 415–417 (1994)

27.3 P. Hinterdorfer, W. Baumgartner, H.J. Gruber, K. Schilcher, H. Schindler: Detection and localization of individual antibody–antigen recognition events by atomic force microscopy, Proc. Natl. Acad. Sci. USA **93**, 3477–3481 (1996)

27.4 G. Binnig, C.F. Quate, C. Gerber: Atomic force microscope, Phys. Rev. Lett. **56**, 930–933 (1986)

27.5 M. Grandbois, W. Dettmann, M. Benoit, H.E. Gaub: How strong is a covalent bond, Science **283**, 1727–1730 (1999)

27.6 T. Boland, B.D. Ratner: Direct measurement of hydrogen bonding in DNA nucleotide bases by atomic force microscopy, Proc. Natl. Acad. Sci. USA **92**, 5297–5301 (1995)

27.7 P. Wagner, M. Hegner, P. Kernen, F. Zaugg, G. Semenza: Covalent immobilization of native biomolecules onto Au(111) via N-hydroxysuccinimide ester functionalized self assembled monolayers for scanning probe microscopy, Biophys. J. **70**, 2052–2066 (1996)

27.8 U. Dammer, O. Popescu, P. Wagner, D. Anselmetti, H.-J. Güntherodt, G.M. Misevic: Binding strength between cell adhesion proteoglycans measured by atomic force microscopy, Science **267**, 1173–1175 (1995)

27.9 U. Dammer, M. Hegner, D. Anselmetti, P. Wagner, M. Dreier, W. Huber, H.-J. Güntherodt: Specific antigen/antibody interactions measured by force microscopy, Biophys. J. **70**, 2437–2441 (1996)

27.10 Y. Harada, M. Kuroda, A. Ishida: Specific and quantized antibody-antigen interaction by atomic force microscopy, Langmuir **16**, 708–715 (2000)

27.11 P. Hinterdorfer, K. Schilcher, W. Baumgartner, H.J. Gruber, H. Schindler: A mechanistic study of the dissociation of individual antibody-antigen pairs by atomic force microscopy, Nanobiology **4**, 39–50 (1998)

27.12 S. Allen, X. Chen, J. Davies, M.C. Davies, A.C. Dawkes, J.C. Edwards, C.J. Roberts, J. Sefton, S.J.B. Tendler, P.M. Williams: Spatial mapping of specific molecular recognition sites by atomic force microscopy, Biochemistry **36**, 7457–7463 (1997)

27.13 R. Ros, F. Schwesinger, D. Anselmetti, M. Kubon, R. Schäfer, A. Plückthun, L. Tiefenauer: Antigen binding forces of individually addressed single-chain Fv antibody molecules, Proc. Natl. Acad. Sci. USA **95**, 7402–7405 (1998)

27.14 T. Strunz, K. Oroszlan, R. Schäfer, H.-J. Güntherodt: Dynamic force spectroscopy of single DNA molecules, Proc. Natl. Acad. Sci. USA **96**, 11277–11282 (1999)

27.15 A. Ebner, P. Hinterdorfer, H.J. Gruber: Comparison of different aminofunctionalization strategies for attachment of single antibodies to AFM cantilevers, Ultramicroscopy **107**, 922–927 (2007)

27.16 S.S. Wong, E. Joselevich, A.T. Woolley, C.L. Cheung, C.M. Lieber: Covalently functionalyzed nanotubes as nanometre-sized probes in chemistry and biology, Nature **394**, 52–55 (1998)

27.17 P. Hinterdorfer, F. Kienberger, A. Raab, H.J. Gruber, W. Baumgartner, G. Kada, C. Riener, S. Wielert-Badt, C. Borken, H. Schindler: Poly(ethylene glycol): An ideal spacer for molecular recognition force microscopy/spectroscopy, Single Mol. **1**, 99–103 (2000)

27.18 T. Haselgrübler, A. Amerstorfer, H. Schindler, H.J. Gruber: Synthesis and applications of a new poly(ethylene glycol) derivative for the crosslinking of amines with thiols, Bioconjug. Chem. **6**, 242–248 (1995)

27.19 A.S.M. Kamruzzahan, A. Ebner, L. Wildling, F. Kienberger, C.K. Riener, C.D. Hahn, P.D. Pollheimer, P. Winklehner, M. Holzl, B. Lackner, D.M. Schorkl, P. Hinterdorfer, H.J. Gruber: Antibody linking to atomic force microscope tips via disulfide bond formation, Bioconjug. Chem. **17**(6), 1473–1481 (2006)

27.20 A. Raab, W. Han, D. Badt, S.J. Smith-Gill, S.M. Lindsay, H. Schindler, P. Hinterdorfer: Antibody recognition imaging by force microscopy, Nat. Biotech. **17**, 902–905 (1999)

27.21 G.U. Lee, A.C. Chrisey, J.C. Colton: Direct measurement of the forces between complementary strands of DNA, Science **266**, 771–773 (1994)

27.22 M. Conti, G. Falini, B. Samori: How strong is the coordination bond between a histidine tag and Ni-nitriloacetate? An experiment of mechanochemistry on single molecules, Angew. Chem. **112**, 221–224 (2000)

27.23 F. Kienberger, G. Kada, H.J. Gruber, V.P. Pastushenko, C. Riener, M. Trieb, H.-G. Knaus, H. Schindler, P. Hinterdorfer: Recognition force spectroscopy studies of the NTA-His6 bond, Single Mol. **1**, 59–65 (2000)

27.24 L. Schmitt, M. Ludwig, H.E. Gaub, R. Tampe: A metal-chelating microscopy tip as a new toolbox for single-molecule experiments by atomic force microscopy, Biophys. J. **78**, 3275–3285 (2000)

27.25 S. Lata, A. Reichel, R. Brock, R. Tampe, J. Piehler: High-affinity adaptors for switchable recognition of histidine-tagged proteins, J. Am. Chem. Soc. **127**, 10205–10215 (2005)

27.26 A. Ebner, L. Wildling, R. Zhu, C. Rankl, T. Haselgrübler, P. Hinterdorfer, H.J. Gruber: Functionalization of probe tips and supports for single molecule force microscopy, Top. Curr. Chem. **285**, 29–76 (2008)

27.27 C. Yuan, A. Chen, P. Kolb, V.T. Moy: Energy landscape of avidin-biotin complexes measured burey atomic force microscopy, Biochemistry **39**, 10219–10223 (2000)

27.28 W. Han, S.M. Lindsay, M. Dlakic, R.E. Harrington: Kinked DNA, Nature **386**, 563 (1997)

27.29 G. Kada, L. Blaney, L.H. Jeyakumar, F. Kienberger, V.P. Pastushenko, S. Fleischer, H. Schindler, F.A. Lai, P. Hinterdorfer: Recognition force microscopy/spectroscopy of ion channels: Applications to the skeletal muscle Ca^{2+} release channel (RYR1), Ultramicroscopy **86**, 129–137 (2001)

27.30 D.J. Müller, W. Baumeister, A. Engel: Controlled unzipping of a bacterial surface layer atomic force microscopy, Proc. Natl. Acad. Sci. USA **96**, 13170–13174 (1999)

27.31 F. Oesterhelt, D. Oesterhelt, M. Pfeiffer, A. Engle, H.E. Gaub, D.J. Müller: Unfolding pathways of individual bacteriorhodopsins, Science **288**, 143–146 (2000)

27.32 E. Kiss, C.-G. Gölander: Chemical derivatization of muscovite mica surfaces, Colloids Surf. **49**, 335–342 (1990)

27.33 S. Karrasch, M. Dolder, F. Schabert, J. Ramsden, A. Engel: Covalent binding of biological samples to solid supports for scanning probe microscopy in buffer solution, Biophys. J. **65**, 2437–2446 (1993)

27.34 N.H. Thomson, B.L. Smith, N. Almqvist, L. Schmitt, M. Kashlev, E.T. Kool, P.K. Hansma: Oriented, active *escherichia coli* RNA polymerase: An atomic force microscopy study, Biophys. J. **76**, 1024–1033 (1999)

27.35 G. Kada, C.K. Riener, P. Hinterdorfer, F. Kienberger, C.M. Stroh, H.J. Gruber: Dithio-phospholipids for biospecific immobilization of proteins on gold surfaces, Single Mol. **3**, 119–125 (2002)

27.36 C. LeGrimellec, E. Lesniewska, M.C. Giocondi, E. Finot, V. Vie, J.P. Goudonnet: Imaging of the surface of living cells by low-force contact-mode atomic force microscopy, Biophys. J. **75**(2), 695–703 (1998)

27.37 K. Schilcher, P. Hinterdorfer, H.J. Gruber, H. Schindler: A non-invasive method for the tight anchoring of cells for scanning force microscopy, Cell. Biol. Int. **21**, 769–778 (1997)

27.38 S. Wielert-Badt, P. Hinterdorfer, H.J. Gruber, J.-T. Lin, D. Badt, H. Schindler, R.K.-H. Kinne: Single molecule recognition of protein binding epitopes in brush border membranes by force microscopy, Biophys. J. **82**, 2767–2774 (2002)

27.39 P. Bongrand, C. Capo, J.-L. Mege, A.-M. Benoliel: Use of hydrodynamic flows to study cell adhesion. In: *Physical Basis of Cell Adhesion*, ed. by P. Bongrand (CRC Press, Boca Raton 1988) pp. 125–156

27.40 J.N. Israelachvili: *Intermolecular and Surface Forces*, 2nd edn. (Academic, London New York 1991)

27.41 R. Merkel, P. Nassoy, A. Leung, K. Ritchie, E. Evans: Energy landscapes of receptor-ligand bonds explored by dynamic force spectroscopy, Nature **397**, 50–53 (1999)

27.42 A. Askin: Optical trapping and manipulation of neutral particles using lasers, Proc. Natl. Acad. Sci. USA **94**, 4853–4860 (1997)

27.43 K.C. Neuman, S.M. Block: Optical trapping, Rev. Sci. Instrum. **75**, 2787–2809 (2004)

27.44 K. Svoboda, C.F. Schmidt, B.J. Schnapp, S.M. Block: Direct observation of kinesin stepping by optical trapping interferometry, Nature **365**, 721–727 (1993)

27.45 S.M. Block, C.L. Asbury, J.W. Shaevitz, M.J. Lang: Probing the kinesin reaction cycle with a 2D optical force clamp, Proc. Natl. Acad. Sci. USA **100**, 2351–2356 (2003)

27.46 A.E.M. Clemen, M. Vilfan, J. Jaud, J. Zhang, M. Barmann, M. Rief: Force-dependent stepping kinetics of myosin-V, Biophys. J. **88**, 4402–4410 (2005)

27.47 S. Smith, Y. Cui, C. Bustamante: Overstretching B-DNA: The elastic response of individual double-stranded and single-stranded DNA molecules, Science **271**, 795–799 (1996)

27.48 M.S.Z. Kellermayer, S.B. Smith, H.L. Granzier, C. Bustamante: Folding-unfolding transitions in single titin molecules characterized with laser tweezers, Sience **276**, 1112–1216 (1997)

27.49 T.R. Strick, J.F. Allemend, D. Bensimon, A. Bensimon, V. Croquette: The elasticity of a single supercoiled DNA molecule, Biophys. J. **271**, 1835–1837 (1996)

27.50 T. Lionnet, D. Joubaud, R. Lavery, D. Bensimon, V. Croquette: Wringing out DNA, Phys. Rev. Lett. **96**, 178102 (2006)

27.51 F. Kienberger, V.P. Pastushenko, G. Kada, H.J. Gruber, C. Riener, H. Schindler, P. Hinterdorfer: Static and dynamical properties of single poly(ethylene glycol) molecules investigated by force spectroscopy, Single Mol. **1**, 123–128 (2000)

27.52 S. Liang, D. Medich, D.M. Czajkowsky, S. Sheng, J.-Y. Yuan, Z. Shao: Thermal noise reduction of mechanical oscillators by actively controlled external dissipative forces, Ultramicroscopy **84**, 119–125 (2000)

27.53 M.B. Viani, T.E. Schäffer, A. Chand, M. Rief, H.E. Gaub, P.K. Hansma: Small cantilevers for force spectroscopy of single molecules, J. Appl. Phys. **86**, 2258–2262 (1999)

27.54 T. Strunz, K. Oroszlan, I. Schumakovitch, H.-J. Güntherodt, M. Hegner: Model energy landscapes and the force-induced dissociation of ligand-receptor bonds, Biophys. J. **79**, 1206–1212 (2000)

27.55 H. Grubmüller, B. Heymann, P. Tavan: Ligand binding: Molecular mechanics calculation of the streptavidin–biotin rupture force, Science **271**, 997–999 (1996)

27.56 G.I. Bell: Models for the specific adhesion of cells to cells, Science **200**, 618–627 (1978)

27.57 E. Evans, K. Ritchie: Dynamic strength of molecular adhesion bonds, Biophys. J. **72**, 1541–1555 (1997)

27.58 E. Evans, K. Ritchie: Strength of a weak bond connecting flexible polymer chains, Biophys. J. **76**, 2439–2447 (1999)

27.59 J. Fritz, A.G. Katopidis, F. Kolbinger, D. Anselmetti: Force-mediated kinetics of single P-selectin/ligand complexes observed by atomic force microscopy, Proc. Natl. Acad. Sci. USA **95**, 12283–12288 (1998)

27.60 T. Auletta, M.R. de Jong, A. Mulder, F.C.J.M. van Veggel, J. Huskens, D.N. Reinhoudt, S. Zou, S. Zapotocny, H. Schönherr, G.J. Vancso, L. Kuipers: β-cyclodextrin host-guest complexes probed under thermodynamic equilibrium: Thermodynamics and force spectroscopy, J. Am. Chem. Soc. **126**, 1577–1584 (2004)

27.61 V.T. Moy, E.-L. Florin, H.E. Gaub: Adhesive forces between ligand and receptor measured by AFM, Science **266**, 257–259 (1994)

27.62 A. Chilkoti, T. Boland, B. Ratner, P.S. Stayton: The relationship between ligand-binding thermodynamics and protein-ligand interaction forces measured by atomic force microscopy, Biophys. J. **69**, 2125–2130 (1995)

27.63 I. Schumakovitch, W. Grange, T. Strunz, P. Bertoncini, H.-J. Güntherodt, M. Hegner: Temperature dependence of unbinding forces between complementary DNA strands, Biophys. J. **82**, 517–521 (2002)

27.64 W. Baumgartner, P. Hinterdorfer, W. Ness, A. Raab, D. Vestweber, H. Schindler, D. Drenckhahn: Cadherin interaction probed by atomic force microscopy, Proc. Natl. Acad. Sci. USA **8**, 4005–4010 (2000)

27.65 F. Schwesinger, R. Ros, T. Strunz, D. Anselmetti, H.-J. Güntherodt, A. Honegger, L. Jermutus, L. Tiefenauer, A. Plückthun: Unbinding forces of single antibody-antigen complexes correlate with their thermal dissociation rates, Proc. Natl. Acad. Sci. USA **29**, 9972–9977 (2000)

27.66 A.F. Oberhauser, P.K. Hansma, M. Carrion-Vazquez, J.M. Fernandez: Stepwise unfolding of titin under force-clamp atomic force microscopy, Proc. Natl. Acad. Sci. USA **16**, 468–472 (2000)

27.67 S. Izraelev, S. Stepaniants, M. Balsera, Y. Oono, K. Schulten: Molecular dynamics study of unbinding of the avidin-biotin complex, Biophys. J. **72**, 1568–1581 (1997)

27.68 M. Rief, F. Oesterhelt, B. Heyman, H.E. Gaub: Single molecule force spectroscopy on polysaccharides by atomic force microscopy, Science **275**, 1295–1297 (1997)

27.69 M. Schlierf, M. Rief: Single-molecule unfolding force distributions reveal a funnel-shaped energy landscape, Biophys. J. **90**, L33 (2006)

27.70 G. Neuert, C. Albrecht, E. Pamir, H.D. Gaub: Dynamic force spectroscopy of the digoxigenin-antibody complex, FEBS Letters **580**, 505–509 (2006)

27.71 O.K. Dudko, A.E. Filippov, J. Klafter, M. Urback: Beyond the conventional description of dynamic force spectroscopy of adhesion bonds, Proc. Natl. Acad. Sci. USA **100**, 11378–11381 (2003)

27.72 M. Evstigneev, P. Reimann: Dynamic force spectroscopy: Optimized data analysis, Phys. Rev. E **68**, 045103(R) (2003)

27.73 E. Evans, E. Leung, D. Hammer, S. Simon: Chemically distinct transition states govern rapid dissociation of single L-selectin bonds under force, Proc. Natl. Acad. Sci. USA **98**, 3784–3789 (2001)

27.74 X. Zhang, E. Woijcikiewicz, V.T. Moy: Force spectroscopy of the leukocyte function-associated antigen-1/intercellular adhesion molecule-1 interaction, Biophys. J. **83**, 2270–2279 (2002)

27.75 B.T. Marshall, M. Long, J.W. Piper, T. Yago, R.P. McEver, Z. Zhu: Direct observation of catch bonds involving cell adhesion molecules, Nature **423**, 190–193 (2003)

27.76 B. Heymann, H. Grubmüller: Molecular dynamics force probe simulations of antibody/antigen unbinding: Entropic control and non additivity of unbinding forces, Biophys. J. **81**, 1295–1313 (2001)

27.77 M. Odorico, J.M. Teulon, T. Bessou, C. Vidaud, L. Bellanger, S.W. Chen, E. Quemeneur, P. Parot, J.L. Pellequer: Energy landscape of chelated uranyl: Antibody interactions by dynamic force spectroscopy, Biophys. J. **93**, 645 (2007)

27.78 R. Nevo, C. Stroh, F. Kienberger, D. Kaftan, V. Brumfeld, M. Elbaum, Z. Reich, P. Hinterdorfer: A molecular switch between two bound states in the RanGTP-importinβ1 interaction, Nat. Struct. Mol. Biol. **10**, 553–557 (2003)

27.79 R. Nevo, V. Brumfeld, M. Elbaum, P. Hinterdorfer, Z. Reich: Direct discrimination between models of protein activation by single-molecule force measurements, Biophys. J. **87**, 2630–2634 (2004)

27.80 R. Zwanzig: Diffusion in a rough potential, Proc. Natl. Acad. Sci. USA **85**, 2029–2030 (1988)

27.81 C.B. Hyeon, D. Thirumalai: Can energy landscape roughness of proteins and RNA be measured by using mechanical unfolding experiments?, Proc. Natl. Acad. Sci. USA **100**, 10249–10253 (2003)

27.82 R. Nevo, V. Brumfeld, R. Kapon, P. Hinterdorfer, Z. Reich: Direct measurement of protein energy landscape roughness, EMBO Reports **6**, 482–486 (2005)

27.83 F. Rico, V.T. Moy: Energy landscape roughness of the streptavidin-biotin interaction, J. Mol. Recognit. **20**, 495–501 (2007)

27.84 P.P. Lehenkari, M.A. Horton: Single integrin molecule adhesion forces in intact cells measured by atomic force microscopy, Biochem. Biophys. Res. Commun. **259**, 645–650 (1999)

27.85 A. Chen, V.T. Moy: Cross-linking of cell surface receptors enhances cooperativity of molecular adhesion, Biophys. J. **78**, 2814–2820 (2000)

27.86 G. Pfister, C.M. Stroh, H. Perschinka, M. Kind, M. Knoflach, P. Hinterdorfer, G. Wick: Detection of HSP60 on the membrane surface of stressed human endothelial cells by atomic force and confocal microscopy, J. Cell Sci. **118**, 1587–1594 (2005)

27.87 T. Puntheeranurak, L. Wildling, H.J. Gruber, R.K.H. Kinne, P. Hinterdorfer: Ligands on the string: single molecule studies on the interaction of antibodies and substrates with the surface of the Na$^+$-glucose cotransporter SGLT1 in living cells, J. Cell Sci. **119**, 2960–2967 (2006)

27.88 E.P. Wojcikiewicz, M.H. Abdulreda, X. Zhang, V.T. Moy: Force spectroscopy of LFA-1 and its ligands, ICAM-1 and ICAM-2, Biomacromolecules **7**, 3188 (2006)

27.89 M. Ludwig, W. Dettmann, H.E. Gaub: Atomic force microscopy imaging contrast based on molecuar recognition, Biophys. J. **72**, 445–448 (1997)

27.90 P.P. Lehenkari, G.T. Charras, G.T. Nykänen, M.A. Horton: Adapting force microscopy for cell biology, Ultramicroscopy **82**, 289–295 (2000)

27.91 N. Almqvist, R. Bhatia, G. Primbs, N. Desai, S. Banerjee, R. Lal: Elasticity and adhesion force mapping reveals real-time clustering of growth factor receptors and associated changes in local cellular rheological properties, Biophys. J. **86**, 1753–1762 (2004)

27.92 M. Grandbois, M. Beyer, M. Rief, H. Clausen-Schaumann, H.E. Gaub: Affinity imaging of red blood cells using an atomic force microscope, J. Histochem. Cytochem. **48**, 719–724 (2000)

27.93 O.H. Willemsen, M.M.E. Snel, K.O. van der Werf, B.G. de Grooth, J. Greve, P. Hinterdorfer, H.J. Gruber, H. Schindler, Y. van Kyook, C.G. Figdor: Simultaneous height and adhesion imaging of antibody antigen interactions by atomic force microscopy, Biophys. J. **57**, 2220–2228 (1998)

27.94 B.V. Viani, L.I. Pietrasanta, J.B. Thompson, A. Chand, I.C. Gebeshuber, J.H. Kindt, M. Richter, H.G. Hansma, P.K. Hansma: Probing protein-protein interactions in real time, Nat. Struct. Biol. **7**, 644–647 (2000)

27.95 A. Ebner, F. Kienberger, G. Kada, C.M. Stroh, M. Geretschläger, A.S.M. Kamruzzahan, L. Wildling, W.T. Johnson, B. Ashcroft, J. Nelson, S.M. Lindsay, H.J. Gruber, P. Hinterdorfer: Localization of single avidin biotin interactions using simultaneous topography and molecular recognition imaging, ChemPhysChem **6**, 897–900 (2005)

27.96 W. Han, S.M. Lindsay, T. Jing: A magnetically driven oscillating probe microscope for operation in liquid, Appl. Phys. Lett. **69**, 1–3 (1996)

27.97 C.M. Stroh, A. Ebner, M. Geretschläger, G. Freudenthaler, F. Kienberger, A.S.M. Kamruzzahan, S.J. Smith-Gill, H.J. Gruber, P. Hinterdorfer: Simultaneous topography and recognition imaging using force microscopy, Biophys. J. **87**, 1981–1990 (2004)

27.98 C. Stroh, H. Wang, R. Bash, B. Ashcroft, J. Nelson, H.J. Gruber, D. Lohr, S.M. Lindsay, P. Hinterdorfer: Single-molecule recognition imaging microscope, Proc. Natl. Acad. Sci. USA **101**, 12503–12507 (2004)

27.99 L. Chtcheglova, J. Waschke, L. Wildling, D. Drenckhahn, P. Hinterdorfer: Nano-scale dynamic recognition imaging on vascular endothelial cells, Biophys. J. **9**(3), L11–L13 (2007)